CRIMINAL JUSTICE ILLUMINATED

The Effective Corrections Manager:
Correctional Supervision for the Future

SECOND EDITION

Richard L. Phillips
Correctional Consultant
Geneva, Illinois

Charles R. McConnell
Management Consultant
The VMC Group
Niagara Falls, New York
Human Resource and Editorial Consultant
Ontario, New York

JONES AND BARTLETT PUBLISHERS
Sudbury, Massachusetts
BOSTON   TORONTO   LONDON   SINGAPORE

World Headquarters

Jones and Bartlett Publishers
40 Tall Pine Drive
Sudbury, MA 01776
978-443-5000
info@jbpub.com
www.jbpub.com

Jones and Bartlett Publishers
Canada
6339 Ormindale Way
Mississauga, ON L5V 1J2
Canada

Jones and Bartlett Publishers
International
Barb House, Barb Mews
London W6 7PA
United Kingdom

ISBN-13: 978-0-7637-3311-7
ISBN-10: 0-7637-3311-3

Jones and Bartlett's books and products are available through most bookstores and online booksellers. To contact Jones and Bartlett Publishers directly, call 800-832-0034, fax 978-443-8000, or visit our website at www.jbpub.com.

Substantial discounts on bulk quantities of Jones and Bartlett's publications are available to corporations, professional associations, and other qualified organizations. For details and specific discount information, contact the special sales department at Jones and Bartlett via the above contact information or send an email to specialsales@jbpub.com.

Production Credits
Publisher—Public Safety Group: Kimberly Brophy
Acquisitions Editor: Chambers Moore
Production Manager: Amy Rose
Production Assistant: Carolyn F. Rogers
Editorial Assistant: Jaime Greene
Marketing Manager: Matthew Bennett
Marketing Associate: Laura Kavigian
Cover and Text Design: Anne Spencer
Photo Research: Kimberly Potvin
Cover Image: photo © AP Photo; column © Ron Chapple/Thinkstock/Alamy Images
Chapter Opener Image: © Masterfile
Printing and Binding: Malloy, Inc.
Cover Printing: Malloy, Inc.

Library of Congress Cataloging-in-Publication Data

Phillips, Richard L.
   The effective corrections manager : correctional supervision for the future / Richard L. Phillips, Charles R. McConnell.— 2nd ed.
      p. cm.
      Includes bibliographical references and index.
      ISBN 0-7637-3311-3 (hardcover : alk. paper)
      1. Prison administration—United States. 2. Correctional personnel—United States.
3. Personnel management—
   United States. I. McConnell, Charles R. II. Title.
      HV9469.P47 2005
      365'.068—dc22
6048
                              2004011541

Printed in the United States of America
11 10 09 08 07   10 9 8 7 6 5 4 3 2

Contents

Part I The Setting

Part II The Supervisor and Self

Part IV The Supervisor and the Task

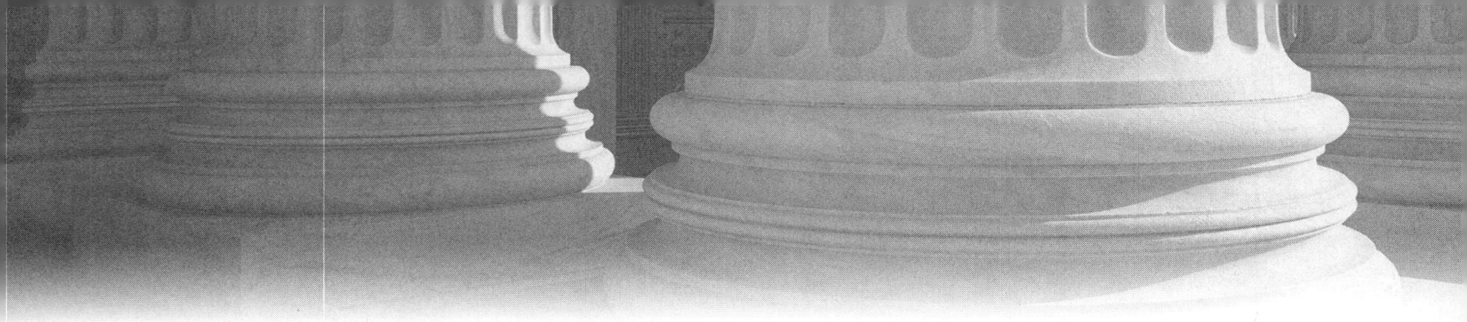

"Give me the right staff, and I can run a maximum security prison in an old red barn."

—Austin McCormick

Writing a second edition to any work brings an author face to face with himself and his chosen topic in an interesting way. Things inevitably change in the time that passes between writing the first edition and contemplation of a revision. Academic theory, professional practice, legal structures, and even the views of the author can look very different five or ten years later.

One might argue, at least to a certain extent, that the quality of a first edition is demonstrated by the amount of core material that remains in the second edition. And for that reason, augmenting the original material in this book has been interesting as a matter of self-criticism. Looking at the original work product—some of which was first compiled almost a decade ago—one finds oneself thinking things like, "Why in the world did I say it that way?" and "How could I possibly not have written about (name any number of things)?" The revision process has resulted in moderate but hopefully relevant changes in the existing chapters, as well as the addition of major material in new topical areas. Complete new chapters have been added to cover the evolution of corrections, ethics, teambuilding, and reengineering.

This is not a book about the security side of correctional management. The reader is not going to find information on key control, tool inventories, inmate accountability, or inmate discipline. These things are important—indeed vital and central—to the operation of most correctional agencies. Every correctional administrator certainly should master the relevant, practical tools and technologies of the trade. But the above quote, attributed to one of the early leaders in U.S. corrections, signals that this book is about managing people in a correctional setting—a setting that could be a halfway house, a probation office, a county work camp, and certainly a maximum-security prison.

In the foreword to the first edition, J. Michael Quinlan talked about contemporary correctional managers operating in "an era of dwindling government resources and a movement to shrink the size of government in the United States." A weak economy in the early part of this decade has meant that most public correctional organizations still are being asked to do more with proportionately less. Inmates are no less difficult to manage than in the mid-1990s, and some would argue that the continued influence of prison gangs and shifting demographics are major contributors to that trend. The litigation and regulatory environments are no less demanding, and probably even more so.

These factors continue to make it important for correctional facilities to be administered in a professional, effective manner. Properly equipping managers for this task is critical, and this publication is intended to serve as a background reference work for those involved in either developing their own managerial skills or teaching others how to be effective supervisors in the correction environment. Effective correctional management is an essential element in the stability and safety of the various components of the correctional system, and an important factor in the operation of this nation's criminal justice system.

Corrections has changed around the margins over the years. New technologies are available and certainly can be of use. New legislation and litigation can alter the context of decisions and priorities over time. However, this book reflects a particular point of view—that corrections is at its heart a people business. Whether discussing public or private operations, confinement or community facilities, the fundamentals of sound correctional management are the fundamentals of managing people.

Introduction

". . . the quality of correctional life depends far more on management practices than on any other single variable."

—John DiIulio

The field of corrections continues to present one of the most difficult managerial tasks society can offer. Many other occupations are challenging, and some—police work, for instance—can be dangerous in much the same way. But today corrections is a part of the collective consciousness of the United States. This nation is experiencing serious public policy (and particularly fiscal) questions about how correctional systems are managed. Long out of sight and mind, corrections is in the open, and correctional management is under the spotlight.

As John DiIulio, a respected scholar and commentator on corrections and the criminal justice scene, said in addressing this issue, "Prison administration in a free society poses important questions of political theory and governmental practice. Correctional managers govern men who are far from being angels. How ought they to govern? What are the ends of good government in the 'society of captives' and how can they be achieved?"

DiIulio, who did extensive research into the workings of the federal prison system, concentrates on the macro issues of correctional governance. But his observation about the importance of management practices applies equally well to the individual supervisor and the day-to-day management of correctional operations at the micro level.

Anyone who has worked in a correctional setting for any length of time can find their own personal examples of how poor management of people or resources at the individual department level has created major organizational problems. In extreme cases, those inadequate decisions and techniques have cost lives. Few, if any, other occupations present managers with such difficult problems on day-to-day basis, and where the stakes are so high.

The unique management challenges of corrections can be distinguished from those of other fields in a number of ways.

- Correctional administrators are asked to protect the public from often-dangerous incarcerated criminals.
- They must operate facilities that are as safe as possible for agency employees.
- They must deal with the problems of institutional crowding in an era of intense fiscal pressures.
- They must supervise and prevent violence among poorly socialized, aggressive inmates—protecting inmates from each other to the extent possible.

- They are obliged to provide drug treatment and literacy programming to poorly motivated subjects.
- They must gainfully occupy large numbers of inmates who have minimal or no work experience.
- They must balance the need to provide a reasonable level of programs and activities for inmates against the public's current sentiment that correctional institutions provide too many amenities and that they are "resorts."

Managers of community-based corrections programs have the additional burden of being unable to control the actual whereabouts of their clientele for large portions of the day, while those individuals are in close contact with the general public.

These factors set corrections aside as a unique field, and many other published works have discussed those macro-level dynamics. Yet at the individual supervisory level, managing in the correctional environment has a great many traits in common with management in other specialties.

- Correctional managers must satisfy their organizational superiors with respect to accomplishing a specific mission.
- They must monitor and control workflow.
- They must develop and stay within budgets.
- They must deal with employee performance issues.
- They ordinarily must work in an organized labor environment.
- They must comply with applicable statutes, regulations, and laws.
- They must see that numerous concrete tasks are accomplished— supplies purchased, grounds maintained, plumbing repaired, meals served, hallways cleaned, and laundry processed.

In those respects, correctional work is not unlike that encountered in other "total institutions," such as the military, hospitals, and residential mental health facilities. True, in correctional settings many tasks are performed by inmate workers who are not highly motivated and who may not even have the requisite skills for the job to which they are assigned. But to a large degree, correctional management at the individual supervisory level has many, if not most, of the same underpinnings of management elsewhere in society.

This book starts from the central premise that many of the insights and skills that have proven so critical to successful management in the private sector and other public organizations are fully applicable to correctional management. Yet one should bear in mind that supervision in corrections has some very different aspects that require adaptation.

One key difference between corrections and other service organizations and agencies is that the population served by correctional staff is held (very loosely in the case of community corrections programs) involuntarily. This produces a work environment that can be hostile and even dangerous at times. It also means that staff at every level must be concerned with security issues that present themselves in very few other occupations. Indeed, this comprehensive overlay of security concerns permeates the correctional environment and has an impact on virtually all management issues. It differentiates correctional work from all other settings, save perhaps the high-security environment of a locked mental health institution.

Add to the security overhead with which the correctional manager must cope the fact that the population in any correctional institution or program generally has no vested interest in participating in the correctional process. Consider the societal mission of corrections—ensuring public safety and equipping (often unwilling, antagonistic, hostile, dangerous, illiterate, poorly socialized) inmates with skills that will reduce the likelihood of future criminality. This lack of inmate interest in participating in correctional endeavors has very clear implications for the staff who try to manage work and other programs in a confinement or supervised community setting.

Paradoxically, it also is true that correctional programs and facilities could not operate in an orderly manner on a day-to-day basis without the compliance of the subject population. But that compliance is superficial at best. It is grounded in (coerced by, it may be argued) a well-defined security and/or supervision structure. That structure is backed up by a disciplinary system that can levy considerable punishment against the noncompliant inmate or community supervisee. Every agency or facility employee is expected to participate in these critical security and control systems. One is hard pressed to find another managerial environment with these overtones, which impact line and supervisory staff equally.

The fact that the physical operation of confinement facilities depends on manpower supplied by the inmate population also produces a bifurcated supervision situation. A line staff member in the official chain of command also can be a first-line supervisor with respect to inmates working in his or her charge. While this book will not dwell on that specific aspect of correctional management, it is clear that many of the skills and personal traits that help managers direct staff activities also can apply to directing inmates.

There is one final difference for those working in a confinement setting. All of the community-like functions of a correctional facility are provided within a security apparatus that can vary from the very low constraints of a minimum-security correctional camp to the extreme restrictions of secure penitentiaries. A manager at a camp will be far less constrained in directing day-to-day operations within the limits imposed by tool control, key control, inmate supervision, and other procedures, than will be the manager in a very secure facility. But both will, to a certain degree, have to take those important security concerns into account.

These factors are important in understanding how the setting in which correctional administrators operate makes their day-to-day management tasks somewhat different than their noncorrectional counterparts. But these differences do not mean that fundamental management principles do not apply to corrections. Rather, it means that in some cases special approaches are needed to their application. Managing a correctional agency or any other enterprise hinges on effectively recruiting, training, directing, and motivating people, and for that reason at the individual supervisory level, corrections and other professions can draw on the same fundamental management principles.

This book is intended to be read and used by first- and second-line institution managers (including those supervising inmates), and mid-level managers at the headquarters level. It also can be of value to those with or without prior formal training in management, and for potential supervisors. It can serve as a refresher text for managers at all levels of the correctional organization. It also is pertinent to many upper-level managers—the people who supervise the supervisors of the supervisors—in terms of lending perspective to the top-down view of what happens at lower levels.

Readers can use this book for general information about correctional management. They can use it also as a reference, seeking out specific topics through either the index or the table of contents. They also can use it as a textbook for supervisory development classes.

This text draws some of its content from the excellent material developed by author McConnell for his work, *The Effective Health Care Supervisor, 5th edition* (Jones and Bartlett Publishers, 2003). But it is in every way keyed to the needs of managers in the correctional setting, whether at a first-line level or higher in the organization. Chapters include corrections-related examples or exercises that illustrate or apply the material they accompany. These may be used for individual reading and study or informal discussion, and are also intended as activities for supervisory development classes.

Readers will find that there is no correct order for the material in this book. Although divided into chapters along topical lines, it is really not possible to deal with any single subject without some awareness of others. "Communication" is a case in point; it is the primary topic of several chapters, yet the principles of effective communication make their presence felt in a dozen or more other chapters. Each chapter is implicitly or explicitly part of perhaps several other chapter's topics, but is written to stand independently.

Because of the way the subject material is intertwined, chapters can be read selectively. But it may be most helpful to begin with the first four chapters for the sake of obtaining an overall perspective. Then, read those chapters on topics of specific interest or that touch on a current problem. For example, if the last meeting you attended was a disaster and you would like to learn about effective meetings, go straight to Chapter 22. Do not worry about skipping chapters that simply do not apply to your situation—just as long as you are certain they do not apply. For example, if you do not have budget responsibility at present, save Chapter 23 until later. Use your valuable reading time for the topics that will do you the most good on the job.

Each of the chapters begins with a short vignette or other illustrative material drawn from the real world. The reader should not presume that these stories are presented in an exact manner; in some cases it was necessary to alter some facts for privacy reasons. But each of them is founded in an actual organizational situation, and is provided to show how the content of the chapter relates to the reality of corrections.

Corrections is a profession that inevitably interfaces with legal issues, and therefore this book touches on a few legal topics. Neither of the authors is an attorney, and the information contained in those sections with legal overtones should be considered as general guidance only, and not dispositive on any matter. Agency counsel, and if necessary private counsel, should be sought in any instance where legal issues arise in the workplace itself.

Whether in a correctional environment or other types of organizations, supervision can be a difficult task. One of the conditions making it so is the fact that in many cases there are no clear solutions to problems. If this book was presenting technical task instructions, it would simply say, "Here's how to do it, period." However, the problems of supervision more often than not are common-sense problems of people, most of whom are unpredictably, but quite naturally, different from each other. When presented with a specific problem, the "correct" answer may be any one of several courses of action, or no action at all, depending on the people involved. The employee involved in technical tasks may spend a great deal of time in a world that is very clearly delineated as to options and choices. There is only one way, for instance, for the records technician to correctly compute and record the terms of an inmate's sentence. But the supervisor has no such fixed guidelines for his or her actions. There is only general guidance on how to suppress a disturbance, make complex personnel deployment decisions, or decide how to distribute scarce budget resources.

Parts of this book are concerned with what are necessarily gray areas for the correctional manager. The book can guide supervisors in making many decisions. It cannot, however, prescribe solutions to "standard" problems, since few such problems exist in the real world of corrections. Moreover, every idea presented herein will not conform to each agency's policy, regulatory requirements, or statutory schemes. But in its totality, it presents a wide array of informational and strategic resources for the new and renewing correctional manager.

Two final comments. The first has to do with the content devoted to quality management and reengineering-related issues. Many current practitioners will recoil from the thought of even so much as reading about Management by Objective (MBO), Total Quality Management (TQM), or any other quality-related management scheme that seems "gimmicky" to them. This will no doubt result from their unsatisfactory experience with other such endeavors. The material presented in this book is not intended to provide the means for a manager to implement any such program on his or her own. In virtually all such cases, these kinds of programs are systemic in nature and require major resources that only can be committed by top agency officials. Rather, by acquainting line managers with these concepts and practices, the authors intend to plant the seeds of an inquiring, productive, quality mindset in new managers and those who seek to upgrade their skills and abilities. The conceptual and practical materials on these topics provide an important foundation for overall managerial practice.

Second, while this book does not dwell on it, the need for a sense of humor in any line of work seems to the authors to be of paramount importance. As serious as corrections is, having a sense of humor is an important trait for a balanced, successful manager. It happens only because of the rules of alphabetization, but it is not inappropriate that Scott Adams' entertaining book *The Dilbert Principle* leads the list of recommended reading. The managerial and bureaucratic foibles Adams chronicles have common threads that weave in and out of corrections, just as they do in the business world. But the real lesson that should be instilled in the reader is to find and take those opportunities to laugh at pieces of our profession when laughter is appropriate, and thus leaven the gravity of our work with healthy humor.

Whatever value this book possesses for corrections professionals lies largely in its potential as a working guide. Use it as particular questions and needs suggest. If it helps on the job in any substantial way, even only now and then, it will have served its intended purpose.

Richard Phillips
Correctional Consultant
Geneva, Illinois

About the Authors

Richard Phillips is an experienced correctional manager with a 35-year career in juvenile and adult corrections, serving in the state, federal, and private sectors. His correctional experience includes assignments in minimum-, medium-, and high-security correctional facilities, as well as in urban detention settings. He has held a variety of field management positions, as well as regional office and headquarters administrative posts. He currently is a private correctional consultant who has provided services to a variety of federal, state, and private criminal justice agencies. He has served as an accreditation auditor for the American Correctional Association (ACA), and in a leadership capacity in a variety of professional organizations. He is author or contributing editor of numerous publications for federal agencies, ACA, and private publishers. He holds a BA in sociology from Northern Illinois University.

Charles R. McConnell
Management Consultant
The VMC Group
Niagara Falls, New York
Human Resource and Editorial Consultant
Ontario, New York

Charles R. McConnell started his professional career in industrial and management engineering, followed by 29 years in human resources management in the health care industry. During those years he served as a senior manager for affiliated organizations of a multifacility health care system based in Rochester, New York, and as a senior consultant with the Management and Planning Services division of the Hospital Association of New York State. He currently is an independent management and human resource consultant who has published 18 books and about 300 articles, and serves as editor of a quarterly professional journal. He holds a BS in engineering and an MBA from the State University of New York at Buffalo, and has served as adjunct faculty at several colleges.

We would like to express our gratitude to the following individuals who, during the development of this project, reviewed the manuscript. Their comments and suggestions were extremely helpful.

Acknowledgments

Dr. M. George Eichenberg
Tarleton State University—Central Texas
Killeen, Texas

Dr. William G. Archambeault
School of Social Work
Louisiana State University
Baton Rouge, Louisiana

The Setting

I

Evolving in a Changing Environment

> ### Chapter Objectives
> - Identify the dimensions in which the correctional manager's work environment is changing most significantly, and develop an awareness of the significant factors contributing to the evolution of the manager's role.
> - Review the principal paradigm shifts that are contributing to major change in the management and delivery of correctional services.
> - Highlight the importance of flexibility, adaptability, and self-motivation as significant determinants of managerial success.

Nothing in progression can rest on its original plan. We might as well think of rocking a grown man in the cradle of an infant.

Edmund Burke

Life is its own journey, presupposes its own change and movement, and one tries to arrest them at one's own peril.

Laurens Van der Post

■ Dimensions of Change

The reality of corrections today is that it is changing. Indeed, first-line managers—those who supervise the people who do the hands-on work—are caught up in a period of bewildering change that some (whether by choice or involuntarily) will not survive. It is a period that will see the managerial role transformed in ways that most of today's working managers could never have anticipated when they entered the workforce. For those entering or progressing through the

Then and Now

It was 1971, and the new case manager walked in the front gate of the penitentiary—looking around with a questioning eye and not a little apprehension. One of the first things he noticed on the high concrete wall was a wire on top, stretching around the entire 26-acre compound. It was the "snitch wire" that would sound an alarm in the control center and all the towers if it was pulled down by an inmate trying to escape. So-called "taut-wire" alarms were the technology of the day when it came to perimeter detection systems. He was identified visually by a tower officer 30 feet above, and was identified the same way as he left. Personal ID methods were all there was.

Fast forward 30 years. The same person walks into a state-of-the-art high-security prison. On the perimeter itself there are motion sensors in the ground and detectors woven into the fencing. Microwave detection devices are seen at gates and other points where vibration-sensitive detectors are not appropriate. Key locations on the perimeter and inside the institution are under closed circuit surveillance (some with circuitry that automatically puts the feed from that camera on a monitor if there is any motion in the camera's field of view). In the gate processing area, he is photographed by an officer in a control center, using the closed circuit camera feed at the checkpoint. The photo will be used to identify him again when he leaves. He is required to provide a fingerprint for a device that will make a comparison with a print provided when he leaves. High technology has reached prisons in force.

ranks of correctional management, the changing interactions of many factors will be constant companions throughout their careers.

Major components of that changing environment include the following issues.

Ongoing Budget Concerns

Budget restrictions in a tight economy continue to exert pressure on correctional agencies to find new ways to perform their mission. Budget pressures have caused states to not open newly constructed facilities due to an inability to afford to hire staff. For instance, in recent years the state of Arizona (to name just one) postponed activation of several facilities due to staffing issues. Budget and recruitment problems combined to prevent their use, notwithstanding the critical population problems in its other facilities. In some cases fiscal cutbacks have caused the closure of correctional facilities that have operated for years in the seemingly secure knowledge that with a growing prison population the state could "never shut us down." One of the authors drives regularly past the town of Sheridan, Illinois, where roadside signs advocate the reactivation of a nearby prison. Its employees, the town leaders, and even inmates no doubt thought this institution would never be closed, but it did close when state budget woes became too severe. Indeed, the facility where that author first served as a department head (a modern institution where a great deal of capital development had taken place in the last decade) was closed in that same budget crunch.

Privatization

Private corrections continues to make inroads into the once-sacred public corrections arena. This book will not elaborate on the argued pros and cons of prison privatization. It is only necessary to note that private corrections firms have survived several shakeouts (political, organizational, and financial) and continue to survive in the "marketplace." While there are continued attempts by organized labor to make inroads in private correctional facilities, for the most part private correctional facilities are non-unionized. And where there are no unions, there are wide open avenues for the kind of organizational change this chapter addresses.

Cost-Effectiveness

Privatization also raises questions in the minds of legislative budget specialists. "If private prisons can be run less expensively, why can't our prisons be run at a lower cost also?" In some cases, state agencies have been forced to submit "bids" for providing correctional services, which have been evaluated against private correctional organization bids.

Technology

New technologies (far beyond those mentioned in "Then and Now") continue at a rapid rate. Any visitor to a corrections convention such as that convened by the American Correctional Association will see a multitude of new technologies that purport to make the correctional manager's job easier. From perimeter security measures to drug testing equipment, from remote monitoring equipment to new restraint methodology, managers have to decide how new technology will fit into their physical plant, programs, and budget.

Overcrowding

Prison crowding rates will continue to be high (and likely will continue to rise). There is little expectation that these pressures are going to abate. As the U.S. prison population grows and public and political pressures continue to advocate for longer and harsher sentences, this factor will impact the job of every administrator in changing ways.

Program Issues

Demands for more programs, particularly in the area of substance abuse, are likely to increase. As the cost of incarceration climbs, it is sound public policy to seek ways to reduce recidivism and the overall cost of repetitive criminal behavior. With so much criminal activity in the United States linked to drug dependencies, it only makes sense to bias programs and staffing in the direction of programs that help offenders in those areas. How to do that in the face of crowding and budget limitations is the challenge.

Workforce Management

Personnel issues will persist, including recruitment and the loss (by retirement or other avenues) of experienced and effective employees at all levels. Not that losing experienced personnel is a new issue. But at a time when offenders are becoming more difficult and the organizational challenges seem to grow daily, the loss of career employees presents an even more acute problem for the remaining managerial personnel.

Inmate Management

Inmates seem to grow more difficult to manage as years pass; in particular the issues presented by prison/street gangs are numerous. But the demonstrated needs of many inmates present a real dilemma—if you don't have the resources you can't give inmates a chance to change, and if you don't provide an opportunity to change, the likelihood grows that the inmates will remain in the system and return to it after release.

Litigation

Legal considerations are more and more of a factor in day-to-day prison operations. It is true that prison administrators have complained about the courts for years. But there is no arguing against the proposition that the cumulative effect of decades of court intervention has imposed a complicated network of restrictions and requirements on prison administrators. And with every new court ruling, day-to-day management becomes more difficult.

These are the major issues fomenting change in many correctional agencies today. They probably reflect what some managers have been going through for some time. They may reflect what is waiting around the corner for the rare agency or institution that has so far escaped such pressures.

■ Paradigm Shifts

The basic paradigms of the generations of line employees who entered the American workforce in the past have come under severe attack. These include such standbys as:

High-security facilities will always be the heart of the system. This is clearly no longer true as the value and cost advantages of inmate classification systems continue to be evident. The realities of the high costs of maximum security confinement have been driving agencies toward increased utilization of medium- and low-security facilities, which offer not only cost benefits but the opportunity to provide improved program and service delivery to offenders.

The way correctional services are currently delivered is the best available. This is self-evident only to those who have decided to not look for other ways. Indeed, private corrections has raised this issue quite forcefully, and research is divided on it; the fact is that the representation of cost savings is a real enough factor that the field has to deal with it.

We work in an essential industry that will never close its doors. Tell that to the former employees of prisons that have been closed or private corrections firms that have not survived for various reasons.

Corrections professionals will (and should) always control the fundamental way the system is structured. This premise has been dramatically impacted by judicial involvement and budget necessities. It is also being challenged by private corrections, in that the top executives of some of the largest private corrections firms have no correctional experience.

Corrections work means reasonable pay and a decent retirement. This is belied by the vulnerability of private corrections employees and managers whose pensions are invested in company stock, and the lack of security in what is generally a non-unionized workplace.

Securing a job with a government agency will lead to employment security. In law enforcement, things have always seemed particularly secure. Serve 20 or 25 years, reach age 50 or 55, and a defined benefits retirement program is waiting for you. Now, think of the ebb and flow of private correctional activities—dependent on contract renewals for sustained operations. Think of the aforementioned prisons that have been decommissioned—their staff forced to either retire early, transfer, or find other work outside corrections.

All of these demonstrable paradigm shifts suggest one conclusion—corrections is no longer a static or stable field. It increasingly requires its management ranks to adapt and adjust to shift priorities and plan for change by being willing to consider new workplace strategies.

For most people, a paradigm can be both a clarifier and an obstacle. Incoming information that fits within an existing paradigm is seen clearly because it confirms expectations. Information that is inconsistent with a paradigm, however, cannot be seen nearly as readily and, in some instances, can hardly be seen at all. The inconsistencies disturb the person's equilibrium with their environment. Possible reactions include fear, uncertainty, frustration, resistance, and the inability to imagine any good resulting from the pressures being experienced.

Correctional managers are prone to resist these paradigm shifts for a number of reasons:

- They are at risk in the process, and this manifests itself as fear and uncertainty.
- They are internal to the organization and cannot step back and objectively view what so intimately involves them.
- They are affected far more than they might ever be able to acknowledge by some long-held paradigms that are presently under concentrated—and largely successful—attack.
- The power of habitual behavior is strong.
- The inertia inherent in any bureaucracy reinforces the natural tendencies to resist at the personal level.

Managers who fear replacement can best ensure their futures by becoming paradigm breakers by refusing to remain satisfied with the status quo for very long. To overcome these forces, it is necessary to start afresh—to work backward from desired outcomes to appropriate processes—ignoring past practices to the extent possible. At times it is necessary to deliberately think along different paths, to deliberately turn away from what is known and follow a line of thought that feels wrong and that causes discomfort.

In some cases, assistance from outside the organization, whether from professional consultants or others, can be helpful. These resources can force those in the organization to get out into the uncomfortable territory where the creative solutions are to be found.

■ Motivation and Empowerment

Motivation is a key element in successful supervisory performance. A great many correctional functions (at both the line and supervisory level) depend on individual initiative and insight into human behavior. But because of the ways in which the supervisory role is changing and because of the dramatic changes in the field of corrections that are altering that role, the average supervisor can be caught in a classic motivational crunch.

Supervisors are susceptible to the same negative morale influences as the nonsupervisory staff—pressures to do more with less, deal with more intractable inmates, manage with the courts looking over one shoulder, and many others. Yet managerial personnel are expected to be sufficiently self-motivated to help lift the line employees' level of motivation.

There is no question that as the one most responsible for the output of the work group, the supervisor can have a significant effect on the group's outlook and effort. It is important that the supervisor does everything possible to be "up" when the group members are "down." This leader must be a cheerleader at a time when the employees might feel there is nothing to cheer about. Surely this seems like one is expected to put up a false front for the employees.

Why, one might ask, should the supervisor not exhibit signs of the same frustration and lack of confidence in the future that the employees feel? After all, in today's fiscal environment the supervisor may be called on to undertake tasks that were never before part of the role. These may include deciding how to curtail programs that heretofore were thought untouchable, or even deciding how to implement reductions in force or facility closures. Their burdens may indeed be far greater than those of the line staff. Well, simply stated, if the supervisor's behavior reflects only the doom and gloom the staff members may feel, they themselves will be dramatically affecting employee behavior—and not in a positive direction.

Improving morale is an uphill struggle in many correctional settings. Morale and motivation are, of course, complex considerations that at any time can depend on a variety of factors. But a great deal of what is related to the supervisor's ability to self-motivate will depend on that individual's personal relationship with the elements of the job. If the supervisor genuinely likes his or her work and finds satisfaction and fulfillment in necessary supervisory tasks, that will serve as a positive example for the group members. However, some supervisors have been lured into the role primarily by title, status, pay, and perks (as opposed to an innate enjoyment of the work involved in supervising others in a particular function). In all probability, such a manager will not rise to the challenges of the shrinking organization and the flattening management structure.

In management circles and in the literature, a recent trend has been to base supervisory strategies on "empowerment." Indeed, empowerment is a key element in obtaining the most from a given group of employees—of maintaining and boosting their morale and fostering initiative. And yet in almost every respect empowerment is no more than that old standby, delegation (much more fully discussed in Chapter 6). The problem has been that most of what has been called delegation was not delegation at all. So delegation—as both a term and an observed management practice—has acquired a tarnish that no amount of polishing can remove. Perhaps that is why empowerment is now used to describe these processes.

What is important is that any group's leader must truly be empowering in relationships with employees—delegating properly and fully to the fullest extent of his or her capacities. In these days when management structures are becoming leaner and leaner, empowerment is essential. Empowerment stands as the only practical way to expand and extend the leader's effectiveness and to pursue the constant improvement that is expected in the contemporary correctional workplace. When it comes to seriously improving the ways in which the group's work is accomplished, empowerment acknowledges the fact that no one knows the details of the work better than the person who performs it every day. This is one of the themes to which this book will return repeatedly.

EXERCISES

Exercise 1-1: Responding to External Pressure

Due to recent developments in the private sector, it is no longer a given that publicly managed prisons are the only option for government. Private corrections has highlighted the issues of cost-effectiveness, quality program delivery, and employee job security. Management in both the public and private corrections arena is impacted by these changes in many ways.

Discuss the following questions:

1. Are there other pressures (beyond those mentioned in this chapter) that create a need to reexamine underlying paradigms, and if so, what are they?

2. How has private corrections impacted the paradigms under which today's correctional managers operate?

3. Why do you believe private correctional facilities may be more cost effective, and if you believe it to be the case, what managerial benefits do private corrections present?

4. In your view, are there ways that corrections can implement new management paradigms without adverse political impact, and if so, how?

Exercise 1-2: Reinventing the Organization

You have been placed on a high-level agency task force that is to develop options for top management and the legislature. Your goal is to develop a comprehensive redesign of the system that will result in greater operational efficiencies across the entire agency. It is understood that you may not propose any options that will reduce public, staff, or inmate safety. You also must take into account the collective bargaining agreements to which the agency is a party, and all court orders and consent decrees to which the agency is subject.

Instructions

In either words or diagrams, or perhaps both, develop an organizational structure for accomplishing the foregoing objective, designating the functions you believe will have to be performed. You can do this individually, but this activity may be more fruitful when undertaken by small groups (perhaps three or four people). Spend 10 minutes or so identifying functions for your redesigned agency structure, and then consider the following three questions:

1. Did you find yourself using the names of so-called "traditional" correctional activities (custody, case management, business office, hospital, etc.) to describe the functions of your revised operation? Why might you have done so?

2. Did you experience difficulty trying to envision new ways of achieving a prison system' desired outcomes, and if so, why?

3. Are there any reasons to think the three safety issues enumerated in the introductory paragraph will be any more or less important in coming years? If so, what might replace or supplement them and how must those elements be taken into account in any agency redesign?

4. Were you able to develop any really new paradigms for delivering the required services to inmates and accomplishing the agency's public safety mission?

Is Corrections Really Different?

The end product of all business is people.

Rensis Likert

So much of what we call management consists in making it difficult for people to work.

Peter Drucker

■ Process Versus Environment

This book addresses a number of issues that distinguish corrections from other professions, at both the macro- and microlevels. And yet the essential premise of this book is that management in corrections is the management of people.

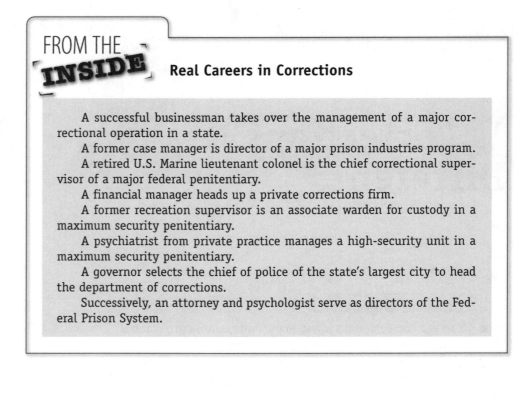

FROM THE

INSIDE   **Real Careers in Corrections**

A successful businessman takes over the management of a major correctional operation in a state.

A former case manager is director of a major prison industries program.

A retired U.S. Marine lieutenant colonel is the chief correctional supervisor of a major federal penitentiary.

A financial manager heads up a private corrections firm.

A former recreation supervisor is an associate warden for custody in a maximum security penitentiary.

A psychiatrist from private practice manages a high-security unit in a maximum security penitentiary.

A governor selects the chief of police of the state's largest city to head the department of corrections.

Successively, an attorney and psychologist serve as directors of the Federal Prison System.

There are certain core skills and traits that managers can use in all types of organizational settings, including corrections. But for the questioning reader, that assertion alone may be insufficient. So the opposing sides of an age-old argument are ready for examination:

It doesn't matter how well it worked anywhere else,
it won't work here—this is corrections.

versus

Good management is good management no matter where it's practiced;
what works elsewhere will work in a correctional organization as well.

Since this book discusses supervision in the correctional setting, it would seem sensible to explain which view governs the approach taken by this text. Should one focus on the management process, thus agreeing that "good management is good management no matter where it's practiced," or should one give the most weight to the environment, agreeing that correctional operations are sufficiently different to warrant a completely different approach to management?

Many correctional managers are clearly divided on the fundamental issue of process versus environment. Often, all organizational considerations are split into these two distinct categories, which are assumed to be mutually exclusive in some way. These considerations can be condensed to corrections versus "industry," with the latter category including manufacturing, commercial, financial, retail, and all other organizations not specifically devoted to correctional operations. Further, in this simplistic comparison, "industry" frequently becomes something of a dirty word. ("After all, we deal in human life.")

This debate has been refocused to some extent in recent years by the advent of private correctional firms, which manage confinement facilities from the perspective of the business world, rather than that of some level of government. Private corrections is not a new phenomenon; for many years private firms have operated community treatment centers housing minimum-security inmates. But there has been continued expansion over the last decade of private firms into more traditional confinement facilities. Private management of low- and medium-security offenders presents a novel situation for public sector correctional managers. For the first time, the precept of uniqueness is being seriously challenged. Privatization of the correctional function, it can be argued, validates at least part of the "process" viewpoint.

■ The Nature of the Correctional Organization

That the process versus environment argument exists is not at all surprising when one considers the evolution and character of the traditional correctional organization. The role of corrections as it is known today is largely a product of the past 100 years. Many correctional agencies of the early 20th century provided only custodial care in a high-security setting. In those days, for all practical purposes, there was only one occupation: the custodial (or security) service. The mission of the organization was security, and essentially the only management activity was the operation of what was in those days called the "guard" force. No classification staff, no education programs, no vocational training, little if any recreation opportunities—in short, hard time.

Some aspects of corrections involved providing inmate labor to the private sector for agriculture or manufacturing employment. However, that practice was discontinued in light of serious abuses that developed over the years. In that early era, for the most part, little thought was given to operating a correctional agency "like a business." In the 1990s, spurred by federal legislation, there was a small resurgence in several states in the use of inmate labor in private industrial work projects. But criticism of this program centers on the assertion that prison labor infringes on the employment opportunities of civilian workers, and this trend is not widespread. Consequently, management of prison factories can generally be considered as a subset of general prison management.

Because prisons are the most dramatic and visible symbol of corrections, the discussion will focus on them for a moment. The modern correctional facility is vastly different from its counterpart of the past century, or even the prison of 40 years ago. The major purpose of a correctional institution used to be custodial confinement with only basic services and very few, if any, self-improvement programs. That is now the primary mission of only a few ultra-maximum security facilities where a relatively small number of extraordinarily dangerous inmates are held. In the last 20 to 30 years, the role of the correctional institution evolved into that of an organization that, while still committed to public safety and institutional security, has an obligation to society. That obligation is to offer programs and services that inmates can use to reach the point where they can live lawfully upon release, if they so choose.

Correctional agencies of the past had a unique mission, which they fulfilled in a simple, one-dimensional manner that had no parallel in other kinds of organizations. (An exception might be the large residential mental health facilities operated by states until the 1970s). Correctional officers supervised food service, laundry, and other functions. Those support activities were seen as an extension of the custodial function, rather than as separate disciplines. The only similarity with the activities of most other organizations was the direct supervision of the correctional officers who supervised the inmates—the basic process of getting work done through people.

However, the modern correctional setting is far different from the one-dimensional situation of the past. In confinement facilities, there are a wide variety of services and programs to be provided, and numerous complex and sophisticated specialized skills are involved. In the community setting, job training and placement, housing, drug treatment, and other programs have to be organized and overseen. Also, a great many "business" functions, which are not specifically part of corrections but which are critical to the functioning of any agency, are present in today's correctional setting.

Correctional industries certainly provide an excellent example of this new situation, and it is clear that in many other respects the correctional institution of today very much resembles a business. Nowhere is this seen to be more true than in the relatively new area of private corrections, to which the management principles of this book are equally applicable. While effective management always was important, it has never been more so, as modern correctional settings are more crowded, more complex, and more subject than ever before to public and legal scrutiny. The argument about corrections versus private industry is frequently articulated along functional lines. One way of understanding how this happens is by looking at the occupational and training backgrounds of those employed in the field.

Many correctional employees did not start out with the goal of working in corrections. They were originally trained in other kinds of organizations or educated in schools where they were concerned with some noncorrectional specialty. These people, essential to the operation of a correctional agency, include counselors, mental health and medical personnel, accountants, personnel specialists, maintenance staff, food service workers, computer specialists, and others. While acquiring their skills in school and perhaps later practicing them in other settings, these individuals may have had no idea of applying these skills in corrections until an opportunity to do so arose. Many workers see their functions as cutting horizontally across organizational lines and applying equally to corrections, hospitals, manufacturing, or any other field. The movement of these specialists into correctional management ordinarily requires them to acquire more general knowledge about the field of corrections, in addition to developing specific management skills.

Other corrections professionals, however, do come into the field as a result of a specific, early career choice. College programs in criminal justice, corrections, and sociology often serve to feed entry-level correctional positions. Courses in management and exposure to correctional management systems through internships may be included in these academic programs. In many correctional agencies, career ladders are established that allow these individuals to build on their academic training. This progression, too, involves the acquisition of specific, separate management skills that are tailored to the correctional environment. This is a process that can produce managers who think their field is unique and requires unique skills.

Part of the process versus environment argument seems to stem from the differing background and experience of these two categories of personnel, as well as the vertical versus horizontal view of organizations. Employees whose careers have noncorrectional origins may have applied their education and training in other lines of work. This can reinforce the horizontal view of organizations and encourage the belief that basic skills are transportable across industry lines. For example, a nurse in a community hospital may easily be recruited to work in a correctional infirmary, and later move to a position in the private sector working for a managed care organization. In contrast, an employee whose education and training was tailored toward the specifics of the correctional environment may have worked in other kinds of organizations, but often in entirely different capacities. Consider, for instance, the person who leaves a job in retail sales to go to college, and obtains a degree in public administration with a concentration in criminal justice. The person eventually takes a position in a correctional setting as a case manager and, over time, progresses into mid- and upper-level management. This path strongly reinforces a vertical view of organizations because the skills involved are specific to that kind of organization and are not readily transportable across industry lines. In looking back at his or her career, this individual likely may not view that early job experience in the retail field as part of their "real career" in corrections.

Certainly there are some differences between management in correctional organizations and management in other organizations. But once the differences have been explored, one finds that at the individual managerial level, the same core skills and talents can be applied in a variety of organizational environments, including corrections.

■ Identifying the Real Differences

Organizations come into being to fill certain needs. Business and government organizations of all kinds—including those with a correctional mission—continue to exist because they provide something that people want or need. This truth is obvious in the business world; food wholesalers and grocery stores exist because people need food. While it may be less obvious, it is no less true that corrections exists because society has a need. It demands that certain types of criminal offenders be separated from other citizens because they are dangerous or because they will not stop committing crimes. Other segments of corrections (such as probation, parole, and halfway houses) likewise fill societal needs that have less to do with security than with providing supervision and structure in the community.

It should follow that if a set of needs can be fulfilled in a number of different ways, the organizations that do the best job of responding to those needs will be the ones most likely to continue to exist. This is especially true in manufacturing, in which competition is keen and the organization that can meet public needs with the best product at the best price will stand the best chance of success. And while among correctional organizations the competition to fill needs is less evident, it is nevertheless there. This is best seen in the ongoing development of various intermediate sanctions (punishments more stringent than probation and less restrictive than correctional confinement). It also is seen in the emergence of private correctional operations, which claim to be able to provide some correctional services at least as well and at less cost than government agencies.

The basic error in considering correctional management as "different" is the classification of organizations by type. Assigning organizations to categories such as government, manufacturing, retail, commercial, financial, and so on, does not necessarily mean that managerial traits are similarly distinct. Indeed, such classification is simply not sufficient to allow one to judge the applicability of supervisory practices across organizational lines. Rather, it is productive to examine organizations for the degree to which certain kinds of activities are present. Disregarding organizational labels, look at the processes applied within organizations and the kinds of actions required to manage these processes. Look not at what managers do but rather at how they do it.

Two Theoretical Extremes

In his book, *New Patterns of Management,* Rensis Likert developed a view of organizations based on how they do the things they do.[1] Likert expressed a great deal of his work in the form of a "scale of organizations," running from one extreme type to another.

At one end of Likert's scale is a type he called the Job Organization System. This system evolved and applies in industries in which repetitive work is dominant, such as the many manufacturing industries that rely on conveyor belts, assembly lines, and automatic and semiautomatic processes. This system is characterized by an advanced and detailed approach to management. Jobs lend themselves to a high degree of organization, and the entire system can be controlled fairly closely.

If, for example, a manager is involved in assembly line manufacturing, it is possible to break down most activity into specifically described tasks and define these tasks in great detail. The manager can schedule output, deciding to make so many units per day and gearing the input speed of all available resources accordingly. A great amount of structure and control is possible. All this calls for a certain style of supervision, a style suited to the circumstances.

At the other end of Likert's scale is the Cooperative Motivation System. This system evolved in work environments where variable work dominates. Management itself is considerably less refined in this system. Jobs are not readily definable in detail, and specific controls over organizational activity are not possible to any great extent.

In many respects a correctional institution is better described by the Cooperative Motivation System, despite its apparent regimentation and reliance on policy. Yes, department-level managers can make reasonable estimates based on experience. Most institutions develop post orders for each post that outline the major duties to be performed on that post. It certainly would be possible to specify some correctional tasks with quantitative measures, such as performing not less than five cell searches each day or writing at least 10 classification reports per week, but it remains difficult to schedule "output." Within the Cooperative Motivation System, close control is a much less prominent feature than in the

Job Organization System. This matches more closely the variable demands and contingencies of the correctional world.

What makes these differing organizational systems work? Likert contends that the Job Organization System depends largely on economic motives to keep the wheels turning. That is, everything is so controlled that the only remaining requirement is for people to perform the prescribed steps. Therefore, what keeps the wheels turning are the people who show up for work primarily because they are paid to do so. These people are not expected to exhibit very much judgment; they need only follow instructions.

In the Cooperative Motivation System, however, there are no rigid controls on activities. Jobs cannot be defined down to the last detail. Activities and outputs cannot be accurately predicted or scheduled. The nature of the work coming into the system cannot be depended on to conform to a formula. In the Cooperative Motivation System it is not enough that employees simply show up because they are being paid. This system depends to a much larger extent on individual effort and motivation to keep the wheels turning. Clearly the Cooperative Motivation System is closer to the situation most correctional managers encounter than the Job Organization System.

Examined in their extremes, therefore, the Job Organization System and the Cooperative Motivation System can be seen to differ in several important ways. However, the most important difference lies in the role of the human element—the part that people play in each kind of system. Under the conditions of the Job Organization System, the system controls the people and essentially drags them along. Under the Cooperative Motivation System, however, the people control the system and keep it moving. Certainly this is true in corrections, where many staff work in locations some distance from their supervisor and are expected to exercise a certain amount of independent judgment.

Regardless of an organization's unit of output—whether automobiles, toasters, or cell searches—one needs to look at the amount of structure that is both required and possible, and at the variability of the work itself. There are few, if any, pure organizational types. As already suggested, an example of the Job Organization System would be the automated manufacturing plant in which every employee is a servant of a mechanized assembly line. At the other end of the scale, an example of the Cooperative Motivation System at work would be the jack-of-all-trades, odd-job service in which any type of task may come up at any time.

Within corrections, the custodial or security department of a confinement facility or the work of a probation officer may very much typify a Cooperative Motivation System situation. In prison, staff in a variety of posts throughout the institution deal with inmates with widely varying needs. They may encounter everything from a drunk inmate, to an attempted suicide, to an escape attempt, to an inmate assault—each requiring a different response. When supervising offenders in the community, an immense degree of latitude must be given to the individual probation officer in determining how to guide and direct an offender on his or her caseload. Even though there are some prescribed tasks associated with each such job, there is a wide range of discretion in how they are performed. Importantly, an even wider range of possible intervening events occur out of sight of any supervisor and require independent judgment and action.

■ The Real World: Parts of Both Systems

Most organizations possess elements of both the Job Organization System and the Cooperative Motivation System. In prison for example, running a cell house is only generally predictable and does not lend itself well to a rigid routine. In contrast, some elements of the inmate booking process certainly are amenable to an assembly line approach. Managing an offender caseload in a halfway house requires far more latitude than managing the personnel office of a large probation system.

So it seems that the organization of the modern correctional system leans toward the description of the Cooperative Motivation System. There are, however, internal exceptions and differences related to size and degree of structure. A small institution, for instance, may demonstrate many features of the Cooperative Motivation System. On the other hand, while security functions involve the kind of remote supervision and independent functioning consistent with the Cooperative Motivation System, a large facility will include some departments organized along Job Organization System lines. For example, the records department of any reasonably-sized correctional agency entails many highly procedural functions. There is a specific method prescribed for calculating the length of an inmate's sentence. Searches of inmates and their property generally proceed in the same fashion each time. The same people repeat the same tasks day after day. Food service in a large correctional facility with many satellite food service locations may actually include an assembly line process and actual meal delivery by food cart. The principles of this kind of work activity are essentially the same as those for product assembly lines in manufacturing. A large prison's laundry certainly will include repetitive tasks that are highly procedural, and repetitive functions may be found in many other departments as well.

Implications for Supervision

Given the above, what are the implications for supervision in the real world?

Environment and Management Style

The concept of Likert's Job Organization System tends considerably toward production-centered management. The essential interest is in getting the work done, and the people who do the work are more or less swept along with the system. This system is rigid, and the people who keep the system going need only show up for work. On the other hand, the concept of the Cooperative Motivation System suggests people-centered management. People—the employees— are needed to do the work, and more is required of them than simply showing up. They have to take initiative, perhaps make individual decisions and render judgments, and in general must accept a measure of responsibility for keeping the system moving. Nowhere is this truer than in prisons, where correctional officers operate semi-independently in supervising large numbers of inmates in housing units or activity areas.

It is perhaps unfortunate that businesses that evolved along the lines of the Job Organization System sometimes tend to overemphasize production while largely ignoring people. Under the Cooperative Motivation System, however, it is not so easy to ignore people (even by default) since the organization may function poorly or, in the extreme, not function at all if people are not cooperative.

To complicate the situation further, criminal offenders are consummately skilled at exploiting differences among staff for illicit, disruptive, or dangerous purposes. This makes it all the more important that staff work cooperatively and within established policies, however broad or narrow they are.

Decision-making can be vastly different for a supervisor in the Job Organization System as opposed to one in the Cooperative Motivation System. In the former, it is more likely to be procedural, with many decisions being made "by the book." In the latter, specific procedures often do not exist (and cannot exist because of the variability of the work), making it necessary to rely heavily on individual judgment. This certainly would be the case for a correctional watch commander, who may be responsible for a sprawling institution and the lives of hundreds of staff and thousands of inmates involved in a variety of activities. Policy and procedure may abound on the shelf back in the office. But when confronted with an angry group of inmates in the dining room, individual judgment (described in the Cooperative Motivation System and leavened by some personal courage) is the operative factor.

Where Does Your Department Fit?

Decide for yourself what kind of organization you work in. Does it look like a Job Organization System or does it more resemble the Cooperative Motivation System? How your department measures up in terms of certain essential characteristics will have a strong influence on the style of supervision necessary to assure proper functioning. Examine the following characteristics:

Variability of Work The more the work is varied in terms of the different tasks to be encountered, the length of time they take, and the procedures by which they are performed, then the more difficult it is to schedule and control. Tasks that are unvarying and repetitive require supervisory emphasis on scheduling inputs and resources. Work that is variable requires supervisory emphasis on controlling the activities of the people who do the work.

Mobility of Employees If all the employees work in the same limited area and usually remain within the supervisor's sight, the supervisor need not be concerned with certain control activities. However, as employees become more mobile and move about in larger areas, there is a need for the supervisor to pay more attention to people who are out of sight much of the time.

Degree of Professionalism There can be a vast difference in supervisory style depending on whether the majority of employees supervised are unskilled, semiskilled, or skilled. Many components of a correctional system are staffed with educated professionals who are able, and expected, to exercise independent judgment. Managing the activities of professionals is considerably different from managing the activities of unskilled workers whose primary responsibility lies in following specific instructions.

Definability of Tasks The more structure possible in work roles, the more rigid the style of supervision may be. For instance, the job of a sorter in a large correctional laundry may be defined in every last detail in a few specific steps on a job description. Since the job is completely definable, the supervisor need only assure that a well-trained worker is assigned and then follow up to see that the work is accomplished. However, any correctional supervisor who has attempted to write a job description for a line correctional officer will tell a different tale.

Because of task variability, the need for independent judgment, and other factors, the job description for the officer is not written as easily as that of the laundry sorter. Similarly, a parole agent in the field much of the time is not easily supervised. Prime duties involved may range from helping an offender with a job interview or giving an alcohol sobriety test, to arresting a parole violator. The duties of these latter two employees are considerably less definable, so there is likely to be more need for the supervisor to provide case-by-case guidance when necessary and also the need to rely on the individual officer's independent judgment.

In general, despite the many rules and regulations involved in day-to-day operations, the organization of the modern correctional facility leans toward Likert's Cooperative Motivation System, since the activity of a correctional organization is quite variable and centered around people. However, elements of the Job Organization System must be recognized as being present in the institution's policies, procedures, and post orders. This suggests that within any particular institution there may be the need for different supervisory approaches according to the nature of the functions being supervised.

A Word about Quality

There is always room in a discussion such as this for the consideration of quality. Consider again the contention that all organizations exist to serve people's needs. It follows that quality should always be a primary consideration regardless of the form of the organization's output. Businesses basically organized along the lines of the Job Organization System tend to have frequent built-in quality checks at points in the process. As many manufacturers have discovered, however, quality must be built into a product—it cannot be inspected into it.

Organizations tending toward the Cooperative Motivation System also have their quality checks, but these are less numerous and less specific. In the kind of organization that relies heavily on individual enthusiasm and motivation, there is considerably more reliance on the individual employee to produce acceptable quality.

■ External Pressure: An Area of Increasing Concern

The "corrections-is-different, period" argument generally does not succeed in differentiating correctional management from management in other disciplines. However, there are some legitimate differences, in the form of outside pressure, that are making themselves increasingly felt in corrections.

This is not to assert that corrections has a monopoly on external pressure. Every work organization that serves people in any way experiences pressure from outside, even if that pressure is as basic as competition from others in the same business. Corrections cannot even claim the burden of maximum external regulation. Other businesses such as hospitals, insurance, banking, and public utilities are highly regulated as well.

But in some cases prisons and entire correctional systems are under strict court supervision that governs many details of facility and agency operations. Growing judicial intervention, increasing financial constraints, and mounting public scrutiny of correctional operations are realities for today's correctional manager. They combine to create a unique, frequently high-pressure work environment. This interventionist environment began to emerge in the mid-1970s and there is every reason it will continue in the early part of the 21st century.

Some undeniable forces have entered the field of corrections and are reshaping the way that supervisors do their jobs:

- The overall cost of sustaining correctional operations continues to rise as the incarcerated population in the United States grows. Correctional budgets challenge (and in some states eclipse) demands by education and other vital public services for scarce funds.
- There is a continued public expectation for high-quality correctional operations and programs (such as drug treatment and literacy training) despite constant pressure to contain or reduce costs.
- Many segments of the public are beginning to voice objections to programs that are believed by corrections administrators to be valuable but which the public views as unnecessary or even as "perks."
- Privatization—once confined only to community corrections—is becoming increasingly accepted for higher security inmates. This places a form of competitive pressure on public sector corrections as well as pressure through employee unions which fear a loss of members to these generally non-union organizations.
- Burgeoning rules and regulations have made some aspects of correctional management considerably more difficult and complex.

The reality is that the desired outcomes—public safety and humane treatment of inmates—often come only by creatively finding a way through these and other obstacles.

■ Your Supervisory Approach

One should not be misled by what seem to be differences between types of organizations. Correctional agencies are indeed unique in terms of the output they produce, but they are not necessarily unique in terms of the management processes employed. Again, examine your own department—how it is put together and especially the variability of the work and the degree of structure required. To a large extent, a manager's approach to supervision will be determined not by the fact that "this is a correctional operation, not a factory" but rather by the kinds of employees supervised and the nature of their job responsibilities.

EXERCISES

Exercise 2-1: Where Does Your Department Fit?

Take a few minutes to "rate" your department according to the four characteristics discussed in the chapter: (1) variability of work, (2) mobility of employees, (3) degree of professionalism, and (4) definability of tasks. Although this assessment will necessarily be elementary, it may nevertheless suggest which end of the "scale of organizations" your department tends toward.

Rate each characteristic on a continuous scale from 0 to 10. The following guides provide the ends and the approximate middle of the scale for each characteristic.

Variability of Work

0 = No variability. Work can be scheduled and output predicted with complete accuracy.

5 = Average condition. Workload predictability is reasonable. Advance task schedules remain at least 50 percent valid.

10 = Each task is different from all others. Workload is unpredictable, and task scheduling is not possible.

Mobility of Employees

0 = No mobility. All employees remain in sight in the same physical area during all hours of work.

5 = Average condition. Most employees work within or near the same general area or can be located within minutes.

10 = Full mobility. All employees continually move about the facility as part of normal job performance.

Degree of "Professionalism" (by virtue of degree, licensure, certification, or some combination of these)

0 = No "professionals" are employed in the department.

5 = About half of the employees are "professionals."

10 = All the employees are "professionals."

Definability of Tasks

0 = All jobs are completely definable in complete job descriptions and written procedures.

5 = Average condition. There is about 50 percent definability of jobs through job descriptions and procedures.

10 = No specific definability. No task procedures can be provided, and job descriptions must be limited to general statements.

Take the average of your "ratings." This may provide a rough idea of whether your department leans toward the Job Organization System (an average below 5) or the Cooperative Motivation System (an average above 5).

Question

Assuming your "ratings" of the four characteristics are reasonable indications of the nature of your department, what can you say about your supervisory approach relative to each characteristic?

Exercise 2-2: Suggestion for Additional Activity

Try this exercise with a small group of other managers (perhaps three or four) who are familiar with your department's operations. Try to arrive at a group rating for each characteristic.

It Isn't in the Job Description

George Morton, the prison's maintenance supervisor, felt a growing frustration with general mechanic Jeffrey Thompson. Morton considered Thompson a good mechanic, and this opinion was continually reinforced by the consistently high quality of Thompson's preventive maintenance work, his success in completing difficult repair jobs, and his ability to supervise his inmate work crew.

Morton's frustration arose primarily from Thompson's apparent lack of motivation. Thompson always needed to be told what to move on to after each job was finished. If he were not so instructed, he would take a prolonged coffee break until Morton sought him out and gave him a specific assignment. Sometimes he would even send his inmate crew back to the housing unit in the middle of the day, which created problems for the unit staff.

Morton's frustration peaked one day when a small plumbing problem got out of hand. He knew that Thompson had to have seen the leaking valve because it was beside a pump Thompson and his inmate crew had been rewiring. However, when Morton asked Thompson why he had done nothing about the valve, Thompson said, "Plumbing isn't part of my job."

"You could have at least reported the problem," Morton said.

Thompson shrugged and said, "There's nothing in my job description about reporting anything. I do what I'm paid to do, and I stick to my job description."

"You certainly do," said Morton. "Jeff, you're one of the best mechanics I've ever seen. But you never extend yourself in any way, never reach out and take care of something without being told."

"I'm not paid to reach out or extend myself. You're the boss, and I do what you tell me to do. And I do it right."

"I know you do it right," Morton agreed. "But I also know that you usually stretch out the work, just like I'd expect an inmate to do. I know you're capable of giving a lot more to the job, but for some reason you seem unwilling to work up to your capabilities."

Again Thompson shrugged. "I stick to my job description and I do what I'm told."

Discussion Items

1. Consider the kind of employees Morton supervises and the kind of work Thompson does, and try to apply the scale in Exercise 2-1 to an assessment of Morton's supervision of Thompson (variability, mobility, professionalism, and definability).

2. Must Morton's approach to supervising Thompson be more people centered or production centered? Why?

3. Do you basically agree or disagree with Thompson's literal adherence to the letter of his job description? Why?

4. Put yourself in George Morton's position and consider how you might deal with Thompson. In sets of written steps or guidelines, describe ways in which you might go about trying to get this employee to perform more in line with his capabilities.

ENDNOTES

1. Likert, Rensis. *New Patterns of Management.* New York: McGraw-Hill, 1961.

The Nature of Supervision: Corrections and Everywhere

3

Chapter Objectives

- Present the two-sided role of the supervisor as both "functional specialist" (worker) and "management generalist" (manager).
- Address the issue of a newly appointed manager's background, suggesting that the best worker may not always become a successful supervisor.
- Explore likely reasons for a supervisor's tendency to emphasize one side of the role at the expense of the other, with special concern for the "working trap."
- Introduce the overall responsibilities of correctional management in general.
- Establish the nature of correctional management as a strongly people-oriented process unavoidably concerned with day-to-day problems.

As we are born to work, so others are born to watch over us while we are working.

Oliver Goldsmith

There are two kinds of people in the world, "doers" and "pointers": which would you rather be?

Victor Phillips

■ Born to Work or Watch?

Managers often create problems for themselves by behaving as though Oliver Goldsmith was correct when he made the statement quoted above. Perhaps one can safely assume a bit of cynicism or resignation in Goldsmith's words, but

He Didn't Know His Place

The warden was always in the lieutenant's office. He felt comfortable there. After all, he came from the ranks of custody. He had risen through those ranks to be one of the senior wardens in the system.

But it gave the captain and the lieutenants fits. It just wasn't easy to have the warden sitting there (supposedly going through the daily log) when they were talking about how to run the shift. And it certainly didn't work well for him to attend all the lieutenant's meetings, and even be involved in the roster meetings. He was a good warden. He understood their problems. But he didn't know his place.

certainly those who work for others (as opposed to working for ourselves) are occasionally inclined to agree. It is easy to think that people in an organization fall very neatly into preordained managerial and nonmanagerial categories.

But regardless of some appearances to the contrary, no one is "born" either to work or to oversee the work of others. In fact, the second introductory quote—one that one of the authors heard regularly from his father—is much closer to the truth. While it isn't preordained in any way, most work in the world is indeed broken down into those who do it and those who direct the doing.

Note that by separating working and watching, Goldsmith conveys the impression that overseeing is not to be considered work. And it is unfortunately true that many people nurture the feeling that supervision is not real work. More commonly however, people who hold this belief are not and have never been supervisors. Spend some time managing in the real world and a person finds that management really just constitutes a different level of "doing."

The facts suggest that supervision is an activity for which a person can exhibit a talent. Many people are called on to become supervisors because they have exhibited talent for certain kinds of work—usually the same kinds of work they are asked to supervise. However, talent for doing manual, technical, or professional work is no guarantee of the presence of talent for supervision. Although the supervisory and nonsupervisory work may both be closely related to the same human activity, doing one well does not guarantee that the other will be done equally well. Consider the occupations of cook and kitchen supervisor, for instance. The two are closely related positions, but it is not necessarily true that the talented cook will automatically be a talented kitchen supervisor.

■ The Supervisor's Two Hats

Most first-line supervisors in correctional agencies function in the dual capacities of worker and supervisor. They are constantly required to fill two roles, which are characterized as the functional specialist and the management generalist.

The functional specialist is the worker who is responsible for doing some of the basic work of the department. The correctional officer, the records technician, the parole officer, the telephone switchboard operator, the secretary, the maintenance mechanic, the accountant, and many others fall into this category. They are required to perform hands-on tasks involved in actual inmate supervision or supporting the organization's security or program operations. Almost everyone in the organization is or has been a functional specialist of some kind. In any given department, the specialist is ordinarily concerned with some function that is unique or very nearly unique to that department.

The management generalist, on the other hand, is concerned with activities that are common to many departments and to most situations in which someone must guide and direct the work of others. It matters little whether the manager began his or her career as a correctional officer, medical technician, accountant, maintenance mechanic, or parole agent. Running a department requires that the manager be concerned with staffing, scheduling, personnel management, budgeting, and other activities that apply to many departments.

The nonsupervisory employee is a pure functional specialist. (This discussion does not address the need in a correctional setting for line employees to supervise inmates, but many of the strategies for first-line employee supervision also are effective with inmate workers.) Only in some instances, however, is the manager a pure management generalist. Certainly there are few if any management generalists among first-line supervisors. Generally, the smaller the organization or department, the more likely is the supervisor to be both worker and manager. The supervisor of a four-person maintenance department will probably be a maintenance mechanic and jack-of-several-trades as well. The manager of a three- or four-person inmate records department will spend considerable time on nonmanagerial tasks. The unit manager in a 30-cell special housing unit may spend more time on hands-on inmate issues than performing classic managerial duties. Only in the upper levels of management is one likely to find a few true generalists. A deputy warden is unlikely to dictate parole planning documents, compute sentences, or search cells.

The first-line supervisor, then, is both worker and manager, and needs to recognize this fact. However, this recognition alone is not enough, for sometimes wearing these "two hats" is fully as difficult as wearing two real hats at the same time. That is because the nature of management itself creates a natural leaning toward one of the supervisor's two roles. Management is not nearly as well-defined in its own right as are many working specialties. Specialties such as plumbing, accounting, nursing, or social work, can be categorized rather neatly according to certain characteristics. Each is associated with an expected amount of education and training and perhaps the awarding of a diploma, degree, or license.

For instance, a psychologist may be fairly well-defined as someone who has successfully completed a certain number of years of higher education. This person would have received a diploma or degree, perhaps completed a clinical internship, and then passed a licensing examination. Especially among the professions requiring higher education, specialties are well-defined in this manner. Even in occupations requiring little or no college-level work, one can find a considerable degree of definition through simple instruction, on-the-job training, and experience.

Management as a separate field, however, defies the kind of definition just described. A few people trained as so-called management generalists perhaps can be identified in the middle and upper levels. But in management's lower levels there are no sound criteria for defining "manager" without instantly raising the question: "Manager of what?"

Thus management is not automatically thought of as a legitimate and separate field in its own right, requiring a certain body of specific skills, education, and training. People often are not willing to accept management as a profession since it is not readily definable and is not restricted by specific qualifications. Since one requires no particular background to be a manager, many people are left with the feeling that "anyone can do it."

Clearly, to be a manager one must manage something. This leads to consideration of the individual's functional specialty. It also leads to another consideration: Who is best qualified to supervise the work of the department? Should this be the role of the functional specialist or the management generalist? Is it more important for the supervisor to be knowledgeable of all aspects of the department's work, or should the supervisor's strengths lie in the general functions of management?

After brief consideration, one might reasonably conclude that the first-line supervisor should be proficient in both management and the functional specialty. Indeed, it is usually the individual who becomes well-rounded in both areas of activity who makes the more effective supervisor.

■ The Peter Principle Revisited

Almost all first-line supervisors have worked as functional specialists. Although it is true that some people begin their working careers as supervisors, this usually occurs in natural "two hat" situations in which the person enters a small department as both supervisor and worker. Usually, however, a person is offered a supervisory position because of past performance in some specialty. Ordinarily it is the better workers who become supervisors. However, the fact that a person is a good worker does not guarantee that this same person will be a good supervisor. It is precisely this dilemma that Laurence J. Peter was concerned with in *The Peter Principle*.[1] Peter's tongue-in-cheek but nevertheless serious commentary contains a great deal of truth. In brief, his "principle" states that "in a hierarchy, every person tends to rise to his level of incompetence." Recognizing that it is the good worker who is singled out for promotion, Peter reasons that the outstanding worker at any level is likely to be promoted to the next level in the hierarchy. This process may continue until the individual reaches a level where performance is mediocre at best. Here all promotions stop and the person is left, perhaps until retirement, a notch above the level of proven capability for good work.

The best workers do not necessarily become the best supervisors. Although promoting outstanding employees will continue to make more sense than promoting mediocre producers, this practice will never guarantee the presence of effective supervisors. A great deal of the problem lies in the individual and organizational attitude toward management being an "anyone can do it" job. Confident individuals step into supervisory positions assuming that all they need to do is begin giving instructions and commands. Likewise, higher managers promote workers and then go about their business as though believing they have created supervisors by simply conferring titles.

The supervisor's hedge against arriving at a level of incompetence is management development. This can be pursued within or outside the organization, through targeted programs or by individual effort, formally or informally. Regardless of how it is done, the supervisor must learn a lot about the second hat before it fits as well as the first hat.

Most people enter their functional specialties sufficiently well-trained to do the jobs for which they have been hired. However, most workers who become supervisors do so with little or no preparation. Management then becomes a sink-or-swim proposition. A few catch on and perform remarkably well after a short period of time. Many generate enough motion to enable them to stay afloat but then continue spending most of their energy simply keeping their heads above water. Some sink quickly, to their detriment (and often to that of the organization as well).

The player-to-coach transition in the world of professional sports offers an appropriate analogy to the worker-to-supervisor transition. Most successful coaches gained experience as players at some level in the sports they coach. However, only a few outstanding coaches were star performers in their playing days; most were infrequently noticed, solid, dependable team players. Knowing how to play the game well is not enough. The differences between working and supervising are as fundamental as the differences between doing and teaching and as great as the essential difference between following and leading.

Dan Reeves, while head coach of the Denver Broncos, is reported to have said, "I was an assistant coach for many years and made a lot of suggestions. You don't realize until you're a head coach that you have to make decisions, not suggestions." Drawing on a nonsporting figure for a further example, Pope John XXIII once said, "It often happens that I wake at night and begin to think about a serious problem and decide I must tell the Pope about it. Then I wake up completely and remember that I am the Pope."

Reeves' and the Pope's careers, of course, are not an illustration of the Peter Principle—quite the opposite. But these statements suggest yet another part of the problem is making the fundamental shift to full responsibility for a department or function. Not everyone is capable of doing that effectively.

■ The Working Trap

The working trap poses a hazard to every supervisor whose job includes the performance of both managerial and nonmanagerial duties. A great deal of the reason why any person is titled and paid as a manager is to see that a certain amount of work gets done through the efforts of other people. This requires that the supervisor primarily perform management tasks.

It is all too easy for the supervisor to feel and behave more as a worker than a manager. This is understandable when half or more of the supervisor's duties may be nonmanagerial to begin with. However, this leads to a state of favoring one hat over the other—spending a great deal of time thinking and doing as a worker rather than behaving as a manager.

Falling into the working trap can leave a supervisor's time and energy stretched too thinly over numerous technical tasks. While this is happening, some of the department's pure workers remain underutilized. The exercise at the end of this chapter illustrates this point very well.

■ Nothing to Do?

The following nine points were excerpted from a list of "supervisory activities" that a noncorrectional manager shared with an instructor and class in a management development program. Their truth and applicability in the correctional environment are poignantly clear.

As everyone knows, the supervisor has practically nothing to do except:

1. Decide what is to be done and assign the task to someone.
2. Listen to all the reasons why it should not be done, why it should be done differently, or why someone else should do it.
3. Follow up to find out whether it has been done, and discover it has not been done.
4. Listen to excuses from the person who should have done it.
5. Follow up again to determine if it has been done, only to discover it has been done incorrectly.
6. Point out how it should have been done and prepare to try again.
7. Wonder whether it may be time to get rid of a person who cannot do a job correctly; reflect that the employee probably has family responsibilities and that a successor would probably behave the same way anyway.
8. Consider how much simpler it would have been to do the job oneself in the first place.
9. Sadly reflect that it could have been done correctly in 20 minutes, but as things turned out it was necessary to spend two days to find out why it took three weeks for someone else to do it wrong.

The foregoing is more than just a tongue-in-cheek recounting of some of the frustrations of supervision. Implicit in the nine points are a number of reality-based considerations that often come into play as part of the overall supervisory task.

Point two suggests that the assignment of work to an employee is more than simply pointing a person at a task and giving the orders. Proper delegation, discussed in Chapter 6, includes thoughtful matching of person and task, thorough instruction, and assurance that the employee understands why the job must be done.

Although follow-up is an extremely important part of all supervisory activity, point three might make one wonder how timely the follow-up was in the situation described, since late follow-up can be as bad as none at all. Point five, again dealing with follow-up, might be a prompt to ask if corrective action has been taken and thorough instructions provided, or if once again the employee was simply "told to do it."

Point six states to "point out how it should have been done." If this is the first time instructions were offered or efforts were made to find out if the job was understood, then the supervisor has far greater problems than even these few voiced frustrations suggest.

The musings of point seven are likely to be uncalled for at this stage. The supervisor must do considerable self-assessment before writing off any employee as incapable. Also, if honest evaluation does lead to consideration of firing, the rationalization that a successor would "probably behave the same way" is completely without foundation.

Point eight simply illustrates the thought processes that allow the supervisor to fall into the working trap, and the "sad reflection" of point nine does not carry far enough—it should perhaps go on to include: "Why, then, can't I get an employee to do a simple job like this within a reasonable time?" Otherwise, it suggests that the supervisor has compounded some personal errors and attempted to rationalize them away by tagging the employee with the failure.

■ The Responsibilities of Correctional Management

Consider "management" as a single composite person responsible for running an institution or agency. This "person" has several major responsibilities.

First, he or she (whether in a community or confinement setting) is responsible to the public in terms of safety and cost-effective use of government resources. This includes an important responsibility to strive continually to upgrade the capabilities of the correctional organization. Doing this successfully requires the manager to adapt as necessary to the constantly changing fiscal and public environment in which corrections professionals operate.

Second, because of the undeniably unique issues relating to working with criminal offenders in any setting, correctional managers are responsible for the personal safety and welfare of their employees. Related to that is the responsibility for recognizing employees' reasonable needs for a sense of accomplishment, fair treatment, and fair compensation for their efforts.

Finally, correctional managers are responsible to the offenders involved in the institution's or program's operation. This can include providing them with safe, humane living conditions, as well as providing programs and services that could improve their ability to function lawfully upon release (if they choose to do so).

For correctional agencies, fulfillment of the foregoing three broad-scope responsibilities usually is couched in terms of a common, overarching objective—the preservation or restoration of public safety. However, there is likely to be conflict in achieving that objective.

To appreciate the implications of such conflict—felt most severely at management's highest level—imagine doing a job while answering to four or five different supervisors. That is the situation that many, if not most, upper-level correctional managers find themselves in. They are caught in conflict situations involving the needs of the public, political superiors, unions, individual employees, and inmates. Resolving those competing demands within the limits of available resources is the day-to-day reality of today's corrections and is perhaps the single greatest challenge upper-level managers in the field face.

■ The Nature of Supervision

Chapter 2 discussed how patterns of supervision may differ from one organization to another according to various characteristics, and pointed out that such differences may occur even among departments in the same organization. Generally, supervisory styles in corrections will be more dependent on people-centered attitudes than on production-centered attitudes. Paradoxically, they also may be far more regimented than many other work settings.

For example, while predictable on a broad-scale basis, the tasks involved in managing a correctional component are more variable than repetitive at the detail level. While managers can tabulate the raw numbers of inmates coming in and out of an institution, units of input and output to and from the institutions' other systems are difficult to define.

Although a manufacturing supervisor has also had a need to focus strongly on people, the repetitive-task environment may favor a concentration on processes and techniques. Conversely, the necessary orientation for the correctional manager is more toward strength in interpersonal skills.

Considering that "the end product of all business is people," corrections must consider who will be the ultimate "customer" for its services. For a mass-production manufacturing enterprise, the customer is likely to be remote, unseen, and unknown. This often is true even though there may be health and safety implications of the product or service. In corrections, however, there is a direct "customer" in the form of the offender group the organization is working with. The service is hands-on and personal, and as such is of extremely immediate importance to the inmate. And as anyone who has already worked in the correctional environment knows, inmates can be demanding, volatile, and at times dangerous. This means that, in corrections, then, much more so than in most other endeavors, quality considerations have an immediate impact.

Successful supervision comes only through conscientious effort. A considerable degree of dedication to the job is necessary, but no one should feel it necessary to become a workaholic. The person who gives everything to the job to the exclusion of all else is most likely using the job as an excuse to fill other needs. Rather, the effective correctional manager is a person who has a reasonable liking for the work, who has a sincere interest in delivering quality services to the offender population involved, and who can bring to the job the ability to evaluate correctional operations in terms of common sense, decency, and humane treatment.

At the end of a management development class in which numerous techniques were discussed, a supervisor said, "I could really get a lot of good work done around here if it weren't for all the problems that pop up every day." When feeling that kind of frustration, consider this: the problems—those nagging, unanticipated, annoying difficulties that seem to spring up day after day—are a large part of the reason for a managerial job's existence. If there were fewer problems, fewer supervisors would be needed.

To a considerable extent the supervisor is a frustration fighter. If the frustrations did not exist, necessary tasks might well be accomplished without supervisory intervention. The day-to-day problems do exist, and hour-to-hour and moment-to-moment operating decisions have to be made. And the person who must make most of these decisions is the first-line supervisor—the final link between the best intentions of the organization and the actual performance of patient care.

EXERCISES

Exercise 3-1: Wearing Two Hats

Divide a sheet of paper into two full-length columns. Label one column "Technical," and label the other column "Managerial."

In the "Technical" column list the tasks you perform either regularly or occasionally in your capacity as a functional specialist. These should be the things you do not do primarily because you are a supervisor but because you are a specialist in performing certain kinds of tasks. For instance, you may be a lieutenant responsible for the activities of sergeants and correctional officers but you still work directly with inmates, or you may be the supervisor of a maintenance crew but you still replace a malfunctioning light switch when you encounter one.

In the "Managerial" column list the tasks you perform as a management generalist. In these tasks you are applying techniques that cut across functional lines (although you are doing them specific to the function you supervise). Such tasks might include departmental budgeting, doing performance appraisals, interviewing prospective employees, and untangling people problems.

You may find it helpful to develop your lists over a period of several days, as you are confronted with a variety of problems and tasks, rather than generating them at a single sitting.

Questions

1. What do your lists tell you about the "two hat" nature of your job?
2. Can you make reasonable estimates as to how much of your time is spent as a working specialist? As a manager?
3. Are there obvious imbalances in the two lists?
4. Do these lists present any clues to functional imbalances in the way the work unit is organized or your supervision tasks are carried out?

Suggestion for Additional Activity

Compare your lists with those of several other supervisors in a group discussion setting, and develop a composite list that is generally descriptive of the roles of the supervisor. Note that the two sides of the supervisor's job represent two distinctly different roles, either of which must be consciously assumed when the occasion demands.

Exercise 3-2: "If You Want Something Done Right . . ."

Samuel Tiggins, chief probation officer in a large urban area, dreaded the one day each month he had to spend doing the statistical report for his department. Tiggins was responsible for thousands of criminal offenders under supervision in the region, as well as the preparation of large numbers of reports to the courts. At one time the report had been relatively simple, but the demands of the courts and budget justification requirements increased the complexity and detail required in this report. To adjust to these changes, Tiggins had simply modified his method of preparing the report each time a new requirement was placed upon him, so there was no written procedure for the report's preparation.

Faced once again with the time-consuming report (and confronted, as usual, with several problems demanding his immediate attention), Sam decided it was time to delegate the preparation of the report to his deputy chief, Angie Clark. He called Clark to his office, gave her a copy of the previous month's report and a set of forms, and said, "I'm sure you've seen this. I want you to take care of it from now on. I've been doing it for a long time, but it's getting to be a real pain and I've got more important things to do than to allow myself to be tied up with routine clerical work."

Clark spent perhaps a half minute skimming the report before she said, "I'm sure I can do it if I start on the right foot. How about walking me through it—doing just this one with me so I can get the hang of it?"

Tiggins said, "Look, my objective in giving you this is to save me some time. If I have to hold your hand, I may as well do it myself." He grinned as he added, "Besides, if I can do it, then anyone with half a brain ought to be able to do it."

Without further comment Clark left the office with the report and the forms. Tiggins went to work on other matters.

Later that day Clark stopped Tiggins in the hallway and said, "Sam, I'm glad I caught you. I've got three or four questions about how to put together that report, mostly concerning how you come up with the estimates on court time per offender." She started to pull a folded sheet of paper from her back pocket.

Tiggins barely slowed. "Sorry, Angie, but I can't take the time. I'm on my way to a meeting." As he hurried past Clark, he called back over his shoulder, "You'll just have to puzzle it out for yourself. After all, I had to do the same thing."

The following day when the report was due, Tiggins found Clark's work on his desk when he returned from lunch. He flipped through it to assure himself that all the blanks had been filled in, then scrawled his signature in the usual place. However, something caught his eye—a number which appeared to be far out of line with anything he had encountered in previous reports. He took out two earlier reports and began a line-by-line comparison. He quickly discovered that Clark had made a crucial error near the beginning and carried it through successive calculations.

Tiggins was angry with Clark. The day was more than half gone and he would have to drop everything else and spend the rest of the afternoon reworking the figures so the report could be submitted on time.

Tiggins was still working at 4:30 P.M. when Mark Signero, the county sheriff, appeared in the doorway and said, "I thought we were going to get together this afternoon and talk about how to speed up getting those violation reports filed faster. What are you up to, anyway?"

Tiggins threw down his pencil and snapped, "I'm proving an old saying."

"Meaning what?"

"Meaning, if you want something done right, do it yourself."

Questions

1. In what ways is Sam Tiggins attempting to function as both a worker and a manager?

2. Identify at least one primary activity pertaining to each of Tiggins' "two hats."

3. How (if at all) do you believe Tiggins is emphasizing one side of his role at the expense of the other?

4. What is the apparent "working trap" in Tiggins' case, and why does it appear to you that he has become caught in that trap?

ENDNOTES

1. Peter, Laurence J. *The Peter Principle.* New York: William Morrow & Company, 1969, p. 26.

Definitions, Titles, and Other Intangibles

4

Chapter Objectives

- Provide a working definition of management and establish management as the generic term descriptive of all persons who run organizations or organizational units.
- Define management in general as *getting things done through people*.
- Relate supervisor and manager to each other and to other labels applied to managerial positions, and identify the need to differentiate generic labels and organizational titles.
- Clearly identify the supervisor as a manager (regardless of conflicting titles within the organization).
- Differentiate between management as a generic term and the various labels used to identify managers in organizations.
- Define "line" and "staff" and differentiate between these two essential kinds of organizational activities.

Treat people as adults. Treat them as partners; treat them with dignity; treat them with respect. Treat people—not capital spending and automation—as the primary source of productivity gains. These are fundamental lessons from the excellent companies' research.

Tom Peters

Those he commands move only in command,
Nothing in love; now does he feel his title
Hang loose about him, like a giant's robe
Upon a dwarfish thief.

William Shakespeare, *MacBeth*, V, ii, 19

Distinct Sets of Duties—A Plethora of Job Titles

Warden. Superintendent. Chief Administrative Officer. Executive Director.

Colonel. Chief Correctional Supervisor. Major. Chief of Security. Captain. Chief, Security Operations.

Case Management Supervisor. Chief of Case Management. Chief, Social Services. Supervisory Case Manager. Social Services Supervisor. Case Management Coordinator.

Same functions, different names—take your pick.

■ In Search of Definitions

Managing people, in a correctional setting or not, is a complex activity. Learning how to "do" management has to begin with knowing what it is. And so the starting point is defining the concept, and the dictionary is as good a place as any to begin. In the *New Shorter Oxford English Dictionary,* management is defined as:

"The act of managing; the manner of managing; the application of skill or care in the manipulation, use, treatment, or control of things or people, or in the conduct of an enterprise, occupation, etc.; the administration of (a group within) an organization or commercial enterprise."[1]

A variety of other sources list the following as synonyms for management: treatment, conduct, administration, government, superintendence, and control.

Management often is "defined" in terms of its root word, *manage.* The word manage comes from the Latin *manus,* meaning hand. This might seem to lead down the right track, since this origin suggests the use of this part of the body in working or doing. However, it doesn't take long to discover that the original English definition of manage is to train a horse in its paces; to cause to do the exercises of the "manege" (defined as the paces and exercises of a trained horse).

It seems, then, that the word "manage" developed from the description of a specific kind of work. However, among the many other definitions of *manage* are the following: to control or guide; to have charge of; to direct; to administer; to succeed in accomplishing; to bring about by contriving; and to get a person to do what one wishes, especially by skill, tact, or flattery.

Some of the words that might be used to describe "manage" have appeared throughout the dictionary definitions, and except for an oddity or two have provided little new information about manage or management—neither the words nor the concept. Unfortunately, it does little good to look up "manager," since that yields the information that a manager is one who conducts, directs, or manages something. However, the list of synonyms for manager is of interest because it includes director, leader, overseer, boss, and supervisor.

This discussion illustrates that at least in some uses the term *supervisor* is the same as manager. The same dictionary defines supervisor as a person who supervises, a superintendent; a manager; a director. Based on proper use of the English language, then, one can say that supervisor and manager are equal in definition: a manager is a supervisor and a supervisor is a manager.

On a working basis, these terms will be used interchangeably throughout this book. However, the idea of supervisor and manager being equal may not be agreeable to everyone. Everyone has ideas of what a supervisor is and what a manager is, and those conceptions are not necessarily the same. Understanding what these terms mean—the positioning in the organization of persons who may be called managers, supervisors, directors, administrators, or whatever—is largely determined by the use of these words and titles.

It is important to realize that the differences in what these terms mean to each person are not absolute. Rather, managers and their organizations have artificially created these differences in meaning. The chief correctional supervisor responsible for hundreds of correctional officers may officially be called the Captain, and a case "manager" may supervise only a secretary. A records supervisor in a large probation office, when asked to consider enrolling in a management training program, said, "No. I'm not a manager, I'm just a supervisor." Differences in how supervisors and their titles are seen in relation to people who run other organizational units interfere with a complete understanding of what is and is not "management."

■ A Practical Definition

Throughout this book, *management* will be used to mean the effective use of resources to accomplish the goals of the organization. In simpler terms, management can be described as getting things done through people. Regardless of title, as long as a manager is responsible for getting work done at least in part by directing the activities of others, he or she is a manager. This applies to the working head of a three-member maintenance crew or the working supervisor of a four-person inmate records department, as well as to the administrator of a small correctional camp, the chief executive officer of a major penitentiary, or even the director of an entire correctional agency. These bottom and top levels both constitute management, just as the people directing the efforts of others at numerous intervening levels also belong to management.

Throughout this book, management will be spoken of in the broadest generic sense, referring to the processes applied and not to particular job titles. In this context, everyone who directs the activities of others is a manager. By that definition, every employee who supervises inmates—from the single inmate janitor in a business office or housing unit to hundreds of inmate workers in a correctional factory—is a manager. And while this book is not primarily directed toward inmate supervision issues, many of its concepts apply equally well to offenders in a variety of supervised settings.

■ Organizational Labels

It is easy to guess how so many different organizational labels developed for managers. It was most likely a matter of organizational convenience, initially adopted to differentiate between managers at different levels or in different roles. It could be quite confusing if all three levels in a particular correctional agency's business office carried the title of manager. Rather, it makes considerably more sense to identify them as, for instance, controller, business office manager, and accounts receivable supervisor.

The use of manager and its synonyms as position titles did not develop uniformly in all organizations. It is more likely to find, for instance, that the inmate records departments of four different correctional facilities in different states are run by a records manager, records supervisor, chief of inmate records management services, and records officer, respectively. There is little overall comparability of titles from one organization to another. In one institution, a "supervisor" may be a low person on the managerial totem pole and in another may be in the middle or upper part of the hierarchy. Certainly the term *manager* is most sensitive to this effect in its use as a title, and it may apply anywhere in the organizational pyramid in almost any institution.

Line Supervisor

Despite this variation in usage, it is probably fair to say that when "line supervisor" is used to describe a working position, one usually imagines a position in the lower part of the management structure. For this reason the term line supervisor is often used to describe the lowest level of management in the organization—the lowest level at which persons manage the work of other persons. Occasionally, the term second-line supervisor will be used, meaning the second level up—the "supervisor of the supervisor." These title conventions refer to staff supervising staff, although in a confinement setting employees in any of these positions could also be first-line supervisors of inmate workers.

Middle and Upper Management

As the label "upper management" suggests, this is the person or people at or near the top of the organization who are responsible for the entire agency or a major correctional facility. Between top management and line supervisors there are (depending on the size of the organization) a varying number and configuration of positions generally referred to as "middle management."

Variations

Organizational size may render a large part of the "middle management" discussion irrelevant. In some organizations—a small correctional camp or halfway house, for instance—it is likely that the first line of supervision is the only line. For instance, in a small facility the person responsible for maintaining inmate records (and who may be the only employee performing that function) could report directly to the facility administrator, so there is no middle management between the top and bottom levels. In a small probation or parole office, the "chief" may directly supervise all line officers.

Line and Staff Functions

In addition to this managerial tier, it is also useful to differentiate as to whether the functions of a particular manager are "line" or "staff." A line function is one that advances the accomplishment of the actual work of the organization. A staff function provides support to the organization in some ancillary fashion.

For instance, in a correctional facility the departments responsible for security, food service, laundry, educational services, and medical care are among those considered as "line" activities. They actually perform the work of the institution. The personnel department and the business office are two examples of "staff" activities. These employees provide support services that enable the line departments to function, but do not do any of the actual line work themselves. The essential difference between line and staff activities is the difference between doing and supporting.

Relating line and staff to managerial titles, a person can be described as a manager of either a line or staff activity. However, whether the overall function of the department is line or staff, the manager within the individual department possesses line authority in the management of the department's employees. Within each function there is a line of authority that extends from the department head down to and including the first-line supervisor.

Look, for instance, at the custodial, or security, department of a correctional institution. The line of authority (viewed from the top down) may be: chief of security, captain, lieutenant, sergeant, and correctional officer. Ordinarily, each person at each level directs the activities of those at the next lower level, relaying instructions down through the entire line of authority. Hence, instructions from the chief of security ultimately result in actions by correctional officers.

However, in large departments, some staff may actually perform separate support functions within their own department. Within a security department, for example, a specific lieutenant may be identified as a special investigative officer, to handle internal investigations against staff or inmates. He or she would report to the chief of security, and even if a clerk or subordinate investigators were assigned to support this function, the investigative supervisor serves in a staff, rather than line, role in relation to the security department.

The workforce mix of every correctional agency or facility necessarily consists of both line and staff employees. Often there is confusion about the degree of authority persons in staff positions are to exercise. Problems sometimes arise from the three-way relationship among a line employee, a staff employee, and the staff employee's line manager.

A great deal of the potential for this problem may be inherent in the way the staff person projects him- or herself into interactions with line employees. The staff employee may appear to be making decisions and following up on them, allocating certain kinds of resources, and even conveying instructions and direction to others. The employee occupied in a pure staff function often appears to be the holder and exerciser of all management prerogatives except the critical one that essentially defines a manager—the authority to direct other people. One might even say that an effective staff person often looks, sounds, and acts like a manager. This frequently causes problems for some line personnel because it creates the impression that a person lacking proper authority is intruding into another's territory.

Sometimes this misapprehension of authority is because of the place the staff position is in the organization, such as an "executive assistant" to a highly placed administrator. Persons in this position have no line authority, but serve in a staff role to their supervisor—the warden, for example. However, to a certain extent the executive assistant carries an aura of authority based on that of the supervisor's position. Line staff may react to the executive assistant's comments or input without regard to their own supervisor. The potential problems of this kind of situation are many.

Often line managers do not know how to make fully effective use of the staff assistance available to them. Some line managers tend to view staff people as regulators or intruders rather than use them as the advisors and helpers they really are. Also, some managers behave as though they believe a request for staff assistance (or even an agreement to accept staff assistance when offered) constitutes a weakness or an admission of inadequacy. In short, the line manager who does not completely understand the role and function of staff personnel often tries to go it alone, attempting to be all things in all situations, operating without the available staff assistance.

As suggested earlier, the functional difference between line and staff is the essential difference between doing and supporting. Looking at the concept from a departmental level, line personnel do, and staff personnel support. But as far as managers themselves are concerned, they are all "line" with respect to the operation of their own departments and in the direction of their own employees. Their departmental activity may be clearly definable as staff functions (like accounting and personnel/human resources [the latter two terms will be used interchangeably throughout this book]). But as managers they are by definition line personnel when operating as managers within their own departmental chains of command.

Many correctional organizations are organized along functional lines, giving rise to another way of grouping activities for organizational purposes. Correctional operations at each level often are structured along a three-way division of functions:

> Custodial—perimeter security, yard patrols, housing unit supervision
> Program—case management, counseling, religious programs, and education
> Operations—food service, laundry, business office, payroll, general accounting, maintenance

The first two of these would be generally understood to have line functions, while the third is a mixture of line and staff. However, within each of these three areas, the departments are organized along traditional lines of authority. A business manager supervising accounting and clerical staff has line responsibilities within a department that is itself a staff function.

As a variant on this model, some facilities organize custodial and program staff into what is commonly called "unit management." Under this system, multidisciplinary employees supervised by a unit manager oversee the operation of one or more housing units. Correctional officers, case managers, education and mental health personnel have varying levels of responsibility to the unit manager, as well as to a supervisor in their specialty discipline.

This more decentralized method of delivering services to inmates is considered by many correctional systems to offer considerable communication, supervision, and program delivery advantages. This is because typically all staff in these job categories assigned to the unit have their offices in the housing area and can be more effective in dealing with a discrete group of inmates.

■ A Title as More Than a Label

Up to this point this has been a discussion of managerial position titles and the various uses of such titles to differentiate levels of responsibility. However, it is also necessary to recognize that titles also represent "status points" or forms of "psychic income." In years past most prisons were run by top managers known as wardens. Now, however, if one looks through a facility directory such as that published by the American Correctional Association, one finds many chief executive officers, executive directors, superintendents, and similar titles in addition to wardens. The functions and responsibilities of a position may have changed little if at all between the days of warden and superintendent, but the latter title may be more impressive or seem more professional to a larger number of people than the former title. It may seem irrational, but differences in title do matter in terms of how some people see themselves and how other persons view the positions of the title holders.

Title differences are significant only to the extent that they may affect the person's view of their position. Avoid falling victim to the attitude of the person who said, "I'm no manager, I'm just a supervisor." Supervisors are managers, and it is important that they see themselves that way, and clearly identify with the collective body known as management.

EXERCISES

Exercise 4-1: Labels and Titles

Identify and list all the managerial titles used within the entire correctional agency and within the function or department in which you work (for instance, security, personnel, inmate records). If not employed in corrections, use the personnel structure of another agency or organization with which you are familiar.

Questions

1. Which titles do you find most descriptive of the positions to which they apply? Why?
2. What is the line of authority (by position title, from highest to lowest) within your function or department?
3. Is your overall function a line activity or a staff activity? What makes the difference?
4. Are there dualities in function (staff/line) that you can identify?

Exercise 4-2: Balancing the Functions

Janet Morgan was associate warden at a large, medium-security correctional institution. She was an outstanding example of someone who had come up through the ranks as a line correctional officer, sergeant, lieutenant, captain, and chief of security. As her career progressed, she worked her way through college, earning a degree in public administration with a minor in criminal justice.

Because of her broad knowledge gained through experience and academic study, she was understandably seen as a key resource person for assuring that the security functions of the institution were in order. Recently, the chief of security, Bill Hallaran, had relied on her to oversee some of the detailed technical aspects of the institution's security program. As this came to be a more routine event, she found herself resisting these intrusions on her wider management role. Eventually, it became clear that her direct involvement in the security program had created a problem.

One day the warden asked, "What's wrong between you and Bill Hallaran? He says you're not giving him the support he needs on revamping the tool control system. He's also upset because he feels the roster management committee has just about fallen apart because you wouldn't take the chair and see that things got done. Is your work piling up to where you've got too much to do?"

Janet shook her head. "No, my own workload is under control. I know Bill has needed some help recently because of the hiring freeze and the fact he hasn't been able to fill those three vacant lieutenant slots. Some of the things he does are hands-on tasks that are pretty technical and detailed, and I always could do that kind of thing easily. Others, like chairing the committee, I can do easily, because everyone defers to me as associate warden, but that means the committee isn't functioning as the independent body it should be. Actually, though, I think the problem really is that I seem to be in a dual role that I'm not comfortable with, and he's become dependent on me to help run his department."

"If that's all there is to the problem, won't it sort itself out when the hiring freeze is lifted and we can promote some people?"

"Yes and no. There's another level to this. Things have reached the point where I don't really know if I'm an administrator, or if I've slipped back to almost being a department head, or neither, or both for that matter. I'm sure I can run that department—you know that very well. But it doesn't—well at least it shouldn't—take a manager at my level to see that tool and key controls are running like they should. I shouldn't be the one to see that we're getting the most out of staff on the roster. Those things are technical details that are Bill's job, and I can't afford to let my other departments go in order to spend half my time running his. I know that, but I also can see that I'm probably doing even more of Bill's job than I really need to be doing."

Janet continued, "The real problem is that I've always believed that the basic job of a manager was to get things done through people. I've tried to practice that ever since I became a manager. But I slipped back into doing the technical side of Bill's job so easily that I'm worried that maybe taking an associate warden's job was a mistake. I guess I really don't know if I'm supposed to be in upper management or in a more hands-on position that has a technical side."

(This example represents a common problem for correctional managers as they move up the organizational ladder. They know their former discipline (security in this case) very well. If a lower level manager in that area needs help or is not performing exactly as the supervisor would have in the same situation, then the temptation is great to take over the function. This restores the upper manager's comfort level about that department, but it takes its toll elsewhere in the organization.)

Questions

1. How is Janet functioning at two organizational levels in this situation?

2. Do you agree that Janet's performance of technical work as described could be a sign that "taking an associate warden's job was a mistake," as she fears? Explain your answer.

3. Describe one set of circumstances under which Janet's involvement in the technical tasks described would be fully appropriate.

ENDNOTES

1. *The New Shorter Oxford English Dictionary.* Oxford: Oxford University Press, 1993.

The Basic Management Functions

5

Chapter Objectives

- Introduce and define the basic essential management functions: planning, organizing, directing, coordinating, and controlling, and briefly discuss alternative approaches to describing these functions.
- Establish the importance of knowledge of each of the basic functions in supervisory practice.
- Describe the relative influence of each of the basic management functions on the roles of managers at all organizational levels.
- Address the sometimes-encountered tendency to discount much "planning" as superfluous because much of the time the results achieved vary from the plan.
- Address "organizing" at both organizational and individual department levels.
- Interrelate directing, coordinating, and controlling as elements of a supervisory activity cycle.

You can map out a fight plan or a life plan, but when the action starts, it may not go the way you planned, and you're down to your reflexes—which means your training. That's where your roadwork shows. If you cheated on that in the dark of the morning, well, you're getting found out now under the bright lights.

Joe Frazier, heavyweight boxer

The charismatic leader gains and maintains authority solely by proving his strength in life.

Max Weber

FROM THE INSIDE

The Development of Substance Abuse Programs

The idea of starting up a substance abuse treatment unit in a maximum-security penitentiary was a novelty not too many years ago. Inmates were sent to the "pen" to do hard time. Somewhere along the line, though, a thought took root. As long as you had them in custody, maybe if you dealt with the problems that got them there that they might be better inmates. And if they benefited (if and) when they ever got out, so much the better.

So how did the agency go about starting up a treatment program in an institution that had little or no tradition of providing more than basic services to its "convicts"? Well, a manager was selected for the task, to include both planning for the unit and managing it once it was formed. He drew up a prospectus (so to speak) and got support in the organization for the staff and money involved. The decision about where to implement the program came next, and depended greatly on pledged cooperation from the warden and other top staff. After negotiating with the personnel department, a staffing pattern was finalized, positions advertised, and staff selected. A target date was set for completing training and for program activation. In that training, the manager learned the actual abilities and interests of the chosen staff. The skeletal program described to top management to build the program was tailored accordingly, and the actual program was initiated.

■ Categorizations

There are several kinds of basic activities that all correctional managers pursue in fulfilling their responsibilities—activities that are critical to success in the day-to-day management of many organizations. Shirk these fundamentals and, as Joe Frazier intimates, the manager, subordinate employees, and the organization itself will suffer.

Management literature offers various categorizations that contain four, five, or even more elements or labels. One widely utilized breakdown is found in the work of Theo Haimann,[1] who refers to five basic management functions. For the purposes of this discussion it is useful to use Haimann's key activities: (1) planning, (2) organizing, (3) directing, (4) coordinating, and (5) controlling.

This five-way breakdown has served for years as a reasonable, if somewhat general, description of what managers do.

Other delineations of the management functions to be found in the management literature include planning, organizing, leading, and controlling; planning, organizing, staffing, motivating, and controlling; and other variations. Even as early as 1916, Henri Fayol, the French industrialist and early management theorist, was basing a great deal of his management approach on the simple four-function breakdown of planning, organizing, leading, and controlling.[2]

It is important to appreciate that none of these lists of functions represents someone's belief that a particular listing is the correct delineation of management functions while the others are lacking. Certainly the various lists of management functions are more similar than dissimilar. As evident in the examples just cited, nearly all such lists specifically cite planning, organizing, and controlling, and all such lists begin with planning.

The differences among the lists are simply matters of semantics and how one views some of the elements of management. What is directing in one approach may be leading in another. What is organizing and staffing in one approach may simply be organizing in another. What belongs under both coordinating and controlling in one approach (the one used in this chapter) may all be encompassed by controlling in another.

Why all of these differences? Are there not clearly definable management functions that can be kept separate? The truth is that it is hard to clearly differentiate among a number of separately defined management functions in a manner that covers all circumstances. Management is a broad pursuit made up of many overlapping and interwoven activities. The management process is a continuum and at the same time is a cycle. All of the business of "defining" management functions is simply a convenience that allows the examination of portions of the management cycle in a way that emphasizes certain kinds of activities.

Regardless of the labels applied, however, it is the concepts that are important. It will be helpful to understand management responsibilities to develop an appreciation of the kinds of activity managers pursue for certain purposes. Later, this chapter will explore how the emphasis on certain of these basic functions differs according to position or level in the agency's management structure. Specifically, it will describe how a manager's organizational position determines to a great extent which management functions are likely to, and perhaps should, consume most of the manager's time and effort.

■ Management Functions in Brief

Planning

Planning is the process of determining:

> What should be done
>
> Why it should be done
>
> Where it should best be done
>
> By whom it should be done
>
> When it should be done
>
> How it should be done

The account at the beginning of this chapter takes the start of an actual treatment program in a maximum-security penitentiary through this process. Needless to say, there was a lot more going on than is described, but in actuality, the steps to actuate that unit followed very closely the six-point process immediately above.

Organizing

Organizing is the process of structuring a framework within which things get done, and determining how best to commit available resources to serve the organization's purposes and carry out its plans. Organizing essentially includes what is often referred to as "staffing" in certain other discussions of the management functions.

Directing

Directing is the process of assigning specific resources or focusing certain efforts to accomplish specific tasks as required. Put simply, directing is running an organizational unit on a day-to-day basis. Directing includes a great deal of leading, yet leading is woven throughout most of the other functions as well. Directing may also be considered to include motivating, and all it implies is getting things done through the organization's employees, yet motivating is certainly a consideration throughout the other functions as well.

Coordinating

Coordinating is the process of integrating activities and balancing tasks so that appropriate actions take place within the proper physical and temporal relationships. Coordinating does not appear by name in a number of other delineations of the management functions, yet in all cases it is directly implied in descriptions of the tasks managers perform.

Controlling

Controlling is the process of follow-up and correction, looking at what actually happened, and making adjustments to encourage outcomes to conform with expected or required results. Controlling best illustrates the cyclic nature of management and the inseparability of the basic management functions. By its very nature controlling requires directing, coordinating, organizing, and revising plans as they unfold in the real world of the workplace.

■ Planning

Planning takes place any time someone looks ahead at what they might be doing sometime in the future. The "future" may be months or years ahead or it may be only minutes away. Whenever there is an attempt to look ahead and predetermine a possible action for a time that has not yet arrived, planning is taking place.

Planning can be high level and far reaching, as when the local institution's administration and agency headquarters staff develop a long-range plan calling for growth and expansion, or other major changes. However, a great deal of planning as it concerns most working managers is short term, and oriented toward near-future applications.

The development of a 5-year plan for a correctional facility or agency is an example of planning, as is the development of a single department's one-year budget. Likewise, spending half a day developing the work schedule for a department's employees for the coming month entails active planning. Even if it's a simple pause at the end of the day to order thoughts, sort out the notes on the desk, and jot down a list of items to take care of in the morning—this is planning.

Of course, the future (even when predicted minutes before the fact) does not always come to pass as envisioned. The military has a maxim that states, "No battle plan survives first contact with the enemy." Planning for many seemingly routine activities follows that pattern, particularly in a correctional environment. Moreover, generally, the further into the future one is projecting, the less accurate planning is likely to be in any case. It stands to reason that full knowledge of the future will not be attained until the future becomes the present, so the process of looking ahead is always being done with less than perfect information.

The imperfect nature of planning suggests that plans should be flexible, intended to be changed and updated as the time to which they apply comes closer. The correctional environment is rife with a wide variety of difficult and sometimes dangerous diversions from planned courses of action. Anyone who has done any roster management, for instance, will appreciate the necessity to revise the roster throughout the period to which it applies. Anyone who has been through an inmate disturbance (whether it be riot, food strike, work stoppage, or others) will appreciate how every planned activity is forgotten in the need to respond to not only the immediate event but also in the aftermath.

Although more will be said later about how much planning a supervisor's job might involve, at this point it seems reasonable to keep planning modest, in terms of how much is done and how long it takes. Planning is essential to effective managerial performance, but it is possible to fall into the habit of "overplanning." Indeed, some people spend so much time planning that they rarely have time to actually do anything.

Although a great deal of supervisory planning need not be formal or time consuming, it nevertheless pays to be sufficiently thorough and organized to commit plans to writing. Often the simple act of putting thoughts on paper will serve to crystallize ideas and help decide on the essentials.

It is important to remember that the plan is not the objective. Everyone knows what happens to "the best laid plans of mice and men." Plans (especially those that look more than a few days into the future) seem rarely to generate results exactly as planned, so one might reasonably ask, "Why plan at all?"

In defense of planning, the importance of having well-defined targets at which to aim cannot be overstressed. Granted, targets often are going to be missed because conditions change between the time the plans are generated and the events occur. This also happens because of weaknesses in the planning process itself. When there is a target, however, then even a miss teaches something. Things like how badly the mark was missed, and perhaps in what direction it was missed, can be useful in assessing both planning processes and work practices.

Consider a simple analogy in shooting an arrow toward a target. If the target is simply a blank circle, it alone is the "mark," and as long as an arrow hits anywhere in the circle, it is hard to discern a lot about aim and accuracy. However, add a bull's-eye and several target rings, and the shooter has a clearer idea of how to adjust shots and hit exactly where intended.

Keep in mind, however, that when plans are not realized it could be for any of several reasons. It is possible that surrounding conditions have changed and what was once a good plan is no longer valid in the light of new conditions. It is also possible that the plan was inadequate to begin with. Also, there is always the possibility that the plan was well conceived and fully adequate, but failed to work because the implementation effort fell short of what was needed.

In any case, whether or not plans work out well, one usually should learn something from the experience. It has often been said that plans themselves are not particularly worthwhile, but that the planning process is invaluable. Indeed, what is truly valuable is the cyclic process of examining needs, setting objectives, making plans to reach those objectives, implementing the plans, and following up on the total effort.

Plans should never be regarded as cast in concrete. People sometimes tend to try bending reality to fit the plan so as to arrive at the projected results. It is true that a certain amount of this kind of effort is called for with some kinds of plans. Departmental budgets, for instance, should be considered as relatively important targets to be met. However, a plan is first and foremost a guide to action. It is not in itself a predestined action.

First-line supervisors may not feel there is a great deal of planning required of them. However, upon examination it is clear that every management position, even one in the lowest levels of management, requires some planning. A certain amount of planning is necessary to help do a job properly, and people who do not run their jobs will find, at least to some extent, that their jobs will run them.

■ Organizing

Sometimes it may seem that organizing, like planning, is not a particular concern of the first-line supervisor. It is true that a great deal of organizing is done in the context of departmentalization—the process of grouping various activities into separate organizational units. A great deal of this departmental planning takes place at high levels in the organization and may not occur very often. However, first-line supervisors still must organize—deciding which people within the department are going to handle certain tasks. In particular, whenever a manager is involved in making decisions concerning division of labor or separation of skills, that person is organizing.

One basic principle of organizing with which managers should be familiar is unity of command. Unity of command requires managers to assure that in all the activities within their area of responsibility specific employees are responsible for specific tasks.

Responsibilities should never be assigned in such a way that employees have room for doubt concerning who is ultimately responsible for any given task. Likewise, unity of command suggests that no function within a manager's scope of responsibility should be allowed to change or shift without being assigned to some specific person.

Another important concept within organizing (and one over which the individual supervisor has little influence) is span of control. An individual manager can effectively supervise only a certain number of workers. This number is subjectively determined by the manager's knowledge and experience, the amount and nature of the manager's nonsupervisory work, the amount of supervision required by the employees, the variability of the employees' tasks, the overall complexity of the activity, and the physical area over which the employees are distributed.

A supervisor can oversee and control more people who do similar work in the same physical area than people who perform variable work scattered over a considerable area. For instance, the working supervisor of a five-member records department (where all three employees work within the same room) has every opportunity for complete control. The supervisor probably knows all the jobs fairly well, and visual and auditory control of the entire department is relatively easy. On the other hand, a working supervisor in a five-member probation department has a much more limited span of control. This department's employees do many different things and often do their work well beyond the supervisor's visual and auditory control. The first supervisor can readily control five employees; the other will have considerably more difficulty controlling five employees.

Delegation is the most important aspect of organizing to a first-line supervisor. Delegation—the process of assuring that the proper people have the responsibility and authority for performing specific tasks—is of sufficient importance to the supervisor to warrant a chapter of its own (see Chapter 6).

■ Directing

Directing consists largely of assigning responsibilities on a day-to-day basis, letting your employees know what has to be done, how, and by when. It is making all the little (but all-important) decisions so necessary in the operation of the department. It is, in fact, the process of steering the department.

Although "team" analogies in management can wear thin at times, the example of the football quarterback is nevertheless appropriate to directing. The quarterback knows the plays and the strategy as a result of prolonged planning sessions. Yet when he goes onto the field he does not know the exact conditions he will encounter. It is only when he sees what happens on the field that the quarterback can call on what he has learned and respond to the conditions of the moment. It is in this way that the supervisor must behave, making the day-to-day and sometimes hour-to-hour decisions necessary in running the departmental team.

Since a great deal of directing consists of giving advice or conveying instructions, directing is, in a mechanical sense, present along with most of the other management functions. That is, since giving an order is a directing activity, it is really not possible to convey any kind of decision without directing. Some might argue that all other management functions are subsets of directing. At a minimum, directing in some way touches every function, process, or technique used to get things done through people.

The foregoing quarterback example should be a reminder that people who are most successful at directing are successful at leading. One can direct without leading by simply giving orders. Managers can fill leadership positions (although perhaps not very well) without being true leaders. However, direction is more successful when employees are led and motivated effectively.

■ Coordinating

It has been suggested that coordinating—the blending of activities and timing of events—might legitimately be considered a part of the directing function. That idea is worth considering separately, if only briefly, out of recognition of its importance to the supervisor.

A dinner of five magnificent courses will not be particularly successful if the courses are scattered over two or three hours, the dessert comes second, and the entree arrives last.

Likewise, simple job performance is not an essential part of many work activities, but rather when they are performed relative to other activities. In many management enterprises, a potentially large number of tasks involve coordination with other tasks.

In day-to-day correctional operations, it is essential that employees, facilities, supplies, and services all be combined. And this combination has to blend the right relationship to each other, at the right time, using the right people, in order for the component to run smoothly. Throughout each department, and certainly between and among the departments of a prison, it is necessary for activities to be coordinated. It would make little sense for the recreation department to schedule a movie for the inmate population if there was no extra correctional officer coverage available for the darkened auditorium. Likewise, would it be wise for the inmate commissary to sell convenience food items packaged in glass containers to inmates in high security housing? Effective coordination is one of the keys to supervisory effectiveness.

■ Controlling

Plans rarely come to fruition exactly as intended, so many moment-to-moment changes are required in pursuit of departmental objectives. The controlling function involves evaluating progress against objectives and making adjustments or new decisions as events proceed. The terms most descriptive of controlling are follow-up and action. Take note of how things are going as compared with how they should be going, and make new decisions and provide new direction to effect corrective action.

Controlling is often the most neglected of the basic management functions, especially in terms of the strength of follow-up on implementation of earlier decisions. The problems of limited or nonexistent follow-up will be examined further during later discussions on delegation and supervisory decision-making.

Emphasis and Priority

The basic management functions of planning, organizing, directing, coordinating, and controlling were presented in a given order for an important reason.

Generally, the first elements of this list should occupy a proportionately larger amount of the time of people in the upper levels of management. In the real world, though, top managers and middle managers are frequently prone to continue behaving in the manner of first-line supervisors. They spend significant amounts of time dealing with day-to-day operating problems. Indeed, managers at all levels in all organizations are frequently prone to "crisis management," expending most of their time and effort in reacting to present events and conditions rather than looking ahead.

Because of the nature of departmental supervision, the first-line supervisor will concentrate more on activities toward the bottom of the list of basic management functions. It is the lower echelons of management who are rightly more concerned with the problems of the moment. Those at the top of the organization should be more concerned with planning where the organization is going relative to its long-range goals, and deciding on courses of action required to support those goals. The correctional officer in the cell house is concerned with bar tapping and counting inmates, while the chief of security is planning next year's budget request and developing a new emergency plan for response to a riot. As the example in Chapter 4 showed, however, even those nearer the top of the organization can fall into the trap of personal involvement in operational details.

As noted, planning, organizing, directing, coordinating, and controlling are all part of every manager's job. In a large correctional organization, top management may spend 70 or 80 percent of the time involved in broad-based planning and organizing. In the same organization, except for the regular practice of delegation (a part of organizing), the first-line supervisor may spend 80 or 90 percent of the time on a combination of directing, coordinating, and controlling (See **Figure 5-1**).

How an individual manager sees his or her approach to these basic management functions will be largely influenced by the approach that person has taken to the job since becoming a supervisor. A great deal of what a manager does has been determined by the concept of management that person held before becoming a supervisor, and the kind of supervisory training he or she received.

Process Versus People

In discussing basic management functions, there necessarily is a focus on a number of practices that are often described as management processes. In doing so there is a risk of creating an impression that management is strongly process oriented. It is tempting to believe that to be successful in management one needs to learn a number of processes and then apply the appropriate processes to circumstances as they arise.

It is indeed true that planning and organizing are processes. But controlling, delegation, and leading (to name some fairly broad functions) are processes also, as are absentee control, scheduling, and interviewing (to name some more narrowly delineated functions). And one could name dozens of other so-called functions or techniques that are processes.

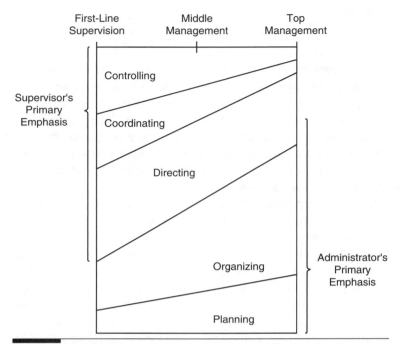

Figure 5-1 Typical shift in emphasis on basic management functions from lowest to highest levels of management.

With all of this apparent emphasis on functions and processes, it is appropriate to remind ourselves that the central focus of management is people. A person might spend a great deal of time learning management processes—most management education is, in fact, heavily weighted toward process—and never become a successful supervisor. In the long run, success at any level of management will depend on ability to work with people.

EXERCISES

Exercise 5-1: The Management Functions in Action

Marcie Denton is chief of maintenance for a relatively new maximum-security correctional institution. Included in her many responsibilities is that of maintaining all electrical and mechanical equipment. Among the prison's equipment are four large air-conditioning units.

When she helped establish the department, Marcie designated three men—one of them a working foreman—to handle all electrical and mechanical repairs and maintenance. Agency policy empowered the foreman with the authority to requisition parts and materials through local sources, up to a limit of $50 per order.

With the foreman, Marcie developed a schedule for preventive maintenance and routine periodic overhaul of the air conditioners. The foreman was to make sure that the schedule was followed.

One warm August day an air conditioner broke down while the foreman and one mechanic were repairing a pump at the power house. The other mechanic was off on sick leave. When word of the breakdown reached Denton, she reassigned the working foreman to the air conditioner and left the mechanic working on the pump. Since the repair to the air conditioner involved a short interruption of electricity to two departments, Ms. Denton arranged a 15-minute break in electrical service for 2:00 P.M. with the supervisors of those departments. She wisely cleared the outage with the chief of security, and notified the associate warden over her department.

The repair to the air conditioner was successful, but unusual wear in the equipment was evident when it was disassembled. To guard against future breakdown, Ms. Denton conferred with the manufacturer's local representative and subsequently increased the frequency of scheduled preventive maintenance for this unit. She also decided to check the other units for similar problems.

Instructions

1. Analyze the case description paragraph by paragraph and identify the basic management functions which:

   a. Were apparently performed sometime in the past (for instance, "Included in her many responsibilities is—" implies that organizing has been accomplished).

   b. Are being actively performed by Ms. Denton (for example, "Ms. Denton arranged a 15-minute break in electrical service" describes an act of coordinating).

   c. May be suggested by these events as being additionally necessary for the future.

2. Be especially sensitive to the "management cycle": follow-up and correction (controlling), frequently involving directing and coordinating, may lead to further planning and organizing.

Exercise 5-2: What a Way to Start a New Job!

Randy Parker was appointed to a position as unit manager of a new 180-bed detention unit being built adjacent to a medium-security correctional institution. The unit was intended to serve the needs of the local court, supplementing bedspace available in the local jail. Parker was selected after serving for 7 years as a correctional officer and 4 years as a unit manager in the main institution. He was selected about 90 days before the unit was scheduled to open, and was tasked with developing a firm budget and staffing plan for the unit.

Parker developed a master staffing plan based on providing each 60-bed wing of the three-wing unit with a core staff in line with staffing of special housing units of a similar size in other facilities throughout the agency. A phased activation plan was anticipated, with a new wing being activated and staffed about every 30 days. As is typical for this agency, a small pool of sick and annual relief personnel were included in the roster to cover a variety of staff vacancies.

One Tuesday morning, the warden of the main institution received word that a riot was in progress at the state's penitentiary, and that the institution should be on standby to provide assistance as needed. One of the possibilities mentioned was that the new detention center would be activated to hold inmates moved out of the penitentiary. About the same time, Mr. Parker received word that one of his key staff members, the lieutenant with whom he had done much of the planning for the unit's activation—particularly the security procedures—had taken seriously ill during the night.

Parker worked to develop a plan that would allow the detention center to open if that became necessary. With the concurrence of the chief of security, Parker began to locate key staff and position them for new assignments. He reviewed the list of available sick and annual relief staff from the main institution (a group already depleted by vacations and illness in that facility). Parker was able to pull one senior lieutenant who was about to be promoted to captain, and who had been a supervisory correctional officer in the county jail before becoming a State correctional officer. He also located two senior sergeants who had worked the main institution's special housing unit recently, and advised them they might be called upon to serve as acting lieutenants in the detention unit. He then made arrangements with the shift captain to cover their regular positions with other sick and annual relief personnel should the move be necessary.

Shift supervisors were needed for the unit; Parker had identified the two senior sergeants as possibilities for these posts. He was tempted to step into the breach himself since he had run a housing unit inside the correctional facility and knew he could run a shift in the new unit. However, he had no idea how long this emergency activation would last and he did not want to spread himself too thin by assuming an additional burden when he might be needed elsewhere. He and the chief of security decided to reassign the three supervisors involved, and to authorize Parker to utilize both sick and annual relief staff and overtime to provide line officer coverage.

The riot was brought under control late that day, and in its aftermath it was determined that one housing unit was too badly damaged to use for at least several months. Parker's warden was advised that 145 inmates would be moved to the detention center the following day. It became necessary to make the changes Parker had planned.

That day, special operations staff from the penitentiary transferred six busloads of inmates to the detention center. These inmates and their personal property were processed and all three wings of the detention unit were activated within an 8-hour span. Asked later if the correctional agency's emergency response plan (a number of elements of which had been put into effect) appeared adequate, Mr. Parker was able to suggest that some procedures be strengthened in specific ways.

(This example is based on several actual incidents, in which inmates displaced by riot or natural disaster were relocated on short notice to about-to-open housing units.)

Instructions

The examples contain numerous descriptions of the basic management functions: planning, organizing, directing, coordinating, and controlling. Identify as many of these as possible.

Note that a single activity often includes elements of two or more of the basic functions in combination, but that on close consideration one basic function usually emerges as dominant. Try to identify the dominant element in the sequences described above.

ENDNOTES

1. Haimann, Theo. *Supervisory Management for Hospitals and Related Health Facilities.* St. Louis: The Catholic Hospital Association, 1965, pp. 10–12.
2. Fayol, Henri. *General and Industrial Management.* New York: Pittman, 1949.

The Supervisor and Self

II

Delegation: How To Form the Habit

6

Chapter Objectives

- Identify delegation and its practice as a major influence on effective supervisory performance.
- Convey the importance of proper delegation in the correctional environment, in terms of its value to supervisor, employees, and organization alike.
- Identify the common reasons behind the failure to delegate.
- Establish a perspective on delegation that will also help the supervisor work under authority delegated from higher management.
- Develop a pattern of steps representing a desirable approach to proper delegation.
- Encourage the supervisor to build positive delegation habits.
- Suggest how the individual supervisor can objectively evaluate his or her own delegation performance.

Surround yourself with the best people you can, delegate authority, and don't interfere as long as the policy you've decided upon is being carried out.

Ronald Reagan

One of modern management's most important functions—effective delegation of work—is a subject for plenty of preaching but not enough practice.

Earle Brooks

Contract-Critical

It's a regional office, and the head of the community corrections division is responsible for monitoring compliance by the contract community treatment facilities the agency uses for releasing offenders. Community corrections officers are delegated the responsibility for on-site monitoring, reporting to their regional superior. These conditions include not only programmatic issues (substance abuse treatment, job assistance, and the like), but also the physical conditions of the facility itself.

An extremely embarrassing series appears in the local paper, reporting on the conditions in the halfway house. It turns out the building has serious code violations. It is not staffed in such a way as to adequately track when offenders are coming in and out. Some personnel are not qualified to actually hold the positions they have. Not all the required programs are being provided. These are all issues that are contract-critical.

When the head of the regional office comes down off the ceiling, he calls in the community corrections administrator and tries to find out what happened. He learns that the delegation was incomplete. There has been no follow-up either at the field level or by the regional administrator. The local community corrections officer had slipped in doing his job, and the regional administrator had as well.

■ Taken for Granted

Effective delegation has long been a subject of considerable exhortation. It is the basis of numerous workshops, seminars, and other educational programs. One need only review a small sample of management literature to obtain an idea of how often delegation is discussed in print. Despite the frequency with which it is mentioned, however, delegation can still be a weak link in the chain of correctional management. In actual practice, it often is treated lightly, if at all—even by supervisors who are genuinely interested in becoming better managers.

Although it is the subject of a great deal of good advice, delegation in its truest sense is frequently overlooked when it comes to conscientious practice. All too often the ability to delegate is unconsciously taken for granted, assuming an expertise that may not be present.

Many people who are responsible for the work of others tend to take delegation for granted because:

- They do not fully understand the true nature of delegation.
- They do not recognize, or at least do not fully appreciate, the extent of the power of their own habits in preventing improvement in delegation.
- They have not yet come to view delegation from the perspective of the employee to whom the work is delegated.

■ The Nature of Delegation

Significant problems are often observed when evaluating the way delegation actually is put into practice as a management technique. Few supervisors believe or will admit that they do not delegate—they believe they should delegate and are convinced they do delegate. However, a great deal of what is called delegation is incomplete and ineffective. Few supervisors delegate significantly beyond simple organizational delegation.

In Chapter 5, delegation was described as part of the basic management function of organizing. Most supervisors are automatically (if passively) involved with organizational delegation. A manager places a person in a job and gives that person a job description that essentially consists of a set of instructions. In doing so, the manager has delegated—charged an employee with carrying out one or more tasks for which the manager is ultimately responsible. The job description is, in a real way, the employee's authority to perform certain functions. Thus, in a very real sense any department consisting of more than one person is necessarily subject to organizational delegation because the normal division of activities requires different people to perform various tasks.

This chapter deals with a major step beyond that type of delegation. That is because each and every task to be accomplished by the department will not appear on someone's job description. The job description for a supervisor (as well as the job description for a partially self-directing professional such as a case manager or probation officer) will include a catch-all statement such as: "performs all other duties as required." The responsibilities of each department include many tasks not specifically mentioned in the job descriptions, located at various levels within the department. But invariably the supervisor is responsible for ensuring that someone does them.

A supervisor's job description may also contain many tasks that rightly belong there, but need not necessarily be done by the supervisor in their entirety. For most supervisors, there are probably a number of tasks that can be accomplished equally effectively by either the supervisor or a subordinate. A great deal of a supervisor's effectiveness is determined by how well he or she utilizes the capabilities of employees to accomplish such tasks, rather than doing the work personally.

Delegation is both a process and a condition. It is, in part, the act of assigning work to an employee, a process that generally is well understood. But consideration of delegation often stops at this point. Effective delegation is achieved by going well beyond the assignment process itself. It requires a thorough, mutual understanding by supervisor and employee of what specific results are expected and how these results may be achieved.

The hows and whys of the delegation process are relatively simple, and it is by understanding and consistently pursuing them that a manager can attain complete delegation. Like any desirable activity, effective delegation must be pursued conscientiously and intentionally, perhaps for a considerable time, before it becomes a habit. Doing this also can require breaking some old habits, because most managers are much more accustomed to doing than to having others do.

◼ What About "Empowerment"?

Already mentioned in Chapter 1, the concept of empowerment rose to prominence in the latter 1980s and early 1990s, partly in conjunction with the TQM movement. It has become a latter-day buzzword of major (some would say fad) proportions. Business has supposedly been discovering that empowerment is "the thing to do" with employees to enable them to make their best possible contributions to organizational success.

Although it is described in a variety of ways, the essence of empowerment as it is used in most organizations is letting the employees solve their own problems and implement their own solutions, or letting the employees decide what needs to be done and go ahead and do it.

What is the difference between empowerment and delegation? Absolutely nothing. That is, there is no difference between empowerment and proper delegation. Even in a standard thesaurus that has been in use for many years, empowerment and delegation are synonyms for each other along with commission, assignment, deputation, and a number of other words.[1]

Although delegation and empowerment mean exactly the same thing as words, in the real world they are regarded quite differently. Empowerment is "in," and a great many people are behaving as though this concept called empowerment is a vast improvement on mere "delegation."

But if the concepts are really one and the same, what happened to delegation? What actually happened to delegation is decades of misuse of the term, decades of regarding delegation as no more than giving someone a task to complete, or as no more than giving someone an order. Proper delegation, however, has always consisted of giving someone the responsibility for a task, an understanding of how to accomplish the task, and the authority required to complete the task. True empowerment is identical to proper delegation. A supervisor gives an employee a problem to solve or a condition to correct, specifies the desired outcome or acceptable range of results, and provides the employee with the authority or whatever other resources are necessary.

Not surprisingly, the problem that was at the heart of most difficulties with delegation is now emerging as a fundamental problem with empowerment as well. It is a problem of management style and approach by the delegating manager, and is a control issue with many higher managers. In brief, these managers cannot let go sufficiently to allow delegation or empowerment to work. Their retained authoritarian management styles send one clear message to the employees: "You are free to make whatever decisions you want, as long as they are the same decisions I would make."

The correctional environment has a particular claim on this problem. That is at least in part because of the high degree of public sensitivity to riots, escapes, crimes committed by parolees, and other incidents involving public safety. Correctional administrators often believe they must keep a personal hand in lower-level decision-making in order to prevent such problems from arising. Whether referred to as proper delegation or empowerment, this process will work as intended only if the manager—the person who does the delegating or empowering—is committed in advance to accepting the decisions of the employees. And for many correctional administrators, that is a difficult and sometimes risky thing to contemplate.

■ Why Delegate?

The first reason to delegate is for yourself. The old adage, "If you want a thing done well, do it yourself," is largely a fallacy as far as management is concerned. One person, no matter how competent, can do only so much and still continue to do things well.

It is not uncommon to see supervisors who try to handle so many activities that they are able to hold things together only loosely at best. They may believe that because they work long hours, take extra work home, and generally try to fill every request that comes along and solve every problem that arises that they are a dedicated, effective, hard-working manager. They may be dedicated and hard working, but if this is the way they approach the job, chances are they are not effective.

Failure to delegate effectively is one of the principal causes of managerial failure and is a leading reason why many people in management do not get promoted to more responsible positions. Failure to delegate effectively will not necessarily halt a management career at the first level. Many persons, by dedication and "workaholism," manage to rise several levels through sheer energy and activity. However, even the workaholic—the manager who continues to do as much as possible alone—will eventually reach a level where the job is too big to handle. Failure to delegate effectively is an even greater hazard for higher-level managers than for first- and second-level supervisors. In fact, delegation failure at all levels of management is damaging.

Managers need to delegate, to assure that some employees are capable of taking care of some of the tasks, problems, and requests that the manager usually has to handle. A manager can achieve a measure of peace of mind by knowing that one or more persons can capably step in when illness, vacation, or other factors require it. Delegation can build a manager's image as a leader with employees and will likely improve his or her standing with higher management. That's because supervisors are properly judged not by what they do but rather by what their department does.

Delegating certain tasks can free up more time to concentrate on true supervisory activities. Rarely is there enough time in the supervisor's day for all the thinking, planning, and communicating needed to maintain and improve a department's effectiveness. Generally, a supervisor who delegates effectively gains a greater degree of freedom from technical tasks and is able to function as more of a manager more of the time. This could also include sufficient freedom to assume greater responsibility, taking on functions delegated by the next level of supervisor.

B.C. Forbes noted that, "The most successful executives carefully select understudies. They don't strive to do everything themselves. They train and trust others. This leaves them with time to think. They have time to receive important callers, to pay worthwhile visits. They have time for their families. No matter how able, any employer who insists on running a one-man enterprise courts unhappy circumstances when his powers dwindle."

The second reason to delegate is for your employees themselves. Managers owe it to their employees to delegate. Supervisors should seek to help subordinates learn and grow, rather than acting in a manner that holds them back. Managers build their department by building individuals.

In the same way that an immediate supervisor can determine a manager's fate, so the fate of subordinate employees is largely in a manager's hands. Managers can guide employees upward in terms of growth and development, or can hold them back. A manager can challenge and interest subordinates by giving them responsibility and opportunity, or can lock them into routine and boredom.

Most employees would rather be stimulated and interested than unchallenged and bored. Most would rather learn and grow than stagnate. Most would rather be serving useful purposes than doing unimportant or inconsequential work. And most would rather work (because they enjoy work, or, at worst, because they wish to pass the time more quickly) than be idle.

When a manager delegates, employees get a taste of greater responsibility and perhaps some decision-making experience. They are likely to take more pride in what they do, reflect higher morale, both individually and as a department, and exercise more individual initiative.

Effective delegation within a department also presents opportunity to the employees. All employees will not react the same way to the chance to assume greater responsibility or do different work; some people do not wish to go in those directions. However, the simple presence of the opportunity is a tonic to all employees whether they avail themselves of that opportunity or not. And the absence of such opportunity will affect many employees in a way that eventually affects the outlook of all.

■ Failure to Delegate

Managers fail to delegate for a number of reasons. Some of these reasons are complex and involve fear of competition from employees or the loss of recognition for task accomplishment. These are factors about which the supervisor may be only dimly aware. It is often difficult for a manager to avoid thinking of being in competition with line employees, since some of them seem to foster an air of competition as they attempt to progress and grow in the organization.

The effective supervisor encourages growth, and a legitimate challenge taken up by an employee can look a great deal like direct competition. The fear of competition from subordinates often goes hand in hand with a degree of insecurity. But the presence of two or three eager promotion-oriented employees with their sights set on the supervisor's job also can be a stimulus that keeps the supervisor in a heads-up and growing attitude.

Although virtually every employee is legitimately concerned to some extent with how they may be doing in their supervisor's eyes, a manager's first concern should not be, "Does my boss think I'm doing a good job?" Rather, it should be, "Is this department seen as doing a good job?" When employees perform well, the credit is the manager's as well as theirs. As the department is recognized, so is the supervisor recognized.

As noted, old habits are a major reason for failure to delegate. That is easy to understand. After all, most managers' earliest working years were devoted to actual performance of line duties. They often fail to deemphasize that habit, even though they may have spent more recent years supposedly directing line activity.

A great deal of sound advice is available concerning delegation. Having said that, though, it is sound advice only if it is followed. But most of the time the advice offered in step-by-step approaches to delegation is not followed. Alternatively, if it is followed at all, it is not conscientiously pursued long enough to become ingrained as a new supervisory habit.

Consequently, in order to try to adopt a new approach to delegation, managers often must change the way they do things. The classic example is the former chief of security who, after being promoted to warden, cannot stop hanging around the shift supervisor's office offering advice (see the story at the beginning of Chapter 3). Many a technical specialist who has excelled to the point of being recognized and promoted to upper management is more comfortable reverting to old tasks than in learning and exercising new management duties.

Anyone who has put forth a conscious effort to change a deeply ingrained habit will fully appreciate how difficult it can be. But it certainly can be done, depending on how much effort is applied, how long this effort is sustained, and how deeply ingrained the old habit is.

Attempts to improve the way delegation is carried out frequently go the same way as attempts to apply information from a time management seminar about improving effectiveness. There is a flurry of initial activity that is followed by stops and starts that diminish, and eventually vanish as old ways take over once again. This ordinarily occurs because the manager is only dimly aware of, or perhaps has not thought at all about, the immense barrier presented by habits—habits that may have become so deeply ingrained through years of practice that they are now second nature.

The key to improving delegation skills lies in the constant awareness of the need to overcome old habits. It is necessary to start on a small, or at least modest, scale, conscientiously applying the new process to one well-defined task or project at a time. Doing so over and over is necessary, until the new way becomes a habit that is strong enough to keep the old habit from returning.

Another common reason for the failure to delegate (one that many supervisors may readily recognize and admit) is the feeling that available employees simply cannot handle greater responsibility. This is a classic situation. A manager hesitates to try someone with more responsibility because of the uncertainty as to whether they can handle it. Yet the only way to find out if they can handle it is to try them.

A manager may feel that they cannot delegate because there is no one in the ranks who can handle additional responsibility. That person should act as if he or she has seriously tried to prepare those involved to assume more responsibility. Again, the correctional environment is prone to foster this kind of problem. The consequences of internal management problems can reach life-threatening levels and many managers place themselves under pressure to personally be sure the institution is running properly.

The most prevalent reason given for failure to delegate is also the most obvious reason: lack of time. Here there is a basic contradiction—managers tend to think most seriously about the need to delegate only when the pressure is most intense. The in-baskets are overflowing. Problems are coming from all directions. There is a realization of the need to delegate to improve effectiveness—or at least to help get caught up. However, delegation takes time. Managers need time to pick people and prepare them; time to prepare the work; time to do any number of other things.

And when the workload is heavy there simply is not that kind of time. A manager may promise to do some serious delegating "as soon as the rush is over." However, when the rush finally goes away so does the feeling of urgency that came with it: "Delegate? The workload is under control, so where's the need?"

The time required to accomplish proper delegation is an investment that must often be made when there is precious little time available. The returns of delegation are not immediate. It often is necessary to expend an additional amount of time and go even further behind for the sake of improvements that will not yield real time savings until weeks or perhaps months have passed. However, this hurdle—the "time trap"—must be overcome before improved efficiency through delegation can become a reality.

Reflect on Example 3-2 for another example. Then, consider the case of the prison education department supervisor who was required to submit a detailed statistical report of the department's classroom and vocational training operations each month. Each time the manager faced the report, which required about 4 uninterrupted hours to complete, this person considered delegating this task to a certain employee who could handle it with proper training. However, the supervisor estimated it would take 8 to 10 hours to go through the report point by point with the employee. So the employee's training was put off until a "convenient time" between reports. However, the convenient time never came, and soon it was time to submit the next report and the pressure was on again.

After rationalizing the way through this "time trap" several times, the supervisor finally recognized the need to invest extra time and effort. When the time came to do the next report, the supervisor and the designated employee closeted themselves and went through the report number by number and line by line. It took an entire day. The next month's report took the two of them nearly 6 hours, still longer than it had taken the supervisor alone. The third report was done in the same way in about 4 hours.

The fourth month's report consumed about 6 hours. However, this consisted of 5 hours of the employee's time, working alone, and an hour of the supervisor's time to audit some important statistics. By the sixth month of the new arrangement, the employee was doing the report in 4 hours and the supervisor was spending perhaps 5 to 10 minutes to review and approve the report. Improvement took a long time, but the early investment of time and effort paid off in later returns.

The impact of the failure to delegate can be significant and can be felt well beyond the individual nondelegating supervisor. Consider the case of the business office manager in a correctional agency's headquarters who was not promoted to comptroller primarily because of delegation weaknesses. Because this supervisor did not delegate, neither of the two assistant business managers in the office had been allowed to demonstrate their ability to handle additional responsibility. This left the department without a potential successor to the manager, had he been promoted. The business office manager had been the only possible in-house candidate for the comptroller's job, so the job was advertised outside the agency and filled through that avenue.

Because the manager lost out on a promotion, one of the two assistants also lost the chance to become business office manager. In addition, two lower-level jobs that could have been opened up by this promotional chain reaction remained closed. Thus, four people missed out on possible promotions simply because one of them had not practiced effective delegation.

A nondelegating supervisor's department suffers in a number of ways. Unchallenged people with time on their hands usually do not exhibit high morale. As morale goes down, productivity and quality are likely to suffer. A department so afflicted can acquire a reputation for discontent and unreliability, and once acquired such a reputation is not easily shed. In the extreme case, such a department will begin to lose its better people as they go elsewhere in search of work environments in which there is opportunity for growth and advancement. Employees who do not consider job interest and challenge as particularly important will remain to drift—and complain.

■ Looking Upward: The Other Side of Delegation

The supervisor who sincerely wants to improve at delegation should look both downward and upward—downward toward the employees to whom he or she will delegate, and upward toward the higher manager from whom delegated tasks are received. Put yourself in the position of your employees and consider how your superior should, or perhaps actually does, delegate to you. It helps to look toward your supervisor regarding delegation, because your employees similarly look toward you.

The motivational forces that cause people to work exist in a complex mix that may vary dramatically from person to person. But the net effect of these motivating forces is often not all that different between a manager and most of his or her employees. For the most part, employees will want many of the same things that their supervisor wants. Many of them aspire to supervision, wishing to rise from the ranks of the work group as perhaps did their supervisor.

Not all employees are likely to aspire to management, though. Not all of them will seek challenge in their work as most managers do. Generally, however, their motives will parallel those of their supervisor, and they can be dealt with in a manner that generally reflects relationships up the chain of command.

A manager therefore is on a reasonable course when trying to pattern some elements of a relationship with each individual employee after an idealized working relationship with their own supervisor. Attempting to do so will put a manager in a position of being able to work on improving delegation both upward and downward. While striving to become better at delegating, a manager can work to achieve a more appropriate delegation relationship with his or her superior.

This is a highly personal approach, because it finds the manager looking out for his or her own best interests (as well as the best interests of line employees and the organization) by improving delegation effectiveness as well as supervisory communications up the chain of command.

The sections that follow address some of the real-world considerations that arise when discussing delegation.

Ideal Versus Real

In an ideal relationship with a manager's supervisor, both manager and supervisor would know where the other stood at all times and on all matters. In the real world, of course, this relationship may be somewhat less (perhaps even significantly less) than the ideal. However, it is possible to use knowledge of needs as a supervisor and an individual manager's concept of an ideal supervisory relationship to:

- Reshape and generally improve that relationship, actually guiding the supervisor along the path toward more effective downward delegation.
- Establish a new and more effective pattern for downward delegation.

Managers have to determine, as realistically as possible, the dimensions of delegation their supervisor must use in order to allow them to fully meet the requirements of the job. After these needs have been defined, the manager must apply the same process to their relationship with each subordinate employee in the work group. By constantly envisioning a position "in the middle," a manager can apply a number of considerations to delegation both upward and downward.

Differing Perspectives

Every person in the chain of command has one primary upward channel of communication. However, every managerial person in the chain of command has as many primary downward channels as he or she has direct reporting employees.

As the person in the middle of three consecutive points in the chain of command, the supervisor needs to appreciate the differences between the upward perspective and the downward perspective. Therefore, the most constructive view to take when looking up the chain of command is, "I am but one of a number of people reporting to this superior, so I need to make it as easy as possible for my superior to communicate with me." Conversely, the downward view suggests. "Each of these people who reports to me is looking to me for guidance, so I must show him or her how to relate to me so that I can meet his or her needs as well as possible."

Whether rank-and-file employee, first-line supervisor, or middle or upper manager, the person looking up the chain of command sees a single point of contact and thus a single and presumably dedicated source of assistance. However, the person looking down the chain of command invariably sees multiple points of contact that, taken together, are capable of making overwhelming demands.

Not Just Problems, but Solutions

A large part of becoming an effective delegator consists of teaching employees to do their homework before bringing problems and concerns up the chain of command. However, the business of homework, or thorough staff work, as it might be called, can be dealt with both upward and downward by the individual supervisor. The reaction of many employees to the presence of a problem is to take that problem to their supervisor and ask for advice and assistance. The supervisor, however, does not need problems—he or she needs solutions.

The way not to communicate with one's superior is to say, in effect, "I have a problem and I need your help." Rather, it is necessary to do some homework first. Analyze and isolate the problem, identify causes if possible, develop alternative solutions, and identify specifically what is needed from the supervisor. Then go to the supervisor and say, "Here's a problem. This is why I think it's a problem and what probably caused it. Here are two or three possible solutions, and here's the solution I like best and why. This is what I think I need from you in order to proceed. What is your advice?"

In such a case, the supervisor often need only answer "yes," "no," "do this," or "do that." Quite often the problem will be nearly self-solving once it has been subjected to analysis. In any event, the manager will have made it as easy as possible for the supervisor to communicate, without intruding on the supervisor's time to analyze the problem and without transferring ownership of the problem to the superior.

In parallel fashion, the supervisor should encourage subordinate employees to do the same kind of homework to the fullest extent of their abilities. It doesn't matter whether it involves dealing with a specifically delegated task or a new problem encountered directly by the employee. The supervisor should instruct the employee in techniques of analyzing and refining everyday problems, offering alternative solutions, and making recommendations.

The employee who does the necessary homework does not always relieve the supervisor of what can sometimes become extensive involvement. The supervisor may have input to offer beyond the reach of the employee. However, the employee who does the necessary homework has accepted partial ownership of the problem and has focused the problem so that it can be dealt with more efficiently. At the heart of effective delegation is the need to teach the employees to do as much as possible on their own before calling for help. One highly successful supervisor expressed management's needs well by saying, "Don't bring me more problems, bring me solutions."

Reasonable Deadlines and Follow-Up

Although superiors may be lax regarding deadlines and follow-up (no level of management is exempt from such weaknesses) there is no need to be similarly lax with subordinates. This is another area of supervisory behavior in which managers can do a great deal of improving in working both upward and downward.

The supervisor who is lax concerning deadlines and follow-up shapes subordinates' expectations accordingly. If the supervisor says simply, "Take care of this when you can," the subordinate may respond immediately, within days or weeks, or never. And if the superior says, "Let me have this by Friday," but the subordinate knows through past practice that the supervisor is not likely to mention it again for 2 weeks, it will probably be 2 weeks before the work gets done. Such managers have, through their actions over time, conditioned their employees to expect this kind of behavior. Add to this conditioning the fact when a delegator uses time-related instructions—"When you get a minute," "Whenever you can," "Sometime soon." These phrases can mean anything from right this minute to never, depending on the employee's interpretation.

No matter how non-urgent or unimportant it may be, any task that is worth assigning is worth assigning a reasonable deadline. Once a deadline is assigned, regardless of the task's significance, that deadline should never be allowed to pass without some kind of closure taking place. If closure does not occur in the form of delivered results, then the manager who assigned the task and set the deadline should immediately follow up with the employee.

If a supervisor does not set deadlines externally, the line manager should set them internally anyway. Commit to such deadlines, if possible by promising results to the supervisor by some given day and time. Do not promise immediate results, but rather allow reasonable flexibility in the self-imposed goal. (If the supervisor is at all fair and effective, he or she will immediately indicate whether too much or too little time has been allocated.) If unable to deliver when promised because of circumstances beyond control, do not wait for the supervisor to follow up; rather, inform him or her of the delay and its causes.

This notion of applying reasonable deadlines and faithful follow-up is even more important in the supervisor's relationship with the employees in the work group. These are also the steps that ensure the thoroughness and effectiveness of proper delegation. One can assign a thoroughly defined task to a well-instructed employee and still get nothing if there is no specific requirement for completion or follow-up. The supervisor who assigns no deadlines projects a casual attitude toward timely accomplishment of work that can permeate the entire workgroup.

Worse still, if the supervisor usually assigns deadlines but lets them slide by without saying anything, the employees will come to see the supervisor as lax, disorganized, and possibly uncaring. It is this latter condition that is most likely to occur. It is easy to set a deadline when a task is assigned, but it is not quite as easy to remember to follow up on that deadline. By failing to follow up on the deadlines they set, many supervisors create credibility problems and other difficulties for themselves.

Assigning reasonable deadlines and following up on them may represent a way in which a supervisor can considerably improve his or her managerial effectiveness with the least amount of effort. The supervisor can simply set reasonable deadlines for every task that is delegated and faithfully follow up every time a deadline is missed.

■ The Nuts and Bolts of Delegation

Selecting and Organizing the Task

Some things should never be delegated and some things almost always are appropriate for delegation. As discussed in Chapter 3, work as a supervisor lends itself to separation into two general categories: technical tasks and managerial tasks. The records office supervisor, for example, still must have technical knowledge about sentence computation to be sure the department's employees are doing their jobs properly. However, the supervisor also must tend to purely managerial tasks like budgeting and scheduling coverage.

Most technical tasks are likely to be subjects for delegation. There may be a few a manager should continue to control personally, either because they are of sufficient importance to warrant personal attention, or because they occur so infrequently that any training time invested would never be paid back. But for the most part, there is considerable benefit in delegating purely mechanical or narrowly technical duties.

Of course, few if any true managerial tasks can be delegated in their entirety. After all, they are among the reasons why the manager was entrusted with supervisory authority in the first place. For instance, specific personnel management tasks relating to hiring, firing, promotion, demotion, criticism, discipline, and performance appraisal cannot be delegated.

Other managerial activities may lend themselves to partial delegation. A manager may obtain staff input and assistance in planning, scheduling, budgeting, purchasing, and other such activities. Nevertheless, the authority to approve, recommend, or implement still calls for the exercise of supervisory authority.

Take the time to make a list of duties you perform that could reasonably be delegated to an employee. Consider each workday for a period of weeks. Write down each such task whenever one occurs. You may be surprised at the significant amount of work falling into the category of tasks that can be delegated. Preparing routine reports, answering routine correspondence, preparing service schedules, ordering supplies, serving on certain committees, performing certain technical tasks, and many other activities may present themselves as candidates for delegation. List them all, and rank them according to two criteria: the amount of time they require and their importance to the institution. In short, establish a priority order of tasks for delegation.

Do not, however, attempt to delegate all these nonmanagerial duties at once. Do not even consider working with just two or three of them at the beginning. This requires establishing delegation as a habit, and new habits are tough to form.

Pick one task to begin with, preferably one that is either of most importance to the operation or takes the largest part of your time, or both. You should plan on delegating a single function, or as much of one as possible, to a single person. This will avoid the situation in which a function is so broken up that no one person is able to develop a sense of the whole job. Also, in considering activities to delegate, concentrate on ongoing functions—on jobs that regularly recur. There is little to be gained by delegating a one-shot activity if you can do it faster and better by yourself.

Determine the specific authority you will have to provide the person to whom you delegate an activity. Plan also on defining the limits of that authority. As suggested earlier, operating instructions themselves are often the full "authority" needed to perform a task. However, there are instances in which the person must be able to call on certain resources necessary to do the job. In all instances the authority given should be consistent with the responsibility assigned. For example, if you make an employee responsible for ordering office supplies you should also be sure that agency policy gives that person authority to sign the necessary purchase requisitions.

Generally, a manager should consider delegating as much technical task authority as possible, within policy. Ordinarily, even some of the routine portions of a few of managerial tasks can be delegated. For example, you can delegate a great deal of the numerical work involved in preparing your departmental budget as long as you maintain final decision-making authority over the complete budget.

Select the Appropriate Person

To emphasize a widely violated precept, someone in the second level of management (supervising one lower level of supervision) should always delegate to an immediate subordinate. The chief of security should not be issuing orders to line correctional officers without going through the shift commander. A chief probation officer in all but the smallest office should work through his or her deputy. To bypass that intervening level and delegate directly to an employee two levels below is to undermine the authority of the first-line supervisor. First-line supervisors, however, will not have this problem since all of their employees report directly to them.

Pick the employee to delegate to by matching the qualifications of available employees with the requirements of the task to be delegated. How well this is done will depend to a great extent on how well the manager knows the employees' strengths, weaknesses, attitudes, and capabilities. If a manager wishes to delegate a portion of the department's budget preparation, an employee who appears to have the aptitude for numerical work would be a natural candidate. If the task to be delegated consists of guiding new employees through a department orientation, other attributes come into play. The manager should be looking for an employee who knows the rest of the employees well enough to introduce new people to them. He or she should know the department's work well enough to describe it to a newcomer. They should be reasonably friendly and adept at conversing with new people.

However, it is not enough to simply make a subjective match between the requirements of the task and the perceived abilities of the person. The manager must consider who likely is ready to assume additional responsibility. Also, it is valuable to concentrate on those employees sufficiently mature and reliable to give the assignment an honest try.

Beware of either over-delegating or under-delegating. When a manager over-delegates, the employee to whom the task is given is clearly not ready to handle it. While a modest amount of challenge is certainly desirable, too much challenge can be overwhelming to the employee. Over-delegation frequently leads to an employee's failure in a first attempt at handling increased responsibility, a harsh beginning that is not easily overcome.

On the other hand, under-delegation (assigning a task to an employee who is overqualified and can obviously handle it with the greatest of ease) can be fully as damaging. Under-delegation is a waste of an employee's capabilities and often results in that employee's boredom and stagnation. Ideally, delegation should provide a modest amount of challenge, modest but recognizable opportunity for growth, and the opportunity for diversification and expanded usefulness. Also, the employee must be able to see the importance of the delegated task.

The manager making a delegation decision must also be reasonably convinced that the employee in mind has the time available to handle the delegated task. Even if person and task are properly matched, problems can result by assigning more work to someone who is already fully occupied. Importantly, in such a situation not only the delegated duties will be poorly executed, but it is likely that the employee's core responsibilities will suffer as well.

Finally, in selecting an employee to take on a specific task, keep an additional factor in mind. If delegation is to serve its proper purpose, the supervisor must be willing to accept the employee's decisions as though they were their own. To do otherwise not only undermines the process but sends a message to the employees that the supervisor really does not have confidence in their abilities.

Instruct and Motivate the Person

One of the most common errors in delegation is turning an employee loose on a task with inadequate preparation. (Another look at the example in Chapter 3 is instructive here.) It is at this point that the pressure of time can set the stage for delegation failure. If the task being delegated is one the manager has previously performed (and very often this is the case) there may be few instructions, procedures, or guidelines existing in writing. It may be that the only available instructions are those in the manager's head. In gathering the information needed to turn over a job, it may be necessary to put those instructions in writing as well as prepare to teach the employee how to do the job personally.

When completely ready to turn a task over to an employee, the manager should be able to provide satisfactory answers to the following questions:

- Are all the details of the assignment completely clear in my mind?
- Am I prepared to give the reasons for the task, fully explaining why it is important and why it must be done?
- If necessary, can I adapt all the instructions and procedural details to the level of the employee's knowledge and understanding, and if necessary should I reduce them to writing?
- Does the assignment include sufficient growth opportunity to motivate the employee?
- Does the employee have the training, experience, and skills necessary to accomplish the task?
- Am I giving the employee sufficient authority to accomplish the results I require?

Assuming satisfactory answers to the foregoing questions, turning a task over to an employee then becomes a critical exercise in two-way communication. When meeting to make the actual assignment, encourage the employee to ask questions. If questions are not readily forthcoming, ask the employee to restate the instructions. Whenever possible, demonstrate those parts of the activity that lend themselves to demonstration and have the employee perform those operations satisfactorily. Throughout the entire process, emphasize two-way communication.

Last in the process of turning over a task, but extremely important, is the necessity for supervisor and employee to achieve agreement on the results expected. The precise methods by which those results are achieved may not be particularly important—several people may do the same task slightly differently. But the anticipated output must be known and agreed upon.

Maintain Reasonable Control

Control of delegation is largely a matter of communication between supervisor and employee. The frequency and intensity of this communication will depend significantly on the manager's assessment of the individual. A manager should know his or her employees well enough to be able to judge who needs what degree of control and assistance. Employees are bound to differ from each other in their response to delegated responsibilities. Some may need to be checked frequently and their activities monitored relatively closely. With others, touching base every few days will be sufficient.

Since the degree of control necessary will vary from employee to employee, the hazards of over-control or under-control are always present. Over-control can destroy the effects of delegation. The employee will not develop a sense of responsibility, and the manager may remain as actively concerned with the task as though it had never been delegated it at all. Under-control is also hazardous in that the employee may drift significantly in unproductive directions, or perhaps make costly or time-consuming errors that could have been avoided.

Having decided the approximate extent of control the individual needs, proceed to set reasonable deadlines for task completion, or for the completion of portions of the task. Prepare to follow up as those deadlines arrive.

Two points to be stressed at this stage are (1) the reasonableness of the deadlines and (2) the timeliness of follow-up. Give the employee plenty of time to do the job, including, if possible, extra time for contingencies. However, when a deadline arrives and the manager has not been presented with results, he or she should take the initiative and go to the employee. There are few, if any, valid reasons for letting a deadline slide quietly past without asking for the results that are expected.

Laziness in enforcing deadlines will cause far more grief than just weakening the effects of delegation. Let only a few deadlines slide by unmentioned and some employees will automatically adapt to this pattern of behavior and assume that deadlines are unimportant. On the other hand, if a manager makes it a habit always to follow up on deadlines, employees will pick up on this pattern and produce timely results.

Throughout the entire delegation process, try to avoid being a crutch for the employee. Regardless of how much guidance and assistance the manager is called on to provide, try to avoid solving problems for employees. Rather, focus on showing employees how to solve their own problems.

Delegation Failure

The responsibility for most instances of delegation failure rests with the supervisor, not with the employee. Some failures are to be expected; delegation is an imperfect, sometimes highly subjective process. However, when delegation problems arise, assess responsibility for the problem with the following questions:

- Did I assign a task only to take it away before the employee could truly demonstrate any competence at the task?
- Did I maintain too much or too little control?
- Did I really provide the employee with authority commensurate with the responsibility I delegated?
- Did I split up an activity such that no single person with some authority could develop a sense for the whole?
- Was I overly severe with an employee who made a mistake?
- Am I giving proper credit to the employee for getting the job done?
- Am I keeping the more interesting tasks for myself, delegating only the mundane or unchallenging activities?
- Have I slacked off in my own work as I delegated certain activities away, or have I used the time saved to increase my emphasis on managerial activities?

U.S. Secretary of State and retired Army General Colin Powell once said on the subject of failure, "The same day I have a disappointment, I try to reflect on it and say, 'What did I do wrong?' I never look for scapegoats. Once you have experienced a failure or disappointment, once you've analyzed it and gotten the lessons out of it—dump it."

Would that everyone could have that attitude, even in circumstances when living with the concrete results of mistakes and the mistakes of subordinates. Just as importantly, would that managers could instill that in their subordinates.

When an employee fails in the performance of a delegated task, the failure is often shared. Sometimes the failure is largely the employee's—for doing something incorrectly or for not following instructions. But many times the fault lies with the supervisor. And since everyone learns from failures far more often than from successes, punishing or overcorrecting an employee who makes a mistake makes little sense.

Peter Drucker takes the position that, "The better a man is, the more mistakes he will make, for the more new things he will try. I would never promote into a top-level job a man who was not making mistakes. . . . He is sure to be mediocre." Criticism should, of course, be delivered when deserved, but it should always be constructive and include the means for correcting the errant behavior and avoiding a repetition of the problem.

In many situations in the world of corrections, there is essentially no room for error without grave consequences. A warden simply cannot afford to allow a death row inmate to escape without paying a significant price in career status and perhaps even in human life. However, in headquarters functions and in many support and administrative activities in the institution itself there is indeed room for errors, and they occur frequently. A tabulation error in a monthly report on consumption of dairy products may cause a budget problem next quarter. But this kind of problem has much less impact than a riot caused by a lieutenant misjudging how to break up an unauthorized inmate group, or a probation officer deciding whether or not to issue a warrant on a violator.

The employee who is soundly taken to task for making a mistake while exercising reasonable individual initiative will likely shy away from exercising further initiative for fear of punishment. The stakes can be higher in correctional work, as noted, and errors are a natural part of career growth. But it is still accurate to say that the employee who never makes a mistake is not growing very much. This employee (and especially the supervisor who never makes a mistake) is the one who never exercises initiative or tries anything new or different.

A decision to delegate is, at best, a calculated risk. Managers should be willing to take that risk. After all, one does not know for certain about an employee's capability until that person has been given a chance with an actual assignment.

Managers should feel a strong incentive to practice and control the process of delegation. As suggested earlier, managers cannot always expect to do everything of importance on their own. Since a manager cannot shed final responsibility for task performance, he or she will be held responsible for any serious mistakes the employee may make. Knowledge of this retained responsibility can be sobering, but functioning as an effective supervisor means recognizing the necessity to delegate to get things done through people.

■ Building the Habit

It already has been noted that delegation (or lack thereof) should be looked at as a habitual behavior. The delegation habit begins by concentrating on a single activity at a time, working with one employee until that activity has been completely incorporated as a part of the person's job. Proceeding in this fashion, one activity at a time, it may take a long time to make significant progress with the list of activities eligible for delegation. However, this methodical approach, conscientiously pursued, is the surest way to learn delegation as a true habit.

Incomplete or improper delegation is an undesirable habit and as such can be difficult to alter. However, the practice of proper delegation, once substituted for old habits, is itself a strong, useful habit that will serve well throughout a management career.

EXERCISES

Exercise 6-1: To Whom Should You Delegate?

You are the controller in the headquarters office of a state correctional agency. Eight of your staff members are briefly described as follows:

1. Frank is technically competent, seems to communicate clearly, especially in writing, and pays attention to details.

2. Marsha has consistently shown good judgment in matters of finance, particularly in analyzing reports and statements for financial implications.

3. Barbara is a dependable employee. Young and fairly new to the operation, she nevertheless appears ready to handle increased responsibility.

4. By both credentials and experience, Tom is your most technically qualified employee.

5. Carlos is low key, polite, and a diplomatic "people person." He is clearly your best letter writer.

6. Paula is an empathic individual. People are generally comfortable with her and inclined to speak freely.

7. Sam has displayed both technical and managerial skills. He has successfully run several special projects.

8. Maria is an organizer. Few if any details of arranging a gathering of people escape her attention.

The functions you can delegate to your employees include the following:

a. Provide technical support on a special system-wide study of the operation of the inmate commissaries in all of your agency's institutions.

b. Manage the study, including responsibility for technical content.

c. Schedule and organize periodic project review meetings.

d. Review and approve all correspondence relating to the study.

e. Analyze and approve all expense reports relating to the study.

f. Write monthly status and progress reports for the study.

g. Answer all inquiries concerning certain activities of your department and the agency itself, with respect to fiscal matters.

h. Locate and screen applicants for potential employment in the department.

i. Requisition standard supplies for the office and assure that the supply room is adequately stocked.

Instructions

Create a possible pattern of delegation by matching employees and assignments in a manner that appears to make best use of each person's capabilities. Note that there are more assignments than people and that any one person may be capable of taking on more than one of the assignments. (Caution: avoid over-delegation—assigning an employee a job for which he or she is probably not qualified.)

_____	1. Frank	_____	5. Carlos
_____	2. Marsha	_____	6. Paula
_____	3. Barbara	_____	7. Sam
_____	4. Tom	_____	8. Maria

Exercise 6-2: A Delegation Key

This is an extension of Exercise 5-1 concerning Ms. Marcie Denton, chief of maintenance. When she helped establish the department, Marcie designated three men—one of them a working foreman—to handle all electrical and mechanical repairs and maintenance. She empowered the foreman with the authority to requisition parts and materials up to a limit of $50 per order.

Questions

1. Since Ms. Denton has already made the foreman responsible for maintenance and repairs, why would she then find it necessary to take specific steps to give the foreman "authority to requisition parts and materials up to a limit of $50 per order"?

2. What does this illustration tell you about responsibility and authority in proper delegation?

Exercise 6-3: What, Me Worry?

Juan Ramirez, manager of the inmate-operated print plant at a large medium-security correctional institution, was about to retire in 6 months. Ramirez was responsible for all production of all printed forms for state agencies. He also supervised the operation of two warehouse facilities staffed by minimum security inmates from adjacent correctional camps. Over time, it had become obvious that the workload for the plant was going to grow to the point where new facilities would be needed. Juan had done little planning for the future because he knew he wouldn't have to deal with the consequences of his delay.

Finally, pressure from the warden forced Ramirez to face the daunting task of preparing a proposal for a new, larger plant. It would have more modern equipment, including an elaborate climate control system that would keep humidity and heat from affecting the print process. The lead time for the entire process would be at least 18 months, and a growing inmate population was going to be a critical factor in about 2 years. To make matters worse, there were only 10 weeks until final budget submissions were due in the legislature.

Ramirez decided he was going to delegate the preparation of the plant proposal to his assistant, Frank Santarelli. He called Santarelli to his office, and said, "Frank, we've got a problem. We need to get a plan together for the legislature to design, spec out, and fund a print plant that will employ twice as many inmates as we're working now. The population is going to be going up so fast with this new 'three strikes legislation'. We are going to be in big trouble if we don't have those new jobs on line within 2 years. This is your baby. Run with it. And by the way, you only have 10 weeks to get it done!"

Santarelli wasted no time in voicing his concerns. "Boss, I'm sure I can do this, but I've never been involved in construction planning, budgeting on this scale, or for that matter, writing anything that is going to go before the legislature. How about let's work on this together, at least at the start until I can see how you want me to do it?"

Ramirez said, "Look, the reason I'm giving you this is because I'm not going to have to live with this new factory, you are. Besides, if you don't have any more initiative than that, I may as well do it myself." He grinned as he added, "Actually, if you really want that promotion to factory manager at the penitentiary, this could be your ticket out!"

About 3 weeks later (after hearing nothing back from his boss), Santarelli stopped Ramirez on the plant floor as the latter was making an impromptu inspection of a big forms job for the Department of Social Services. He said, "Juan, I need to talk to you. There is no way I can pull this all together in the next 2 weeks."

Ramirez replied, "What do you mean 2 weeks? You should have 7 weeks to go."

Santarelli responded with, "What you didn't know was that the warden needs a week to review it and then headquarters wants 3 weeks to give it a once-over before it goes to the legislative budget committee. And it has to be to the committee a week sooner than you thought so their analysts can figure out where to get the money. I just found all this out today. We're in big trouble here."

Instructions

Ramirez committed several significant errors in "delegating" the preparation of this proposal to Santarelli. Identify at least three such errors in the case description.

Using as many steps as you believe necessary, describe how this instance of delegation might have been properly accomplished.

ENDNOTES

1. *Roget's Thesaurus of Synonyms and Antonyms,* Halo Press, 1990, sections 755, 759, and 760.

Time Management: Expanding the Day Without Stretching the Clock

Chapter Objectives

- Place "time" in perspective as an unrenewable resource that influences all managerial activity.
- Identify the common time-wasting practices encountered in organized work activity.
- Address the problems presented by excess planning and organizing, or, commonly, "overplanning."
- Identify delegation and planning as key considerations in the manager's effective use of time.
- Offer practical suggestions for analyzing one's use of time and improving one's effective use of available time.
- Isolate the sources of time-wasting pressure inherent in the organizational environment and suggest the manager's appropriate response to these pressures.

Time is fixed income and, as with any income, the real problem facing most of us is how to live successfully within our daily allotment.

Margaret B. Johnstone

Whether these are the best of times or the worst of times, it's the only time we've got.

Art Buchwald

Making It Happen in a Crisis Situation

It was a riot/hostage-taking situation in a major institution. There was a media frenzy that attracted national attention. The director of the agency had to brief his higher-ups in government every day at 11:00 A.M. This time drove every stage of the bureaucracy's preparation for this important event. Working backwards, it involved:

By 10:00 A.M., a draft of the briefing materials had to be ready for the director's review.

By 9:00 A.M., the preliminary briefing content had to be ready in the assistant director's office for review.

By 8:00 A.M., all reporting departments had to have their raw input to the assistant director's staff for collation.

By 7:00 A.M., the on-site commander (the warden in this case) had to report in as to the current situation.

By 6:00 A.M., all tactical and response elements at the riot site had to have reported to the warden on their current status.

Thus, under the best of circumstances, the prepared briefing materials were 5 hours old when they were presented to the director's superiors. Of course, if events eclipsed this process, that information was presented ad hoc. But the formal report prepared every day of the crisis was a result of this kind of rigid time structure. There was no plan for this; it simply happened in response to the needs of the agency during an emergency. Staff at every level calibrated their schedules, activities, information-gathering actions, and resources in order to make it happen.

■ Time and Time Again

To the corrections professional, time is not only a measure that offenders mark very carefully. It often is a critical factor in fulfilling the agency's or institution's objectives to ensure public safety and provide humane conditions of confinement for inmates. Indeed, for everyone, time is a resource of considerable value. Moreover, time is always moving, always going forward regardless of whether it is put to some specific use. Effective time management is critical to being a good manager.

In addition to being self-consuming, time is a fixed resource. There are just so many minutes in an hour and so many hours in a day. But paradoxically, while available time is indeed limited, for all practical purposes the demands on a manager's time are open-ended.

It is common to hear people assert that "time is money." And to many people in business, time well spent is the difference between loss and profit. But to providers of all kinds of services, including corrections, wasted time represents wasted cash resources, since salaries and expenses go on even when service is not being rendered. This chapter begins with the assumption that few people—no matter how organized and efficient—could not improve their use of some of the working time available to them.

Thankfully, correctional work seldom is driven by the kind of situation mentioned in the opening vignette. But even though managers usually have more discretion in how they budget their time, they still often are ruled by the demands of the job. It may seem as though the work and the hours rarely come out even, there being considerably more of the former than the latter. Managers react to this disparity in various ways. At one extreme is the supervisor who continually stays late in an effort to get "caught up," a state that seems never to arrive. This approach is not the answer, and neither is the answer found in the other extreme—walking out the door at quitting time with many problems remaining to be solved. Certainly there are days when extra time and effort are the only answer. But when this is the regular solution it simply turns a full-time job into a full-and-a-half-time job without truly improving the way the department runs.

With occasional necessary exceptions, the place to begin improving performance is in the efficient use of regularly scheduled work hours. Managers owe it to themselves, and certainly to their agency, to get more out of each day by putting more of the proper effort into each day.

However, a little candor is in order here. A great deal of "working" time goes to other things. Late starts, slow starts, long breaks, long lunches, and meetings of questionable value (and especially periods of social conversation) are part of many people's working style. These nonessential activities can drain supervisory and nonsupervisory productivity. Personal considerations and social relationships are important in a work environment. However, few people realize the extent to which these things cut into the time for which they are being paid to perform certain duties. Many a manager has stretched a coffee break into a half hour while complaining of how much work there was to be done.

Aside from out-and-out nonproductive activity, though, a great deal of time is wasted in the ways managers approach supposedly legitimate tasks. The following section will briefly highlight the readily identifiable time wasters, then offer suggestions for incorporating time savers into individual behavior.

■ The Time Wasters

A number of common time wasters are important to comment on at this stage.

Failure To Delegate

Failure to delegate in a thorough and effective manner is one of the greatest time wasters to which a supervisor can fall victim. The cautions of Chapter 6 need not be repeated, but it is important to remain open to ways to improve in this area.

Failure to Plan and Establish Priorities

A 1959 Walt Kelly "Pogo" cartoon shows two characters leaning against each other. One says, "MAN! You know, it ain't the work so much what tires you, it's the planning." The other character responds with, "WHOO! If you got the strength you can say that again." Planning takes effort. And often it is an unwillingness to put effort into planning that catches up with a manager.

You may have heard the term "traffic-cop management" or perhaps "firefighting management." These describe an all too common approach to supervision: very little looking ahead and not much consideration given to deciding which of a number of tasks is most important. It is usually the problem of the moment that gets the attention or the people making the most noise who get heard.

Corrections in general, and prisons more specifically, are a particularly fertile environment for this management style. Confinement facilities have their never-ending series of minor crises and the occasional major crisis, and community-based correctional agencies are similarly vulnerable. Each emerging situation is the perfect rationale for abandoning structured management activities. A manager can be extremely busy practicing this form of management but can nevertheless be wasting significant amounts of time. Without some rationale for approaching work in a manner consistent with its necessity, and for dealing on an ad-hoc basis with only real crises, tasks of lesser importance consume disproportionate amounts of time and not enough time is devoted to more important, but nonemergency situations.

Overplanning and Overorganizing

Overplanning and overorganizing are two common traps that many people fall into when making what they believe are conscientious efforts to improve their use of time. Because they hear so much about the necessity of planning and the importance of being organized, some supervisors tend to overcompensate for their self-perceived weaknesses by going overboard. They rely on plans, task lists, schemes of objectives and subobjectives, and open files, active files, tickler files, dead files, and so on to compensate for other, more fundamental problems.

It would be easy to insert here a psychological treatise exploring the possible reasons behind the tendency to overplan and overorganize. These reasons would probably include avoidance of unpleasant tasks (planning is "good," and while one is busy planning one does not have to be doing). They also might include a sense of insecurity, uncertainty concerning how superiors will deal with mistakes, and a fear of failure. Whatever the reasons, a manager's personal planning and organizing become time wasters when they occur to excess.

What is excess planning and organizing? Only the individual manager knows for certain how much is personally adequate and when overplanning and overorganizing are prevailing. But a few examples can illustrate the problem.

Consider the manager who maintained his calendar in duplicate—one a full-size loose-leaf binder and one a version that would fit into a pocket. When asked bluntly by another manager why he maintained duplicate calendars, he could say only that he especially liked the features of the desktop book but that it was too large to carry around conveniently. Thus nudged into thinking about what he had been doing, the manager realized that the time and effort of maintaining duplicate calendars was not worth the minimal benefit gained. A second manager (representing the opposite of the spectrum) literally worked his entire department from a set of 3 × 5 note cards carried in his shirt pocket. When questioned about the efficiency of his system, he said that he had started keeping notes about his work this way when he began as a correctional officer, and thought that it still was adequate to his needs—a doubtful proposition.

Many managers engage in activities that are not worth the benefit or convenience gained. It's fine to update a "To Do" list daily. But if every day sees the To Do lists (plural because there are lists A, B, C or 1, 2, 3) revamped, reworked, and reordered, then time is being wasted. And if a manager is "organized" to the extent, for example, that a task is pulled from the "To Do" folder, stuck in the "Active" or "Open" file, and advanced through three or four other stages, each with its own name and filing system, then organizing has become overorganizing.

Planning and personal organizing are surely important. However, both of these virtues can become excesses. Quite simply, even the noble pursuit of planning should be kept consistent with the expected results. That is, never pour a dollar's worth of effort into assuring a two-bit payback.

Face-to-Face Contacts

Face-to-face contacts represent the essence of a supervisor's job. However, one-on-one situations readily get out of hand and waste significant amounts of time. Aside from social conversation, which should be controllable within reason, managers experience numerous interruptions during the workday. Handling these interruptions will influence the degree of effectiveness with which available time is used.

For instance, think about the last time a vendor or a counterpart in another agency called to say he or she "was in the neighborhood" and wanted to see you. Did you drop something else in your schedule to see that person? Or the last time you were talking business with one of your employees did you allow another employee or fellow manager who wanted your attention for a moment to interrupt? In short, do you usually give in to the pressures of interruptions and unexpected visits? Other people will readily waste your time if you allow them to do so.

The Telephone and Other Technology

One of the most useful communication facilitators ever devised—the telephone—can also be one of the greatest time wasters. The telephone can be of considerable value when effectively controlled, but often the tendency is to allow it to control us.

Modern phone and computer messaging systems or e-mail may help in this regard. These systems can help cut back on "telephone tag" by enabling callers to relate complete thoughts on a subject, and to receive feedback on that issue via the same technology. But e-mail and other Internet-based communication systems can introduce a major opportunity to lose precious time as well. Whether these systems are a real aid to managers depends on how the organization structures them, as well as the expectations the agency sets for managers to use them.

Meetings

Like the telephone, the meeting is another means of communication that is misused and abused, to the extent that its effects are often the reverse of what was intended.

Sometimes it seems as though the modern correctional agency runs on meetings. However, close analysis of many (if not most) meetings will reveal significant wasted time. This is not to say that most meetings are unnecessary, although some certainly are. Rather, the majority of meetings are too long relative to the results they produce and too loosely managed to be truly effective. The advice offered in Chapter 22 is applicable here.

Paperwork

Today, many correctional managers say that their institutions seem to run on paper. Many must feel as though the in-basket is like the magic pitcher in the children's story of many years ago. No matter how much was taken from it, it was always full. Many managers seem never to be caught up on paperwork, and the supervisor who conscientiously tries to keep the in-basket current usually discovers that this cuts into time needed for more important tasks.

Like many other activities, paperwork can be essential as well as wasteful. It becomes a time waster when some things get done at the expense of others, items of minor consequence get attention while important items remain in the stack, and items that should be ignored receive attention simply because they are there.

Personal Habits

Whether a manager wastes significant amounts of time on the activities just mentioned will depend largely on the habits that have been formed regarding these activities. It is possible, for instance, to allow meetings to be loose and rambling simply out of habit—the way it has always been, following the path of least resistance.

Beyond a manager's personal approach to specific activities, however, there is an overall pattern of behavior to consider. Are you perhaps a slow starter, entering the day with an hour's worth of coffee and procrastination until you take hold and begin to produce? Are you in the habit of jumping from task to task, starting many jobs but completing few? When you reach a point in a task when you need information and assistance from others, do you simply stop and wait for them rather than fill your waiting time with productive effort? For good or for ill, unless a manager exercises conscious control, his or her approach to daily duties will be governed largely by personal habits.

■ The Time Savers

To avoid the productivity penalties caused by the time traps just discussed, there are a number of things to consider that can improve one's ability to get the job done. Some of them can turn the aforementioned time traps into net-time gainers.

To get control of time, a manager must first determine what must be done and in what order things should be accomplished. The process of planning and setting priorities, along with the practice of proper delegation, is a major force in the productive use of time. Volumes have been written about planning, but for these purposes some simple suggestions will provide an effective basis for a solid start in managerial planning.

Determine How You Really Spend Your Time

Make the decision to analyze critically your use of time to determine if you are really using most of your time wisely. To perform this analysis honestly, a degree of commitment is necessary, and it may also be necessary to accept the likelihood of sacrificing some old habits for the sake of self-improvement.

Check Up on Yourself

For some period of time (at least 2 weeks but preferably 3 or 4 weeks) keep a record of how you spend your time. This need not be an elaborate scheme in which you write down every 30-second task or record your activities down to the minute. Rather, reasonable entries might look like: "weekly staff meeting— 90 minutes"; "interview prospective employee—one-half hour"; and "prepare monthly statistical report—3 hours."

When you have sufficient information to work with, sort the results and determine the approximate portion of your time spent on routine tasks, unscheduled tasks, and tasks that could be considered emergencies.

Routine tasks are those you do on a regular basis (for instance, daily, weekly, or monthly staff meetings). They are routine in that you know what they are and approximately when they occur.

An unscheduled task is one for which you know the "what" but not the "when." For instance, you may know there will be times when you have to write a report because of an inmate-related incident or other unusual event, but you have no way of knowing when this will occur.

An emergency task is one for which you know neither what nor when. You simply know that unplanned events occur and you must respond. Responding to an escape plot or investigating an assault would fall into this category.

When you have finished analyzing your time record, you should know most of the things you do and the approximate portion of your time spent on routine, unscheduled, and emergency tasks. Further, you should be able to tell which of your tasks are likely candidates for delegation.

One of the first useful pieces of information gained from this process is an appreciation of how much of your time on average goes to what you categorize as emergencies—tasks for which you are aware of neither the "what" nor the "when" in advance. Correctional life is full of such episodes, and the average supervisor (if there is indeed such a person) will find a surprisingly high percentage of time consumed by "emergency" tasks.

Of course, in making your analysis, you also should be sure that all of the incidents categorized as emergencies really are. Although you cannot ordinarily avoid genuine emergencies, especially those falling clearly within your responsibility, you can learn enough to enable you to allow for them in your planning.

For example, suppose your three-way analysis of several weeks' activity suggests that you spend an average of 35 percent of your time on so-called emergency tasks. Then on average you should leave that much of your time flexible to meet emergencies rather than scheduling yourself to the hilt with known work.

When you have eliminated from consideration those tasks you can delegate, you are left with a number of activities—routine tasks as well as unscheduled tasks—that you must do. The next stage of your planning involves determining how you will approach them.

A Personal Planning Gimmick

You may have heard the following suggestion before, but its enduring applicability makes it worth repeating here.

Of the tasks you know you must personally perform within the coming few days, write down in list form the four or five you consider most important. Then spend a few minutes putting these tasks into their order of importance; that is, task number one should be the first task you must accomplish. The result should be a list of the several most important jobs you must do and the order in which they must be done.

The approach is simple. Go to work on number one and stay on it until it is done. Granted, you may be pulled away for an emergency. However, when the emergency is over, get immediately back to task number one. Turn away all other interruptions, and do not allow your efforts to be diverted. When task number one is completed, move on to task number two and tackle it the same way.

These need not be only desk-bound activities. Suppose you are a factory manager of a correctional industries operation. It would not just be desirable, but critical, that you allocate a significant amount of time out to being on the production floor, talking with civilian and inmate workers. The supervisor of a high-security detention unit needs to be out touring the housing unit, talking to staff and hearing inmate complaints and concerns—not just reviewing incident reports and policies. A probation supervisor should plan on spending a certain amount of time watching line personnel in court or in actual offender-related contacts.

You may not get through your entire list as quickly as you would like. Rest assured, though, that you will have accomplished more than if you had attacked these several tasks in any other manner. Also, when working this way you are always at work on what you deem to be the most important task of the moment.

Do not forget to allow yourself some planning time. The planning process is essential, urging you to think ahead and establish priorities and determine direction before proceeding. A "plan" as such need not be elaborate—perhaps a list on your calendar of four or five items to tackle tomorrow. However, the time you devote to planning will ordinarily pay itself back in improved efficiency several times over.

Set Objectives and Work to Deadlines

It always pays to be working toward a specific objective, whether that objective is the completion of your most important open task or simply the completion of a particular part of a larger task. Any objective, whether organizational, departmental, or personal, should consist of what is to be done, how much is to be accomplished, and when it should be completed.

As a unit manager of a housing area, one objective may be to complete the performance appraisals (what) for all five counselors in the unit (how much) by the end of Wednesday (when). In the case of the correctional industries manager, objectives can be quite clear—the production of a certain number of desks or brooms every month at a certain cost. For the chief of probation, the objective may be to assure that no more than a specific number of unevaluated cases remain on the caseload at the end of each week without resorting to overtime on the part of the probation office staff. For the business manager, it may be that all financial statements must be completed and reconciled with headquarters records within 5 working days after the end of the month.

Related to the "when" portion of any objective, the time boundaries of the task may be externally imposed, as in some statutory reporting requirement. They may relate to internal institutional needs, like a goal to keep fewer than 35 inmates in a prison intake/orientation unit over a weekend, due to staffing constraints. But even if there are no outside time limits, be sure to set a deadline for yourself. As for any deadline for an employee, make self-imposed deadlines reasonable and stick to them.

Write It Down—and Change It

Even someone who is used to doing day-to-day task planning mentally should commit plans to paper. The plan may, and usually should, be extremely simple— a few lines on a desk pad, perhaps a few entries in a pocket notebook. Make sure that tasks, priorities, objectives, and deadlines exist in writing in a place where they will be seen frequently.

Of course, simply noting 2 or 3 days' planned activity in writing does not guarantee being able to follow the plan as written. Each time some unforeseen event intrudes on an activity and upsets the plan, take the new requirement into account and revise the written plan. Get in the habit of spending a few minutes at the end of each day planning for the following day, revising the plan to reflect current needs. A computer, using nothing more elaborate than a word processor, can enable quick updates on these planning documents. A few minutes at the end of each day making necessary changes yields a fresh list, and also can allow old list items to form the basis for a record of past activities.

Face-to-Face Contacts

Face-to-face contacts were described earlier as often being significant time wasters. They need not be an encumbrance to effective time use if approached properly.

The biggest waste of time in face-to-face contacts is the tendency to engage in excessive small talk and social conversation. The images of the crowd around the coffee pot or a group of correctional officers hanging around the yard shack are based in reality. Regrettably, managers also are prone to the same kind of socializing on the job. Social conversation should not be taboo—a certain amount is essential to morale and supportive of good interpersonal relations. However, business can also be conducted on a friendly level, and beyond minor social pleasantries, a manager should make an effort to stay on business in contacts with employees and coworkers.

A great deal of the manager's most important work takes place within the context of the one-to-one relationship with each employee. Consequently, developing a relationship with each employee is recognized as one of the supervisor's most important tasks. This is not to suggest that supervisors minimize contacts with employees or that each contact must necessarily advance the business of the department. Listening to a troubled employee may take an hour without apparently advancing the work of the department at all. But the time is well spent if it results in upholding the morale and productivity of the employee. On the other hand, talking about fishing for an hour with an employee who happens to be an enthusiastic angler and who brings up the topic at every turn is a clear waste of time.

Just as a good manager does not allow employees to waste his or her time, neither should a manager waste their time by delaying them with irrelevancies and nonessentials simply because of being the boss. One of a manager's primary functions is to enhance employees' ability to accomplish their assigned tasks, and wasting their time will not further that goal.

Another form of face-to-face contact that frequently wastes time consists of visitation by others from outside the operation. On first glance, this may seem like no problem at all in a correctional facility—it is a closed environment, after all. But on examination, many institution departments are subject to this phenomenon. The supervisor of education may be approached on a regular basis by textbook vendors or persons who want to demonstrate the "latest in programmed learning" computer software. The hospital administrator may be besieged with calls from salespersons seeking to come in to promote new pharmaceutical products. The chief of maintenance and business manager probably have more than their fair share of inquiries about how to get on the institution's procurement list, with offers to come in and demonstrate their product's superiority.

A manager who is involved in a significant number of such distracting contacts from outside the institution will find that productivity eventually suffers. Once reinforced positively in this behavior, visitors from outside will tend to adopt the practice of calling to see if they can "just drop by." However, it takes little effort to refuse politely to agree to allow a visitor to come into the institution on short notice. And once the recipient of such calls begins to demonstrate a more discriminating response, these visits are likely to diminish in frequency or cease entirely.

The Telephone

How managers make or take telephone calls will determine the extent to which the telephone rules daily activities and consumes excessive amounts of time. Granted, there are many times when a manager cannot avoid a telephone interruption. However, many calls are controllable with a little effort.

When it is possible and when interruptions would be a hindrance, have incoming telephone calls held. This is especially important when busy in important personal contacts. Stopping to answer the telephone wastes someone else's time as well as possibly the manager's. If possible, have calls screened and "sorted out," with whoever is answering the calls taking messages for later action, and putting through only genuinely urgent calls. As a general rule, if there is not a staff member to perform this screening function, do not use an inmate. (Those instances where an inmate may be used to answer a phone should be very carefully controlled, and limited to phones that do not have a connection to an outside line).

Likewise, do not interrupt something important you may be doing to make or return a call simply because you happen to remember it just then. Note a reminder of the necessary call and take care of it later.

When making telephone calls that respond to all but the simplest messages, it is helpful to spend a few seconds organizing relevant thoughts before dialing. Jot down the points to cover, and have these and any necessary references and note-taking materials handy. Otherwise, the call may be prolonged unnecessarily, only to remember after ending the conversation that there was something else important that should have been covered.

Upon returning to the desk at various times during the day and finding calls to return, do not always feel it is necessary to return every call then and there. Urgent calls, of course, should be returned at once. However, routine call-backs are most appropriately handled if there is one particular time of day to do most of this kind of telephoning. Then, go through all possible calls at a single sitting.

For someone who works the day shift, for example, the best time is usually from early or mid-afternoon until the end of the workday. Bear in mind that this time may be a bit more restricted for some staff (such as shift supervisors). This is because many supervisory posts have a log, shift report, or other paperwork to contend with as the shift ends. Conversely, the noon hour is probably the worst time to return telephone calls because many employees (in and out of an institution) are eating lunch.

Use the telephone wisely and it can be one of the best time savers available to you. However, let the telephone use you and you will suffer through lost efficiency and reduced effectiveness.

New Technologies

Many, if not most, agencies are using some form of messaging services and e-mail technology to facilitate contacts that otherwise would be handled by telephone. This approach works particularly well for noninteractive messages, such as daily updates throughout a large organization. It also is effective for disseminating bulletin-type information, such as changes in routine or scheduled activities. Phone messaging can be quite usable for transacting a great deal of business that otherwise would be slowed down by managers playing "phone tag." However, these communication methods do not allow a fully efficient, free flow of interaction between two parties on a complex subject or range of subjects. Successive computer-recorded or e-mail messages to cover the same material take more time than a personal call covering those topics.

These methods also create a reduction in what might be called "communications bandwidth." In relying solely on this type of communication, there is a greatly diminished ability to pick up nonwritten communication cues, such as tone and inflection in a phone call, and body language in personal sessions between two people. The successive exchange of e-mails of phone/computer messaging can be useful, but managers using them should be aware of these shortcomings. Relying solely on them for truly important decisions would seem to be unwise.

In particular, it is important to remember that these impersonal methods of communication are a poor substitute for the personal touch that can enhance a manager's relationship with employees. One of the authors was notified by the agency's automated phone messaging system of a promotion and transfer. (In all fairness, the author was aware that the personnel action was pending, but it still had a far different impact than hearing the news in person.) In another instance of note, an organization used e-mail to notify all employees at once of large monetary awards granted to individual staff members. This method was used to tell the recipients as well, rather than supervisors telling the employees individually before disseminating the information. In each case like this, the supervisor missed an opportunity to give positive feedback and attention to an employee—something that no manager should ever miss.

The use of personal digital assistants, "Blackberries," and other communication devices is taking hold in the private sector and in upper management echelons in corrections. Mid- and line-level managers can also use this technology to good effect. However the possibilities of privacy act violations and compromise of confidential information should be in mind when doing so.

In a related concern, agency legal counsel will no doubt have specific guidance on the subject of how e-mail records should be maintained. In general, however, it is wise to consider that anything written in an e-mail is likely to be discoverable in a matter that is litigated. In thinking about what you commit to emails, remember the old saying of Washington, D.C. bureaucrats, "Don't commit anything to writing that you wouldn't want to see published on the front page of the *Washington Post*."

Meetings

The subject of meetings is of sufficient importance to warrant a chapter of its own (see Chapter 22). At this point it is sufficient to say that meetings should be held only when necessary; planned and organized with specific purposes in mind; started on time whenever possible; kept to the subject; and allowed to consume only the amount of time required to accomplish their specific objectives.

Paperwork

It was noted earlier that it is all too easy to allow the in-basket to control a great deal of a manager's time. Although all managers receive a great deal of material deserving attention and action, not everything that comes in is of equal importance.

Consider the problem of confronting several days' accumulation of incoming material (and for some managers this can be a significant amount of paper). Rather than simply working on the basket from the top down (or bottom up, for that matter), first sort all the paper into three categories.

The first category should consist of those few items that genuinely deserve immediate attention and those that, although not urgent, can be disposed of by noting a quick word or sentence in response. These might include memos from superiors on current issues, and routine approvals requiring no research or consideration.

The second category should include items that will take some time to resolve because they require research or more extended effort, as well as non-critical items that can be resolved at leisure. They might include policies to review, memos from subordinates, proposed activities requiring coordination with other departments, and similar items. Some of the items in this category may be routine or unscheduled tasks, and should be worked into priority planning. Also, a few items in this pile may be clear candidates for delegation. Coming across these should result in tentative decisions as to who might be able to handle them.

The third category of incoming material, and often the largest stack, consists of items that can safely be discarded. Resist the temptation to file every information memo or advertising brochure that comes in. This kind of material, especially information on routine inmate programs or vendor brochures, rapidly becomes out of date. In most instances, when information may be needed in the future about a specific inmate program or a certain line of supplies or equipment, the institution department with the primary interest in that item of information (the education supervisor's office or purchasing agent, in this example) should be able to come up with current information. Simply screen all this material for topics of special interest, pass along things that may be of interest to someone else, and throw away the rest. Rest assured, it is more than 99.9 percent certain that the memo or brochure in today's mail will never be more than office clutter or file cabinet filler should it be kept.

■ How to Respond to Time-Wasting Pressures

Time-wasting pressures close in on managers from every direction. The choice is two-fold. You can give with pressures and bend in the direction they are going. You can become a positive force, refusing to bend, and influencing others in productive directions. And indeed the choice is up to you.

Time-wasting pressures are likely to come from the sources described next.

Your Supervisor

At times it seems as though higher management is insensitive to the demands on a line manager's time. There is no end to the reports that must be produced and the meetings that must be attended. It seems this way especially when the workload already is high and the supervisor comes along with something to add to the pile. However, the manager who places more work with an already overburdened supervisor is not necessarily being insensitive. Remember, the next level supervisor sits at a different organizational level and often does not have full knowledge of everything his or her subordinate managers have to do.

Most supervisors would probably say that they know the full extent of their subordinate manager's workload (a dubious proposition). However, the manager personally is the only one who can truly say if a new assignment is too much. Depending on the relationship the manager has with the superior assigning the work, perhaps not accepting the assignment is an option when other duties already fully obligate available time and resources. Speak up. Let the supervisor know about the current workload level, and that there may not be enough time to do it properly. Talk it over; two-way communication is the key here. Chances are something can be worked out—a compromise, perhaps, or a reordering of priorities that will serve the needs and purposes of both levels of management. There may be many times when the supervisor will not really know how much a subordinate manager has to do unless they are told.

The System

Inefficient practices prevail to some extent in most workplaces, and government organizations and correctional agencies are no exception. Being surrounded by people who are disorganized, careless, and always late says something about the organization itself, not just its managers. Unfortunately, if the organization tolerates or even fosters that behavior, then the manager will be under pressure to behave in the same manner.

However, no organizational system was ever made better without someone taking the initiative to start improving one small section first. And today few, if any, organizations will resist well-reasoned initiatives for improving operations. More than ever in this day of citizen concern over government activities and the resources devoted to sustaining them, bureaucrats and politicians see the benefit of facilitating changes that benefit the public through cost-effectiveness and increased productivity. A few managers who are determined not to give in to the time-wasting pressure of the system can begin to apply positive pressure that will eventually be felt beyond their own departments.

Your Employees

Time-wasting pressure will frequently come from employees, especially those who may be insensitive to the true requirements of their supervisor's position. The manager is there to help employees and "run interference" for them. However, some staff may see their supervisor only in terms of their individual needs. Some supervisors have found they could spend most of their time in unnecessary handholding if they gave way to the pressure. (However, in all fairness to the employees who seem to require handholding, if this situation is widespread in your department then many elements of the manager's approach to supervision need examination.)

■ The Unrenewable Resource

Time is a resource that cannot be replaced. Once a given amount of time is used, it is gone forever. If it is not used, it is still gone. Managers who do not spend time effectively are not simply wasting their own time. In most instances they are also wasting the time of their employees. Conversely, any time a manager saves may be effective in saving time for some or all of the employees in the department.

Do enough planning to assure that employees do things right. Be constantly awareness of priorities. Ensure that both you and your employees are doing the right things, as well as doing things right.

EXERCISES

Exercise 7-1: The Time-Use Test

This exercise may provide you with useful information about the way you utilize your available time. It is suggested that you time yourself to determine how long it takes you to complete the exercise.

1. Read everything before doing anything.
2. Write your name in the upper right-hand corner of this page.
3. Circle the word "upper" in instruction 2.
4. Draw four small squares in the upper left corner of this page.
5. Mark an X in each square.
6. Draw a circle around each square.
7. Sign your name immediately below the title.
8. To the right of the title write "Yes, Yes, Yes!"
9. Circle each word in instruction 6.
10. Mark a large X in the lower left-hand corner of this page.
11. Draw a triangle around the X you just marked.
12. Somewhere in the margin, multiply 7,125 by 306.
13. Draw a box around the word "page" in instruction 4.
14. Speak your first name out loud.
15. If you have followed instructions to this point, say "I have."
16. Add 7,840 and 8,740.
17. Draw a circle around the answer to 16.
18. Draw a square around the circle you just drew.
19. In your normal voice, count out loud backward from 10 to 1.
20. Now that you have finished reading carefully, do only steps 1 and 2.

Exercise 7-2: The Time-Wasting Pressures

For a period of at least one full work week, keep track of the occasions when you felt your time was wasted. Make a note briefly describing each such occasion and also note the approximate amount of time involved.

When you have a week's worth of notes, review them and attempt to relate each occasion to a source of time-wasting pressure. As discussed in the chapter, the sources of time-wasting pressure are your supervisor, the "system," your employees, and yourself.

Considering your position relative to the sources you noted, next arrange your notations into three categories according to how much or how little control you can exercise in such situations. Use (1) full control, (2) partial control, and (3) little or no control.

If you have made realistic, objective assessments throughout the exercise, the resulting category listings should suggest several areas where you can apply concentrated effort to improve your use of time.

Exercise 7-3: Ten Minutes to Spare?

You are a unit manager at a medium-security correctional facility.

This morning you return to work following a week of annual leave to find your in-basket overloaded and your desk littered with telephone message slips. You are greeted by your secretary, who informs you that you are expected to substitute for your supervisor, the associate warden, at an outside meeting today. You have to leave no later than 9:30 A.M. to make the meeting on time, and you know you can plan on being gone the rest of the day.

You are left with 1 hour in which to begin making order out of the chaos on your desk before leaving for the meeting. True to your usual pattern, you set about reviewing all the items on your desk—telephone slips as well as in-basket items—and sorting them into stacks according to their apparent order of importance. You feel that perhaps you can at least get sufficiently organized to be able to begin work the following day with emphasis on your most important unfinished tasks.

About halfway through your hour of organizing your secretary enters to say, "Mr. Wade (the business manager) is here asking to see you. He says he wants 10 minutes of your time to discuss a minor problem with our fourth quarter budget request. Shall I tell him you'll call him about it? Or that maybe he should send you a memo?"

You cannot help feeling that the last thing you need during this hour is an interruption, especially for a non-urgent reason. It occurs to you that your secretary has briefly suggested two alternatives; to these you add a possibility of your own so that you see three choices:

1. Say that you cannot see Wade right now but that you will call him the following day.

2. Ask for a memo detailing the problem so you can look into it at your convenience.

3. Agree to the request for a meeting then and there, and try to limit the discussion to 5 to 10 minutes.

Instructions

1. Enumerate the advantages and disadvantages of each of the foregoing three alternatives.

2. Indicate which alternative you would most likely choose, and fully explain the reasons for your decision.

Exercise 7-4: The Visitor

As the door closed behind her departing visitor, shift supervisor Janet Mills glumly reflected that she had just lost an hour that she could ill afford to lose. To complete the roster she was working on, she was either going to be late for an upcoming labor-management meeting or have to work some extra hours tonight.

It had been a busy day anyway, and the hour had been lost because of a visit from a pleasant, but overly talkative, state police investigator. It seems he was more interested in trading "war stories" about prison cases he had investigated (when he was assigned as investigative liaison officer to the state's main penitentiary) than he was in discussing current cases. As was her practice, Janet agreed to see him when he stopped by her office to discuss a recent stabbing, although she resented the intrusion and he had not called in advance. As the hour went on, she knew she was losing time, but didn't want to alienate the investigator, since she knew she would be working with him on other cases in the future.

This incident, occurring on a Friday, made Janet realize that she had lost time to four such drop-in visits this week alone. This was the only noncorrectional staff member who intruded on her workday routine to this extent. But she generally did not like the idea of simply saying no or otherwise trying to avoid people who wanted to see her. This episode made her more aware that her work is beginning to suffer because of such demands on her time.

Instructions

Develop some guidelines that might help Janet and other supervisors deal with the problem of drop-in visitors.

Self-Management and Personal Supervisory Effectiveness

8

Chapter Objectives

- Round out the review of "The Supervisor and Self" by supplementing delegation and time management with important personal considerations.
- Highlight the key influence of individual initiative on supervisory effectiveness.
- Review the principal barriers to effective performance.
- Address the relationship of personal organizing skills to personal effectiveness.
- Examine the relationship of stress to personal supervisory effectiveness, identifying sources of stress and suggesting how to recognize them in one's own behavior and one's employees' behavior and briefly suggesting how to best cope with supervisory stress.
- Provide suggestions and guidance for organizing for effective performance.
- Provide guidelines for assessing one's suitability for a supervisory role, with implications for successful self-management.
- Establish the importance of self-discipline in achieving and maintaining personal effectiveness.

Technical training is important but it accounts for less than 20% of one's success. More than 80% is due to the development of one's personal qualities, such as initiative, thoroughness, concentration, decision, adaptability, organizing ability, observation, industry, and leadership.

G. P. Koch

The spirit of self-help is the root of all genuine growth in the individual; and, exhibited in the lives of many, it constitutes the true source of national vigor and strength.

Samuel Smiles

FROM THE INSIDE: Self-Management on the Job

The social worker was a fairly good writer, but she had other notable job skills as well, and was promoted to a position as executive assistant to the warden. There, she not only was able to use her communications abilities, but also was instrumental in orchestrating a number of changes in the case management system at the institution. Eventually she suggested, and was given, the seemingly dubious honor of being the only executive assistant in the entire agency who was also the case management supervisor for the institution.

This situation continued for about 2 years. During that time, she suggested and the warden approved a number of changes, including a shift to unit management. This entailed a significant reordering of case management services. While she was managing this transition, the institution experienced a number of major incidents that entailed investigating and writing lengthy, detailed, after-action reports. The executive carried out these duties well—something that was not unnoticed elsewhere in the organization. She was promoted to a more responsible position elsewhere in the agency.

■ It Starts with You

Managers who want true control do not look to sophisticated management theories or techniques, and do not rely on gimmicks. They don't expect higher management to wind them up and point them in productive directions. The unfortunate reality of corrections, like many other professions, is that although a certain amount of upper-management direction is essential, it will not always be there when it should be, and it will not always take an ideal form.

For a solid foundation on which to base development as a supervisor, begin with yourself. Do not presume to manage anything else until you have captured and controlled the essentials of managing yourself. Every employee is a manager of sorts. Even the worker with a few simple tasks and no subordinates is partly a manager of time, methods, and supplies. Since all resources apply together to influence output, the individual worker has at least a limited amount of flexibility in managing output. For supervisory personnel this flexibility is much greater. Performance expected of managers is more results oriented and less methods oriented than what is expected of nonsupervisory personnel, so for many managerial tasks any of several approaches can be taken as long as certain results occur.

Conscientious self-improvement in the use of personal resources will improve the effectiveness with which you manage yourself. The opening vignette in this chapter represents only one of a multitude of such cases in which a capable person who shows initiative is recognized by the organization.

Individual initiative will facilitate managing the efforts of others more effectively, whether the staff involved are correctional officers in a cell house, a plumber supervising an inmate work detail, or an accountant working independently in the business office. These personal resources include initiative, organization, and time.

The effective use of time was covered in Chapter 7 and many of the implications of organization were discussed under delegation in Chapter 6. A few pertinent remarks will round out this discussion of "The Supervisor and Self."

■ Initiative

The lead vignette illustrates how an individual with initiative can showcase and advance his or her own skills. It is true that managers cannot always exercise initiative in the way they might want to, because of limits on their authority. However, when it comes to self-improvement the only constraints on initiative are those placed there by the person. Even in the area of task performance closely circumscribed by institution or agency policy, a great amount of initiative can be possible.

Some first-line supervisors may counter this claim by pointing out that their managers do not seem to encourage initiative but rather seem content to function as "wheelbarrow pushers" rather than as true leaders. This may be a fact of life at times, but the higher manager who is operating as a wheelbarrow pusher is guilty of misguided performance. What do you suppose will happen when the administration decides to eliminate a wheelbarrow pusher (or that person retires) and you suddenly find yourself reporting to a manager who expects you to be self-propelled?

The ability to exercise initiative in self-improvement will depend largely on attitude-related factors that often suggest one critical question: how well do you like your job? There are probably not many supervisors who truly enjoy every minute of every day; jobs are mixtures of things we like and things we do not like, and one can only hope that the former usually outweigh the latter.

If people honestly dislike most of their jobs most of the time, there is little to be done for them. Dislike of one's job is usually reflected in a poor attitude and, in turn, by the absence of initiative. It also has the effect of hurting both the individual and the organization. After all, you can put a square peg in a round hole if you pound hard enough, but in the process, both the peg and the hole will be damaged. One would hope that supervisors who find themselves in a job that is such a totally bad fit will at some point see the problem (if their supervisor already doesn't see it and act on it) and find a new job or a different niche in the organization.

Setting aside those instances, people will never improve as supervisors unless they get moving, driven by the determination to do it on their own. The occasional swift kick administered by higher management has only a temporary effect and usually generates resentment and resistance. Each person has the source of all learning and growth buried within him or her; no one but that person can tap this source.

People are all occasionally haunted by the realization of the need to do more useful things and do a better job with the tasks they now perform. However, most also are always waiting for the "right" opportunity. A thousand declarations have been made about what will be done "when next year's budget kicks in," "as soon as count goes back down," "when I get more staff," or "after the first of the year." However, true initiative says the time is now, not later, and the first places to go to work making initiative count are in personal organization and the use of time.

■ Barriers to Effectiveness

Before any further discussion of personal organization, it will be useful to examine some of the traditional barriers to effectiveness (other than those presented by the poor use of time) that can have a bearing on how a manager approaches work. A few of these may be reflections of specific personalities. Some, perhaps, are so deeply ingrained that they never may be corrected in every respect. However, the mere awareness of them can be valuable in helping understand managerial behavior. This understanding can, in turn, lead to ways of compensating for what may be seen as shortcomings.

Fear of Failure

One major barrier to effectiveness is the fear of failure. Perhaps people shy away from taking calculated risks or making certain decisions because of fear of losing or simply afraid of being wrong. This certainly is a prominent factor in correctional management because the human stakes for mistakes can be so high. However, fear of failure generally leads to procrastination and inaction, which in turn lead straight to ineffective performance. Recognize that there is considerable risk and uncertainty involved in management at all levels. Were this not so, far fewer managers would be needed.

The Search for Perfection

Another barrier to personal effectiveness that is occasionally encountered is perfectionism. This can show up as excess time and energy poured into an undertaking, or the drive to continue seeking the "ultimate solution" rather than solving one problem and moving on to the next. A manager should always strive to do the best job possible under the circumstances. However, a person lured by the prospect of perfection (or driven to it by a perfectionist supervisor) is bound to discover that perfection is rarely, if ever, attainable.

Temper

Another common barrier to supervisory effectiveness in day-to-day working situations is temper. Almost without fail, temperamental behavior impairs interpersonal communication. Generally, as temper increases, true communication decreases, and personal effectiveness suffers.

These barriers to effectiveness were discussed after the section on initiative for a particular reason. The degree to which managers encounter these barriers may be largely a reflection of personalities—of whom and what each person happens to be. However, the awareness of these barriers is what often opens the door to possible change. And this change cannot come from outside; it must be self-inspired.

■ Organization

For many years a popular cartoon has hung on countless office walls. It pictures two little men facing each other across a table. Both are leaning back in swivel chairs and both have their feet on the table. The area around them is in general disarray. The caption is, "Next week we've got to get organized." The cartoon touches on the biggest problem in the business of personal organization—although it is frequently thought about, it is usually put off until some more convenient time.

Of course, organizing is one of the basic management functions, the process of building the framework needed to accommodate the work of the department involved. However, a manager thinking about personal management issues will start by organizing things in the immediate environment that are relevant to the work process.

People vary greatly in how they relate to some degree of order—or lack of order—in their surroundings. Some people are meticulously organized, and others seem to function well in the midst of clutter. However, just about everyone can reach an indefinable point beyond which clutter becomes confusion.

Not everyone can be like a certain institution's records manager who—incredibly for someone in his position—never used desk drawers or file cabinets. The horizontal surfaces in his office were piled high with ragged towers of seemingly unsorted documents ranging from one-page letters and thick presentence reports to court documents and memos. Remarkable as it seemed, he had the uncanny knack of being able to reach into a stack at the right place and pull out the necessary document at any time. This practice worked well for many years, at least when he was there, because when he was on leave or sick, it was virtually impossible for his subordinates to find anything. He also failed to realize that the visual effect of the turmoil in his office was to reduce the confidence of other staff that things were really under control in the records department.

Eventually, events overtook the nonsystem. The institution was expanded to include four new large housing units, the scope of responsibility of the records department increased, the amount of paper in the office expanded accordingly, and the disordered stacks became truly overwhelming. Things began to get lost and stay lost. His supervisor intervened (as probably should have been done long before) and the manager was compelled to put things in order. His staff rejoiced, efficiency in the department increased, and confidence in his effectiveness rose as well.

It is not only the material lying about in the open that reduces effectiveness. A great many things out of sight in desk drawers and file cabinets also can breed confusion and delay. Someone once referred to a refrigerator as a place to keep leftovers until they are old enough to qualify as garbage. Desk and file drawers often are used much the same way.

Let's look at one proven method for restoring control to office paperwork. Go through desk drawers and file cabinets and clean house thoroughly. Keep as few files as necessary, and consider arranging these according to their importance.

First and most accessible should be those things currently being worked on, perhaps a number of one- or two-page items in an "open items" file backed up with a few folders for open tasks of greater size.

Next keep a few folders devoted to items pending, on hold, or likely to become active in the near future.

A third personal file section would reasonably consist of a limited number of files whose future necessity are a manager's judgment call. This could be a key memo authorizing an action or program. It could be critical copies of reports that may be needed someday in connection with litigation.

Frequently used reference material can be kept in a separate file as well.

Least important in the personal filing system are those things that potentially may be useful or helpful someday. These are the items referred to seldom if ever, and they are the things that create most office clutter.

Many supervisors struggle with the pack-rat syndrome—something that is reinforced in the work world by the way a bureaucracy encourages staff to retain documentation in order to "CYA." They hesitate to discard anything because they feel it may be needed some day. They are likely to keep outdated computer printouts, notes of meetings and incidents long past, old magazines and professional journals, and suppliers' catalogues, brochures, and price lists.

For those inclined to be a "collector," consider this: most of the items that have been saved so faithfully will never again really be needed by your institution or agency. The problem, of course, is guessing which things are likely to be important so the unimportant ones can be thrown away. Since there is no sure solution to this problem people tend to save everything—a familiar story to one of the authors, who struggled with this throughout his professional career.

Overcoming this begins with an honest effort to clean out the office. Go through the clutter and throw out or shred everything that hasn't been used in the past 2 years and is not likely to be used in the foreseeable future. Be mindful of any regulations the agency has about preserving key records, and be sure to retain any material that may have a bearing on any potential litigation. But for most people, this can be a productive exercise. Certainly a few things may be discarded that might have been helpful some day, but what of it? No person can possibly cover all anticipated information needs with his or her own resources. In such cases, the best bet lies, rather, in knowing where to go for information when a specific need arises.

Go through the office at least twice each year, purging material and condensing files, again being mindful of litigation and other regulatory concerns. Doing so will produce two distinct advantages: limiting the amount of material kept and providing a memory refresher about what is being retained and where it is kept.

Do not allow material to pile up on desks, tables, or any other surfaces in the office. Some managers strive to keep the material on their desk limited to an amount that can be put into a drawer at the end of each day. If this isn't possible and is personally impractical for some reason, at least once a week review the things that have accumulated and either file them in appropriate places or get rid of them. Keep frequently used reference materials near the usual work place, and consider using a desktop organizer to keep things straight.

Avoid becoming a generator of the worst kind of clutter—unsorted and undated notes. Some supervisors generate many pages of handwritten notes each week. This practice is itself no problem, and often it is better to err on the side of too much documentation rather than too little. However, this applies only when the problem or activity is current or the nature of the subject suggests that all documentation should be preserved. When the immediate need has passed, notes should be sorted down to essentials, assembled in order, properly identified, and filed.

As far as note writing is concerned, there is one small rule to follow that can vastly improve the usefulness of informal documentation: whenever putting pen or pencil to paper, first put the date on the page.

■ Stress and the Supervisor

The classic definition of stress, attributed to Dr. Hans Selye, probably the world's leading authority on the subject, is the nonspecific response of the body to any demand made on it. In 1914, years before Selye's work, Dr. Walter B. Cannon defined stress as the body's ability to prepare itself instantaneously to respond to physical threat. The latter definition describes the oft-cited "fight-or-flight" response.

We now know, of course, that the "threat" Cannon referred to can be emotional as well as physical. Indeed, the demand that triggers stress in the body can be purely physical, purely emotional, or a combination of the two. And managers should all be more than passingly familiar with the fight-or-flight response—increased heart rate, faster breathing, tensed muscles, flowing adrenaline, and other signs that the body is ready for action.

The correctional environment provides more than enough stimulus for this phenomenon. But even in the normal ebb and flow of life, we cannot avoid experiencing a certain amount of stress. It is inextricably related to change. Stress is a common human response to change, not through conscious actions but involuntarily, both physically and emotionally.

Often, stress originates in a feeling of loss of control. When some outside influence—some change—disturbs their equilibrium, people react involuntarily in ways that suggest they no longer have the measure of control they need over events and circumstances.

Stress can be both positive and negative. It is positive when a person is "up" for something, prepared and alert and determined to regain control. People who seem to perform at their best when under pressure usually do so out of response to positive stress. Consider, for example, a key presentation that must be made to top management. The future of the department, as well as the presenter's career prospects, may depend on how well this is done. The person feels "up" for the occasion—tense, perhaps anxious, maybe experiencing butterflies in the stomach. Knowing how much is riding on the presentation usually will motivate the person to be thoroughly prepared and determined to do a good job. This is taking control, reacting positively to positive stress.

A large portion of stress, however, can be negative. And no one can say with any certainty which events constitute positive stress and which constitute negative stress. People vary greatly in their ability to cope with stress in general and to perform well under pressure in a job situation. Whether stress is positive or negative depends largely on how the individual reacts after the stressful event has passed. Positive stress, or "good" stress, is invariably followed by relaxation. Negative, or harmful, stress is not followed by relaxation, and the person continues to experience tension, anxiety, and the like.

Sources of Stress

Stress emanates from three general sources: one's personal life (life outside work), the total job environment, and the nature of the person him- or herself—personality, inherent capabilities, and approach to daily living whether at work or away from work.

Stress arising in personal lives is quite likely to influence job performance in some way. Everyone has known people who seem incapable of leaving the problems of the job at work, instead carrying their frustrations home and allowing them to affect their personal lives. Corrections, and law enforcement in general, is known for high rates of divorce, alcoholism, and other dysfunctional employee behaviors, many of which are thought to be related to job stress. Just as well known are employees who regularly bring their personal problems to work and allow them to affect their performance. Some of the most frequently encountered employee problems a supervisor faces arise with employees whose performance and interpersonal relations are adversely affected by personal difficulties.

A good manager will recognize that it is not possible to separate the person on the job completely from the person off the job. People vary greatly in their ability to keep the work side of life from influencing the nonwork side and vice versa. To some people home is a welcome respite from the problems of the job; they can literally leave their worries on the doorstep. To others, work is a refuge, an escape from a chaotic personal life. To a great many people, however, trouble at work usually means trouble at home, and trouble at home usually spills over into their employment in some way.

The total job environment can induce stress in a number of ways. Inmate contacts and confrontations, of course, are prominent factors. But things as simple as physical working conditions—heating, lighting, furnishings, space, noise, and the like—can create stress. So can organizational policies and practices that are inconsistent or unpredictable. These are common sources of stress in supervisors and line employees alike.

A supervisor's feeling of having less than total control over the work situation can induce stress, especially when the supervisor has total responsibility for a given situation without having full authority over all the elements that must be brought to bear to address the situation. It is common among supervisors to find that they have responsibility—at least implied responsibility—but that they have not been given authority consistent with that responsibility.

A great deal of supervisory stress comes from negative practices of higher management. "Bossism"; management that pushes rather than leads; and management that is authoritarian, unreasonably demanding, or fault finding all create supervisory stress. And although not necessarily negative itself, a change in management that leaves a supervisor reporting to a new superior can be a stress producer. Frequent changes in the chain of command or organizational structure that leave the supervisor reporting to a new manager every few months are virtually guaranteed to produce considerable supervisory stress.

Finally, a major potential stress producer for supervisors is work overload—the fact or at least the perception of having too much to do, not enough time to do it, and not enough resources for its accomplishment. Today's crowded prisons are a classic example of this situation. Many supervisors have learned the hard way that they cannot be all things to all people in a finite amount of time.

As functions of personality, capabilities, and approach, the stress producers that can be at work within ourselves include:

- Self-doubts; a lack of confidence in one's own abilities.
- Lack of personal organization.
- Inability to plan out work and to establish priorities and address them appropriately.
- Perfectionism; setting excessive and unrealistic demands.
- The inability to say no to any request or demand.
- The tendencies to take all problems as indications of personal shortcomings and to take all criticism personally.

A person who is never stressed on the job may have too little to do, and little or no true responsibility. Also, someone who is never stressed by the demands of the job probably falls short of doing his or her best work. Positive stress—urging performance under pressure—produces learning and growth.

To a considerable extent, stress goes with the supervisory territory. But a person who is always stressed—chronically on the verge of anxiety, depression, or panic—can be leaning toward personal ineffectiveness and ultimately to physical or emotional illness.

Coping with Supervisory Stress

To succeed on the job over the long run, supervisors must gain as much control over both themselves and the work environment as is possible. Approaches suggested for the supervisor to apply in combating stress include:

- Learn to say no, or at least to speak up, when that last request or demand finally adds up to too much. The supervisor is only human and is likely to be as stressed as you are, if not more so. Most supervisors will understand, especially if you can suggest alternatives or offer to reorder priorities to serve a pressing need.
- Don't let your pile of accumulated work grow until it becomes totally uncontrollable. Take time to plan. Establish priorities. As suggested in earlier sections of this chapter, tackle one important task at a time and do it completely. Avoid piecemealing tasks, intermittently doing a little on each of several items because "they're all important."
- Delegate. Take Chapter 6 ("Delegation: How to Form the Habit") seriously. And delegate in anticipation of stressful times to come. Do not wait

until the pile is so high that you can no longer see over it to think about the need to train others to do some of the work.

- Vary your pace. Intersperse short, quiet tasks among the more hectic, tension-producing contacts required of you. If you have been in staff meetings all morning, try to allow yourself an hour or two of solitary work in the afternoon. Once in a while reward yourself by working on something that you especially like to do.

- When the going gets rough, take a few minutes to relax. Stretch. Breathe deeply. Take a few minutes to walk to a fellow manager's office instead of calling on the phone. Go out on the yard and talk to the staff or inmates there. A few minutes spent clearing your head by doing something different will pay themselves back in efficiency many times over.

As for managing stress outside of work, there is still no advice better than the time-honored standards: proper nutrition, proper rest, and regular exercise.

Just as problems are a part of work that must be managed, so too is stress. Accept stress as part of the supervisory job, above and beyond the normal levels encountered in a correctional setting. Recognize stress when it strikes. By taking control of the situation and consciously managing the stresses that otherwise seem overwhelming, a person will accomplish more in terms of both quantity and quality, take more enjoyment from the work, and reduce the potential for stress-related illness.

■ Effective Use of Time

We have probably all said to ourselves about learning how to make better use of time: we want to do it, but we do not have the time to learn how. We might just as readily admit that we are too busy doing things inefficiently to learn how to do them more efficiently, or simply that we are too busy wasting time to learn how to save it.

This topic is so important that an entire chapter is devoted to it. Refer to Chapter 7 for a complete discussion of time management.

■ How Well Are You Suited to the Supervisory Role?

Not all persons who work in supervision are equally effective at all parts of the job, and not all supervisors enjoy all parts of the supervisory job equally. The relative effectiveness of many supervisors in the work force can be directly related to how well, personally and temperamentally, they may be suited to the role of supervisor.

How well does any active supervisor fit into the supervisory role? It is possible to examine a few facts and conclusions about yourself and look at the way you relate to the job so you can decide: How well do I fit the supervisory role? And, what can I work on to improve the way I fit this role? This can be accomplished through a thoughtful examination of both personal orientation and performance orientation.

Personal Orientation

To a considerable extent managers will unconsciously approach everyday job activities in the same manner they approach tasks and activities in their personal lives. Most people are very much the same whether on or off the job. Unless people make a conscious effort to behave differently in one or another area of life, they are likely to be governed by the same tendencies in all that they do.

Personal orientation can be illustrated in simplified form as a graph with axes representing ranges of capabilities or tendencies (**Figure 8-1**).

The vertical axis of personal orientation is focus, ranging from totally internal at the bottom of the graph to totally external at the top. Focus represents the extent to which one is affected by or actively, emotionally involved in activities or events to which one is exposed or is a party. External focus is typified by detachment; internal focus is represented by involvement.

As with any depiction of human tendencies or characteristics, this focus axis, as well as other supposed scales discussed here, represents a range with an infinite number of gradations possible between extremes. Rarely do the extremes apply in full. Rather, anyone might show a tendency toward one end of the scale or the other.

The individual with an external focus is not personally affected by matters in which he or she is involved, is witness to, or is otherwise party to. An external focus suggests the presence of the ability to keep from taking things personally, and it also suggests the ability to better cope with events and demands that are stressful in some way.

Internal focus is of course the opposite. The person who is internally focused is personally affected by most of what goes on. All events that the individual is party to are internalized, and the individual finds it difficult if not impossible to remain unaffected or unchanged by events.

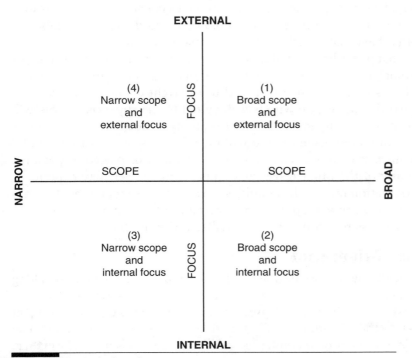

Figure 8-1 Personal orientation.

As far as the role of the supervisor is concerned, an individual who is totally externally focused may be able to cope rationally with a great many difficult matters, but the result may be a "supervision machine"—a decision-maker operating on logic and fact to the exclusion of intuition and all other human consideration. At the other extreme, a supervisor who is totally internally focused probably will not survive for any appreciable time. Considering all of the difficulties that arise in managing people, and recognizing the correctional environment as one that presents particular difficulties, it is easy to accept the internally focused supervisor as a likely candidate for early burnout.

The most desirable focus orientation for the supervisor lies above the origin of the graph, a tendency toward an external focus and thus toward the ability to deal with the various events that come with the supervisory task and to cope with the many problems and other unpleasant situations that are part of the job.

The horizontal axis of personal orientation is scope. Scope simply refers to the kind or amount or variety of activity that an individual can best handle according to temperament or individual ability. The person with a broad scope can "juggle," coping with a variety of tasks in bits and pieces, moving from task to task and back again, and solving unanticipated problems as they arise. The broad-scope individual can function in firefighting style without experiencing undue stress and without becoming overwhelmed by a seemingly unending stream of problems or by the inability to truly "finish" most of what is started.

At the opposite end of the horizontal axis are persons of narrow scope, those who cannot readily function in the scattergun fashion of the broad-scope person but who are most comfortable with and whose talents are best applied to one specific task at a time.

The chances of long-run success in the supervisory role are enhanced if one is possessed of a broad scope and is generally able to constantly reorder priorities and cope with a continuous series of unanticipated demands. Combining the two axes that have been used to describe personal orientation, it is reasonable to suggest that one is best suited to the supervisory role if one's personal orientation lies somewhere in the upper right-hand quadrant of Figure 8-1—above the origin as far as focus is concerned, and to the right of the origin as far as scope is concerned. The person who is well suited to the supervisory role will tend toward being externally focused, with the ability to deal with most matters at a reasonable emotional distance, and will be of broad scope, with the ability to take what comes as it comes. The supervisor who may be placed by personal orientation in any of the other three quadrants of Figure 8-1 may experience problems of orientation ranging from mild, such as having to remember to avoid giving in to a narrow-scope tendency and staying too long on a pet project, to severe, such as a stress reaction to the internalized pain of others.

Performance Orientation

Performance orientation can also be described by a pair of perpendicular axes along each of which there can be an infinite number of gradations (see **Figure 8-2**).

The vertical axis of Figure 8-2 presents emphasis on a continuum from totally personal at the bottom to totally functional at the top of the graph. The higher the degree of functional emphasis evident in a supervisor's performance,

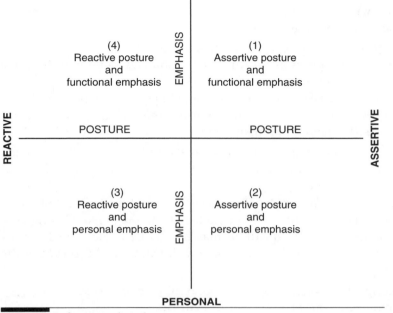

Figure 8-2 Performance orientation.

the more visible will be that supervisor's degree of attention to activities of highest priority. The supervisor having a strong functional emphasis will ordinarily be attending to the most important task at any given time. On the other hand, the more personal a supervisor's emphasis becomes, the more likely that supervisor is to be giving attention primarily to those tasks or concerns that he or she prefers or enjoys. This may be all well and good when it happens that what the supervisor likes to do and what most needs to be done coincide. However, most of the time the tasks the supervisor favors and the tasks that most need to be done are not the same.

The horizontal axis of Figure 8-2 represents posture along a continuum ranging from completely reactive to fully assertive. The supervisor who tends toward the assertive end of the scale ordinarily initiates action, advances ideas or solutions, and moves to resolve difficulties in their early stages and head off problems before they occur. The supervisor who tends toward the reactive end of the scale ordinarily waits to be told what to do, procrastinates on important decisions until there is no more room for delay, and waits until a problem can no longer be denied before attempting a solution. The assertive supervisor deals in innovation and prevention; the reactive supervisor simply responds when pushed by circumstances.

The effective supervisor tends more toward functional than personal in emphasis and more toward assertive than reactive in posture. However, emphasis and posture cannot be considered separate from each other; together they exert combined effects that can be described through association with each other in the four quadrants of Figure 8-2.

In the first quadrant (upper right), tendencies toward functional emphasis and assertive posture describe a supervisor who is generally in control of the job and who operates with a noticeable degree of autonomy. In this, often viewed as the "best" of the four quadrants, the supervisor:

- Concentrates on high priority tasks most of the time.
- Tackles the important problems in timely fashion even though they may be difficult or unpleasant.
- Experiences job satisfaction from accomplishment and achievement.
- Is meeting the needs of the organization in doing what he or she is expected to do as a member of management.

A supervisor functioning in the first quadrant is applying the right approach (assertive posture) to the right tasks (functional emphasis), moving in the right direction without being told.

The second quadrant (lower right) suggests the combined effects of a tendency toward an assertive posture and a tendency toward a personal emphasis. A supervisor functioning in this quadrant may frequently be described as pursuing personal preferences or being out of step with reality. In this quadrant the supervisor may:

- Be unwilling to delegate, personally monopolizing preferred activities, or unable to delegate out of a sense of insecurity.
- Take frequent refuge in preferred tasks of low priority as relaxation, self-satisfaction, or escape.
- Experience goal displacement, being unable to reconcile organizational goals with personal goals.
- Have little appreciation of job purpose, failing to see how this specific supervisory position is supposed to mesh with others as part of the organizational whole.

A supervisor functioning in the second quadrant is generally applying the right approach to the wrong tasks (personal emphasis), moving without being told but going in the wrong direction.

The third quadrant (lower left) is readily describable as the "worst" of the four general situations. The supervisor functioning in the third quadrant may:

- Appear to be running fast and working hard while going nowhere and accomplishing little.
- Exhibit an inability to control interruptions.
- Be unwilling or unable to say no to any request that comes along, no matter how unreasonable or intrusive it may be.
- Avoid that which is difficult, complicated, unpleasant, or potentially stressful whenever possible, taking the "easiest" route through most problems.
- Become disillusioned with the job and the organization.

A supervisor functioning in the third quadrant is ordinarily taking the wrong approach (reactive emphasis) to the wrong tasks (personal emphasis), moving only when pushed and then going in the wrong direction.

Because a person falling into the fourth quadrant (upper left) will usually be doing most of the right things, the combination of emphasis and posture evident in this quadrant might present no problems at all if this discussion concerned people employed in any number of nonsupervisory capacities. However, the supervisor falling into the fourth quadrant may:

- Experience considerable frustration with a nearly complete lack of autonomy.
- Feel "out of control," reacting to the seemingly unending and unpredictable stream of demands made by others.
- Use time ineffectively, moving from task to task and crisis to crisis as directed, having no time left to plan and organize.

A supervisor functioning in the fourth quadrant has the proper functional emphasis but is applying an improper reactive approach. As suggested, this may be acceptable in some nonsupervisory employees; although assertiveness is to be desired in many employees, supervisory or otherwise, there are still many jobs in which an employee may best serve by performing as directed. But although a reactive posture may be appropriate—or at least tolerable—in many nonsupervisory employees, such a posture is contrary to what is expected of an effective supervisor.

■ How Well Do You Fit?

Many people who did not fit well into a supervisory role have either voluntarily abandoned supervision or have failed as supervisors. Many others who also do not fit especially well into a supervisory role have nevertheless remained supervisors, with widely varying degrees of success. It is highly likely that absolutely "perfect fit" supervisors are a minority of the supervisory population of any organization. Therefore it is just as likely that most supervisors have weaknesses that can potentially affect their performance and perhaps ultimately determine whether they succeed or fail in the long run.

Simply identifying one's own weaknesses is a large part of the struggle for improved personal effectiveness. Once these weaknesses are identified, the next essential process is recognizing the difference between those things about one's self that one can actively change and those things about one's self that one cannot readily change but must compensate for or otherwise work around. The final step in the process is acting to remove or guard against one's own weaknesses.

Consider how you believe you must place yourself in regard to personal orientation. You might find, as would most people, that you have more control over your scope than you do over your focus. Although both scope and focus are rooted in personality, focus is much more emotionally based and is therefore much more difficult to alter artificially. Regarding focus, for example, if you are strongly internally focused and truly sensitive to the point of creating intolerable stress within yourself by having to apply firm disciplinary action, you are more prone to minimize this aspect of the supervisory role and thus impair your effectiveness in handling problem employees. Since your focus is emotionally based, unless you are sufficiently near the middle of the graph to allow you to accommodate the occasional emotional stresses of the job, you may be consigning yourself to a working life of strain and unhappiness by remaining in supervision.

Since scope is less emotionally tied than focus, managers can do more to control this dimension of personal orientation. Broad-scope people are team members; narrow-scope people prefer to function as loners. The supervisor needs to be a team member. However, it is acceptable to give in to a tendency to enjoy working as a loner once in a while—as long as one is conscious of the need to be a team member and to willfully function as a team member whenever circumstances demand (which, for the typical supervisor, is most of the time).

Aspects of personality ordinarily have a great deal of influence on both major dimensions of one's performance orientation. Of the two, posture is probably more deeply rooted in personality than is emphasis. Surely there are some whose posture is essentially reactive because they are self-doubting and insecure. Just as surely, however, there are some who are reactive because they lack goal orientation or because they have never been given any clear idea of what is expected of them. Perhaps one's insecurities can be overcome with a great deal of effort and the right kind of assistance and support. However, not nearly so difficult as overcoming insecurity is the deliberate development of goal orientation—which can surely be done if someone wants success strongly enough—and one can pointedly ask for higher management's expectations.

Although unavoidably influenced by aspects of personality, emphasis is more readily altered by the individual than is posture. The key to emphasis is self-discipline, the ability to make oneself give the most attention to the highest priority tasks at hand. The supervisor who can enter each work day by asking, "What is the single most important task I need to accomplish today?" and plan to get that task done and proceed to do it, is on the path of effectiveness. Even the supervisor whose workload may loom as overwhelming is making progress if he or she is always at work on the highest priority task of the moment. To self-discipline one need only add consideration of the need to plan and organize, and to control interruptions while concentrating on priorities, to develop a true functional emphasis.

The self-discipline referred to in the foregoing paragraph is crucial to the improvement of one's personal orientation and performance orientation. To develop an improved personal orientation and to adopt a more appropriate performance orientation is to put oneself in the position of becoming a more effective supervisor. Self-discipline as applied by the effective supervisor is self-management, and self-management comes before the management of others. The person who would aspire to manage others must first become proficient at self-management.

EXERCISES

Exercise 8-1: The Effectiveness Checklist

Provide honest, self-searching responses to the statements listed below, using these responses:

> U = usually (or always)
> S = sometimes
> R = rarely (or never)

(Be candid. No one needs to see your answers unless you choose to share them.)

_____ **1.** I put my objectives—and plans for reaching them—on paper.

_____ **2.** My objectives are expressed in specifics: what, how much, and when.

_____ **3.** For sizable tasks, I use checkpoints or subobjectives so I can assess progress along the way.

_____ **4.** I break large jobs into smaller, more manageable pieces.

_____ **5.** I set deadlines for myself and hold myself to them.

_____ **6.** I use written reminders of what must be done today or tomorrow.

_____ **7.** I avoid thoughts or circumstances that might sidetrack my efforts.

_____ **8.** I know my limitations; I do not set objectives I know I cannot achieve or make promises I know I cannot keep.

_____ **9.** I use positive motivation by reminding myself of the benefits I expect from the completion of a task.

_____ **10.** When facing a disagreeable or difficult task, I am able to distinguish between "I can't" and "I don't want to."

_____ **11.** I am willing to take risks, to try new ways of doing things.

_____ **12.** I allow myself the freedom to fail, to make mistakes and learn from them.

_____ **13.** I keep my personal work area organized and under control.

_____ **14.** I recognize conflicts for what they are and do not back away from making decisions.

_____ **15.** I have a sense of priority that allows me to distinguish between what must be done and what I would like to do.

If you gave yourself U on all 15 statements perhaps you had better go through the list again. There are few managers who do not rate S on at least a few items. Each S or R represents a clear opportunity for self-improvement.

Exercise 8-2: "Where Does the Time Go?"

Bill Martinez, supervisor of education, decided he had to get organized. Recently his work days had been running well beyond quitting time, cutting noticeably into time with his family; but instead of going down, the backlog of work was growing.

Inspired by an article he had read about planning and setting priorities, Bill decided to try to plan each day's activities at the end of the previous day. Monday, he came to the office with his day planned out to the last minute. During the morning he had to complete a report on a recent learning-needs analysis, write the performance appraisals of two part-time instructors, and assemble the balance of materials for a 2-hour in-service training session he was scheduled to conduct that afternoon. After lunch he had to conduct the class, complete the schedule of the next 3 months' training activities (now 10 days overdue), and prepare memos—which should be posted that day—for two upcoming inmate classes.

Bill got off to a good start; he finished the report before 10:00 A.M. and turned his attention to the performance evaluations. However, at that time the interruptions began. In the next 2 hours he was interrupted six times—three telephone calls and three visitors. The calls were all business calls. Two of the visitors had legitimate problems, one of them taking perhaps a half-hour to resolve. The other visitor was a fellow supervisor simply passing the time of day. Neither performance evaluation was completed, and the training materials were assembled in time only because Bill put them together during lunch while wolfing down a sandwich at his desk.

Bill's afternoon class ran 20 minutes overtime because of questions and discussion. When he returned to the office he discovered he had a visitor, an investigative supervisor who stayed for almost an hour talking about several inmates who were in educational classes but who were also under suspicion of involvement in an escape plot. This interview was necessary, since one possibility was that the inmates might try to attempt their escape during an evening school session, but it was time-consuming.

After the investigator left, Bill spent several minutes wondering what to do next. The performance appraisals, the three-month schedule, the class notices—all were overdue.

Deciding on the class notices because they were the briefest task before him, he dashed off both notices in longhand and asked the department secretary to type them, run them off, and post them immediately. He then tackled the training schedule.

When Bill again looked up from his work it was nearly an hour past quitting time. He still had a long way to go on the schedule and had not yet gotten started on the two performance appraisals. As he swept his work aside for the day he sadly reflected that he had not accomplished two-thirds of what he intended to do that day in spite of all his planning. He decided, however, to try again; when he could get a few minutes of quiet time late in the evening, he would plan his next day's activity.

On his way out of the education building, he happened to glance at the main bulletin board. The small satisfaction he felt when he saw the posted class notices vanished instantly when he discovered that both were incorrect—the dates and times of the two classes had been interchanged.

Questions

1. What errors did Bill commit in his approach to planning and the establishment of priorities?

2. In what respects could Bill have improved his use of time on the "blue Monday" described in the case?

The Supervisor and the Employee

Interviewing: The Hazardous Hiring Process

Chapter Objectives

- Stress the importance of supervisory involvement in the hiring process.
- Examine the relative advantages and disadvantages of promotion from within the organization versus external hiring.
- Offer advice on how to prepare for an employee selection interview.
- Present guidelines for interview questioning, specifically identifying kinds of questions that should be avoided and suggesting appropriate lines of questioning.
- Review the significant legal and personal sensitivities occasionally encountered in selection interviewing.
- Describe a recommended general approach to the supervisor's conduct of the actual employee selection interview.
- Describe desirable follow-up action to conclude the interview cycle effectively.

The best man for the job often is a woman.

Anonymous

I have a predilection for painting that lends joyousness to a wall.

Pierre Auguste Renoir

What Would You Do?

The management team of a new institution was in the middle of a hiring push. The facility was to open in 10 weeks, and it looked like enough correctional officers had not yet been hired to get them through training in time for activation. Teams consisting of a correctional supervisor, unit manager, psychologist, and personnel specialist were set up to interview applicants.

In something of a first, identical twins presented themselves in sequential interviews for correctional officer positions. They interviewed well, were well qualified, and in virtually every other way seemed to be good candidates. Then the psychologist posed a situation and a question.

"Suppose you are the correctional officer in the main control center, controlling the entrance to the institution. You, and only you, are responsible for seeing that no one but authorized personnel leave the institution. One of your key post orders tells you that a person taken hostage (no matter what position) loses his or her authority, and that under no circumstance will you open those gates when a person is under duress. Not the warden. Not the director. Not the governor. Now my question to you is simple. What if you and your brother are hired, you are working in the control center, and an inmate brings your brother to the gate with a knife at his throat, threatening to kill him if you don't let him out. What would you do?"

The candidate thought a moment and replied that he would have to release the inmate to save his brother.

Neither he nor his brother was hired. The psychologist discerned and correctly posed a question that went to the heart of the unique relationship between identical twins. And the response (to his credit, an honest one) disqualified an otherwise high-potential job candidate.

■ The Supervisor and the Interview

Oh, that everyone could be like Renoir—so well fit for his chosen profession that he or she simply exudes joy, projecting it onto every project. Indeed, it is this process of finding a good fit between people and their work that is at the heart of the employment interview and the hiring process as a whole.

Actually, there is an immense irony to many contemporary hiring practices, in that the personal interview—usually the heart of the process—often is not a particularly reliable means of finding good employees. Even though many correctional and other government agencies have specific interview procedures, there is no guarantee that the process will always generate positive results. No matter how good a manager is at interviewing or how much he or she knows about the jobs he or she is attempting to fill, it frequently remains impossible to separate true ability in a job candidate from the ability simply to "talk a good job." However, in spite of its shortcomings and weaknesses, reliance on face-to-face employee selection interviews is most often used for locating new employees simply because there is no other practical means available for approaching the task.

Institutions may vary considerably in the extent to which supervisors become involved in the hiring process. In some correctional agencies, major departments or divisions recruit and screen their own job applicants, usually through some specifically delegated authority from the state's personnel department. In most agencies and prisons, however, the employment needs of all departments are served to a more or less equal extent by the personnel department.

Make it a point to know specifically how much of the employment process is vested at the managerial level and how much is done by personnel. Ordinarily a manager can expect to look to the personnel department for locating a number of candidates who generally fit the requirements of the open position; that is, the personnel staff find people who have the appropriate academic or other credentials, minimum required experience, and who otherwise fit the hiring criteria established either in the department or by agency policy. In most cases the personnel staff will screen applicants to locate generally qualified candidates and will arrange personal interviews for the manager who ultimately will supervise the employee.

A very few organizations still may follow the age-old and highly undesirable practice of having a very few people, perhaps only one or two, doing all of the actual employee selection for the entire organization. This outmoded practice is contrary to the fundamentals of supervisory responsibility; a detached party cannot simply "give" an already-chosen employee to a supervisor and expect to facilitate the establishment of the proper employee-supervisor relationship. If the supervisor is to be responsible for an employee's output, then the supervisor must be allowed a consistent amount of authority in selecting that employee. The supervisory role should include, subject to carefully drawn ground rules, the authority to hire and fire. In the past, the authority to hire sometimes was retained by "higher ups" (although the supervisor may have been left to do the firing when things did not work out). As supervisors, however, recognize that although hiring is a sometimes difficult and time-consuming process, it is nevertheless an essential part of the job. To be blunt, if we are not capable of hiring people then we should not be supervisors.

■ Candidates: Outside and Inside

Job applicants often come to the organization in response to employment advertising specific to certain personnel needs. Many also come not in response to specific recruiting efforts, but unsolicited; they fill out employment applications in the hope of finding work. Both solicited and unsolicited applicants reach the attention of the interviewing supervisor following screening interviews in which the personnel department determines that the applicants are qualified to fill some specific need. This determination is most often made in accord with broad qualifications established by the state's or agency's personnel department.

Most correctional agencies operate job-posting systems and have policies governing employee transfers. This means that a number of candidates for any particular position may also come from within the present workforce. As should be the case, upward mobility from the correctional workforce is a priority for many agencies. However, the more specialized a need (electrician, pharmacist, power plant engineer, and so on), the more likely the supervisor is to see a preponderance of external candidates. The more generalized a need (case manager, records technician, and so on), the more likely the supervisor is to see the applicant pool consisting of internal candidates. In more than a few cases, in filling a particular position the manager will be able to choose from among a pool that includes both external and internal candidates.

Most organizations likewise espouse a philosophy of development from within the organization by way of either lateral or promotional transfers. The employee selected from within the agency brings with him- or herself a fund of essential knowledge about the institution and the agency in general. This is information that would require a great deal of time to instill in an outside candidate—time that is lost to productivity on the job. It should be stressed that in some agencies this notion of development from within is more likely to be philosophy than policy. No astute top management official will absolutely require the supervisor to select an internal candidate over an outside applicant in any particular situation. To do so would be to put the supervisor in a position of having sometimes to select a candidate who is less qualified than the one who appears to be the best available. But even though the notion of development from within may not be inviolable policy, a manager should nevertheless incorporate it into practice whenever reasonably possible, for both organizational and practical reasons.

Most supervisors undoubtedly make most of their hiring decisions with immediate departmental needs in mind. However, hiring should also take into consideration the longer-range needs of the organization and its employees. These needs suggest that a certain portion of a department's personnel needs should be filled from within the organization. Neither "always outside" nor "always inside" is appropriate. However, when presented with two or more equally qualified candidates for a given position, there is every good reason for giving preference to the internal candidate.

The existence of real in-house opportunities for varied work experience, promotion, and professional growth will foster employee retention. If lower-level employees continually see all the "better" positions (primarily filled through what staff see as promotional opportunities) going to external candidates, the organization will gradually lose those lower-level employees who are capable of growing and who are interested in promotion.

■ Preparing for the Interview

The first step in preparing for a selection interview should consist of careful review of the individual's employment application and/or resume. In the case of the former, standard agency information will be on the form, which will facilitate comparison of qualifications of various candidates. When a resume-based interview is involved, the interviewer may need to be ready to ask additional questions to ensure that sufficient detail about that candidate is developed.

In any event, the interviewer should have the application materials available in advance of the interview. Never be in a situation where it is necessary to read the application for the first time while the applicant sits waiting. Discourage the personnel department or other referring office from sending applicants with applications in hand. Take a few minutes (preferably longer) to become familiar with information about the job applicant. Otherwise the applicant will certainly be uncomfortable when reading is taking the place of interaction. This also makes it possible that the interviewer, in haste, may miss something important in the application materials and ask questions that have already been answered on paper.

In reviewing this information, remember that regardless of whether a standard application or resume is used, there is a great deal the employment application will not reveal. Employment applications of years past called for a great deal of information that now cannot be legally requested. Assuming little or no control over what is asked on the application, there is little use in going into detail concerning what can or cannot be requested on an employment application. However, a later section in this chapter will discuss some kinds of questions that cannot or should not be asked during an interview.

Likewise, these same questions cannot be asked on an application. Many of today's employment applications are necessarily sketchy compared with those of years ago. In fact, chances are that if an employee has worked for an employer for as few as 2 or 3 years, the application they originally completed might today be considered illegal or at least questionable in some small way. For a person hired before 1980, the application they originally submitted would today almost certainly be illegal in a number of respects.

After reviewing the written material about the candidate, have a definite plan in mind when approaching the interview. Know everything possible about the job being filled, and back up this knowledge with a copy of the job description or at least with a fairly complete list of the duties of the job. Be especially aware of any unique aspects of the job or any unusual requirements the applicant should be made aware of. Be prepared to tell the applicant precisely where the job fits in the total operation of the department and the overall role he or she would play within the organization.

It may also be useful to have a few sample questions prepared in advance of the interview. And while the most valuable questions often emerge while in conversation with the applicant, some starter questions may be useful in getting the conversation going. Be prepared to guide the applicant and listen, never losing sight of your basic purpose—to learn as much as possible about the applicant.

Make sure the interview takes place in private and in relatively comfortable surroundings. An interview in the manager's private office is ideal, unless doing so may involve constant interruptions. If necessary, borrow someone else's office, make use of an open conference room or other available space, or use interview space in the personnel department. The interviewer's ability to learn about the applicant is severely impaired by interruptions, and interruptions can be unsettling to the applicant.

Be aware of the likely state of mind of the job applicant. Even though the person is not yet and may in fact never become an employee, the interviewer is already in a position of authority relative to the person looking for a job. To a potential employee the interviewer is an employer, who enters the interview situation at a definite psychological advantage.

Look at the interview situation from the applicant's point of view. The prospective employer automatically has the upper hand, yet the applicant has far more at stake. The applicant is looking for a job. The interviewer not only has a job but is an authority figure as well. Often the applicant, if truly serious about finding work, will be determined to make a good impression and may be nervous.

Insofar as possible, be familiar with affirmative action, equal employment, and other legal considerations (see Chapter 29) that apply to the employing agency. Use this familiarity to guide interview behavior, but resist the temptation to "play lawyer" and make individual interpretations in questionable areas.

When possible to do so, get advice from the personnel department and other knowledgeable people in the organization. An interviewer who follows the questioning guidelines provided in this chapter should encounter few difficulties. However, if serious doubts emerge about a particular question that could be asked in the interview, do not ask it.

■ Guidelines for Questioning

Although common courtesy should prevail in interviews at all times, courtesy itself is not enough. The interviewer must be constantly aware of questions or comments that could be taken as discriminatory in some way, although not intended in that manner. An applicant may volunteer information related to the following areas, but the interviewer may not ask for it.

Questions to Avoid

Personal Background

Any question that requires the applicant to reveal race, religion, or national origin should be avoided. Direct questions on these subjects should be easy to avoid. Watch out, however, for indirect questions through which a person can claim the interviewer was "fishing" for specific information. For instance, an impermissible question is one such as: "I'd say you were from [name of neighborhood with a population predominantly of one ethnic group], perhaps?"

Age

Age considerations present an interesting situation. Some correctional agencies are subject to statutory limitations on the age at which their employees must retire. Ordinarily, this means that to qualify for a pension, new employees must not have attained a certain age before being hired. In those cases, questions about age are permitted. However, even in those agencies, all positions may not be subject to those statutory provisions, and thus the following discussion is important to bear in mind.

In general, age has become a particularly sensitive area in recent years, calling for heightened awareness on the part of the supervisor. Although the Age Discrimination in Employment Act (ADEA) has been in place since 1967 (see Chapter 29), it was given added scope and influence by the Age Discrimination in Employment Amendments Act of 1986, effective January 1, 1987. Although mostly the concern of the personnel department, this law nevertheless has implications for the supervisor.

The 1986 act prohibits mandatory retirement for most employees and removes the age 70 limit on ADEA protection. As just noted, there are statutory exemptions for some correctional and other law enforcement agencies. However, unless your agency is clearly exempted, stay away from all questions related to an applicant's long-term intentions, such as, "How long would you plan on working before thinking about retirement?" because of the age-related inferences that one might draw.

Except where age-related issues are permissible under statute, an interviewer may not ask any obviously older applicants if they are receiving Social Security benefits. There have been cases of older persons applying for part-time work, with some employers discriminating in favor of applicants receiving Social Security payments. Their rationale apparently was that a person receiving a combined income may be more likely to remain on the job longer than one to whom the part-time job is the sole means of support.

Except as noted above, there are no safe questions (direct or indirect) that a supervisor can ask about age. In most situations it would be permissible to inquire whether an apparently young person is of legal age to enter full-time employment under the circumstances required of the job in question. This would be particularly true in the case of a position involving the use of firearms or deadly force, for which there may be statutory age restrictions.

Also, the agency must assure that all the qualifications sought are truly related to the job. Evaluate individual applicants on their individual capabilities and qualifications, not on general beliefs or personal preferences of the interviewer. It is not defensible to assert, "An older person just couldn't keep up," or "Someone her age would resist new technology," or "I want someone who's more likely to stay 10 or 20 years").

Disability

Similar to concerns about age, there are few if any safe questions the supervisor can ask about disability (whether or not a job candidate's disability is evident). The Americans with Disabilities Act (ADA) (1990) provides a national mandate barring bias against persons with disabilities. This act calls for supportive and accommodating behavior by employers in maintaining disabled persons in the work force. Again, an individual agency may have very clear statutory authority to depart from the strict provisions of the ADA. In those instances, legal counsel and personnel staff should have developed guidelines for all supervisors that ensure any departures from ADA requirements are negotiated very carefully.

There still is a great deal of litigation on the issue of how this recent legislation applies in certain work settings. However, it is likely that an agency could establish bona-fide occupational qualifications (BFOQ), for instance, for a correctional officer that would preclude a wheelchair-bound individual from such a position. One generally would assume that a BFOQ could be constructed for persons with severe sight problems if the job required a great deal of reading.

However, sometimes apparent disabilities are not a real bar from employment. Early in his career, one of the authors had a secretary who was totally blind, and yet who did a fine job, using Braille and other aid methodologies. Her hard-working, cheerful attitude and solid skills more than made up for the occasional problems her sight problem created. On another occasion, when interviewing correctional officers for a new institution, an applicant with rather prominent movement limitations in one leg (the aftermath of childhood polio) was being closely considered over the objections of other managers involved in the hiring process. They believed that the leg problem would predispose the candidate to linger on the main floor of the four-level housing areas, rather than circulate throughout the units. Noting the fact the applicant had engaged in many athletic endeavors over the years and his aggressive, no-nonsense approach to questions on this issue, he was hired. He was one of the most effective, active officers that institution had. These examples do not mean that every person with a disability can work in the correctional setting, but they do illustrate why a manager will give full consideration to such individuals.

During the employment selection interview, the supervisor should ask questions about the applicant's qualifications and experience that focus strictly on the essential functions of the job. The job's peripheral activities that have no bearing on its essential functions—for example, occasional filing or report delivery related to the position of a billing clerk—are subject to "reasonable accommodation" by the organization and have no bearing on the person's capabilities as a billing clerk. More about the ADA appears in Chapter 29.

References
You may not ask if the applicant has a recommendation from a present employer. This may be taken as discriminatory, since it may be difficult for the applicant to secure such a recommendation because of reasons other than job performance (e.g., race, religion, political affiliation).

Relationships
At this stage, the interviewer can no longer ask the identity of the person's nearest relative or "next of kin," even for the simple purpose of having someone to contact in case of illness or accident. This can be taken as probing into the existence of spouse or family, which is not permissible. Even asking about "the person to be notified in case of illness or accident" is risky until the applicant is actually employed.

Citizenship
It is generally permissible to ask if an applicant is a U.S. citizen or is legally eligible for employment. However, this should not be an active concern of the supervisor if the personnel department is fulfilling its proper role in complying with the Immigration Reform and Control Act of 1986 (IRCA). This law requires employers to hire only U.S. citizens and lawfully authorized alien workers. It provides penalties and sanctions for employers who knowingly hire or

continue to employ illegal aliens or who fail to verify legal eligibility for employment (see Chapter 28). Also, most states have statutory requirements in this area, which again should be part of the screening process conducted by the personnel department.

It is worth noting that there are some jurisdictions that have anomalous policies on this subject. An agency functioning in such a setting will develop internal guidelines to balance any competing laws.

In compliance with IRCA, the personnel department must require an applicant to produce certain proofs of identity and employment eligibility within three working days after an offer of employment is made. To help the organization comply with IRCA, the interviewer should resist the temptation to prevail on the personnel department to allow a much-needed new employee to begin work before the necessary proofs are produced.

Immigration reform has presented employers with a set of risks and pitfalls of a kind never previously experienced. These risks and pitfalls cannot be avoided by backing away from the problem. The supervisor who might seek to avoid IRCA problems by not considering an applicant because of foreign appearance or language can face discrimination charges as defined by other laws.

Military Service

In asking about the applicant's military service, an interviewer may inquire only into the general training and experience involved. Depending on state regulations, it may not be possible to ask the nature of the person's discharge or separation. Likewise, questions about whether the discharge was honorable, dishonorable, general, or otherwise may be privileged information, which the applicant may reveal voluntarily but which may not be requested.

Marital Status

You may not ask the marital status of the applicant at the time of the interview. Particular sensitivity has developed along these lines in recent years. Many women have been able to claim, with considerable success, that they were denied employment because of marital status. Employers often proceeded on the difficult-to-substantiate assumptions that:

- Young women recently engaged often quit shortly after they get married.
- Young, recently married women may leave after a year or two to begin families.
- Unmarried women with small children tend on the average to have poorer attendance records than other workers.

Even if an applicant, male or female, has willingly revealed marriage on a resume, an interviewer is not permitted to ask what the spouse does for a living. In the case of a married female applicant, it is not permissible to ask her maiden name. This question can be interpreted as probing for clues to national origin.

Material Issues

An interviewer may not ask if the applicant owns a home or a car. This may be interpreted as seeking to test affluence, which may in turn be taken as discriminatory against certain minorities. However, it is possible to ask if the applicant has a driver's license, if driving is a BFOQ. Generally, the interview is on soundest footing when all questions of a personal nature relate to BFOQs.

Financial Matters

In general, an interviewer may not ask if the applicant's wages were ever attached or garnished. Credit information ordinarily is privileged information. In most situations it may be volunteered by the applicant but not requested by the prospective employer. Some agencies do, however, use "integrity interview" procedures. These may include developing information on applicant financial status in order to determine whether that person is susceptible to corruption by inmates by reason of excessive personal debt or demonstrated poor financial responsibility.

Height/Weight

Direct questions about an applicant's height or weight are not permitted. Neither of these factors should have a bearing on the applicant's suitability for the job unless there is a specific BFOQ on these criteria that is uniformly applied to all applicants for that job category. In most such cases, there is a performance-based criteria (be able to run one mile in 8 minutes) against which all applicants are judged at some point, usually during training.

Testing

Beware of requesting that an applicant take qualification tests not approved at the agency personnel department level. Preemployment tests have long been under fire as discriminatory in matters of age, race, and economic background. Such testing is best left to professionals who design tests that are specifically related to the requirements of the job. They must be statistically validated as nondiscriminatory, administered consistently and in good faith, and evaluated impartially. If the agency has such a test, in all likelihood it has been validated statistically. It may even have been subject to review through the labor-management structure of which the organization is a part. In any case, ad-hoc or locally-developed tests would be subject to question and should not be used.

Education

An employer may require specific educational levels when these are BFOQs directly related to job performance. Otherwise, though, asking irrelevant questions may make it possible to show a pattern of employment discrimination. The hiring agency can require, for instance, a social worker to possess a diploma or degree and a state license. These are essential to the performance of the job as it is structured. However, under most interpretations of the law, a food service foreperson or plumber may not be required to have a high school diploma. That is because it is possible to demonstrate that people in these job categories may perform equally well with or without a diploma. Effective personnel department procedures should take most of the burden off the interviewer in this matter.

Union Membership

You may not ask whether an applicant is or has been a member of a union or has been involved in organizing or other union activities.

There is a final, ironic, note to this subject. A great deal of the information not available to the interviewer may ultimately come to light in the process of background investigations performed by many law enforcement agencies. When that happens, agency procedures ordinarily are in place to deal with any adverse background factors, independent of the supervisor's view of the applicant. This, of course, is the heart of the argument to avoid at all costs actually committing to hire someone until the full background investigation is complete.

Questions to Ask

The questions that may be asked are broad and in many instances open ended. This is as it should be. After all, the interviewer is interested in learning as much as possible about the applicant in a limited amount of time, The way to do this is to listen to the person talk. Try these in conversation with the applicant:

- What are your career goals?
- What would you like to be doing 5 or 10 years from now?
- How would you like to spend the rest of your career?
- Who have been your prior employers, and why did you leave your previous positions?
- What did you like or dislike about the work in your previous positions?
- Who recommended you to our institution?
- How did you hear about this job opening?
- What is your educational background?
- What lines of study did you pursue? (Be careful, however, of attempting to delve into specifics as cautioned in the list of questions to avoid.)
- What do you believe are your strong points?
- What do you see as your weaknesses?
- Have you granted permission for us to check references with former employers (see following section)?

If the personnel department has not already done so, ask the applicant to explain any gaps of more than a few weeks duration that appear in the resume or application. Some applicants will omit mention of unsuccessful job experiences or other activities they believe might reflect negatively on their chances of being hired. The hiring manager should be in a position to make a fully informed decision based on more than just selected pieces of a candidate's background. Also, there is a very clear responsibility for ensuring that the organization is protected from possible allegations of negligent hiring, should an employee who was hired without a reasonable amount of background verification cause harm to others during the course of employment.

Employment References

The whole topic of employment references represents a quagmire of potential legal traps for both the supervisor and the personnel department. Regulations vary greatly from one jurisdiction to another.

Ideally, the personnel department should have secured the applicant's signed permission to check references as part of personnel's screening interview. This usually involves obtaining a specific signed release for each former employer. For good reasons, many applicants indicate that their present employers not be approached for reference information in the preliminary stages of the application process. They ordinarily grant permission to check with most (but not always with all) previous employers. No reference checks should be performed by anyone in the absence of an applicant's signed authorization to do so. Even if a signed release is presented, reference checking should be left to the personnel department. Unless specified in agency policy, interviewing supervisors should not endeavor to check an applicant's references personally.

References must be checked systematically in accordance with strict guidelines governing the kinds of information requested and the kinds of responses to expect. In these litigious times, many unsuccessful applicants are quick to charge that they were not hired because of defamatory information acquired from previous employers. It is therefore important for all references to be checked by persons who do so regularly enough to be sensitive to the legal pitfalls that can be encountered. For example, an applicant might claim to have been defamed in being labeled on a reference check as "uncooperative and unreliable." This person might succeed in pursuing such a claim if there is insufficient information in the individual's past employment record to "prove" such contentions.

Reference information used in making a decision not to hire needs to be substantiated in the past employment record. For instance, to stand up as valid if challenged, a reference of "poor attendance" should be backed up (in the personnel file maintained by the employer supplying the reference) with attendance records or with records of disciplinary actions dealing with attendance.

You might be tempted to use personal contacts such as counterparts at other institutions to check on potential new employees applying from those locations. However, this information is not considered valid in making an employment decision, and it should not be used in the selection process.

■ The Actual Interview

Assume that as an interviewer, you have reached the point where you are well aware of questions you should avoid and those you should and may ask. In addition, assume that you and an applicant are now face-to-face in private. You remain mindful of the edge you hold in this interchange and of the possibly uncomfortable position of the applicant. Having reached this stage, proceed according to the following guidelines.

Put the Applicant at Ease

At the beginning of the conversation, put the individual at ease and instill a degree of confidence. Try several different topics at the start of the interview—for instance, the weather, the ease with which the person may have found the interview area, or an invitation to enjoy a cup of coffee—to get the person talking and move toward conversational rapport. Whatever opener that is employed (and usually it need be only brief) it will get an interview off to a far better start than the shock of something like, "Good morning. Why do you want to work in a prison?"

During these first few critical minutes try to avoid making judgments and freezing a picture of the applicant in your mind. First impressions are often difficult to shed, and when formed while the person is not yet at ease they can be unfair.

Avoid Short-Answer Questions

Avoid asking questions in such a way that they can be answered in one or two words, and especially avoid questions that can be answered simply "yes" or "no."

For instance, a question such as: "How long did you work for the Sheriff's department?" might simply be answered: "Three years." This gives very little information. Rather, a request on the order of: "Please tell me about the work you did at the Sheriff's department," requires the person to use more than just a couple of words in response. The purpose here is to learn about the applicant, and that can be done only by getting the person to talk.

However, avoid permitting the applicant to wander off the subject for minutes at a time. When this occurs, interrupt (as politely as possible) with another question or a request for clarification intended to draw the conversation back to the focus of the interview.

Avoid Leading or "Loaded" Questions

Avoid questions that lead the applicant toward some predetermined response. Take, for example, a question posed along these lines: "You left the Sheriff's department because the pay raises weren't coming along, is that right?" This is channeling the applicant toward a response that the interviewer may have already decided is correct. It is far better to ask, "Why did you leave the Sheriff's department?"

Leading questions can be a particular hazard when talking with someone who is shaping up favorably as the interview progresses. In the process of unconsciously deciding they like this person, the interviewer may begin to bend the rest of the questions in a fashion that calls for desirable answers. Most people are sensitive to leading questions. Depending on the nature of the questions, they will feel either forced or encouraged to deliver the answers that seem to be wanted.

Ask One Question at a Time

Avoid hitting the applicant with something like, "What kind of work did you do there and why did you leave?" This is in fact two questions, and although they may be properly asked one after the other, the person should be given the opportunity to deal with them individually. Combining or pyramiding questions tends to throw some people off balance; a person prepared to deal with a single question is suddenly confronted with two or three at once. It is always preferable to limit questioning to one clear, concise question at a time.

Keep Writing to a Minimum

Unless the agency has certain forms that must be filled out during the interview, try not to take extensive notes while the prospective employee is talking. This can be disconcerting to the individual, as note-taking creates the impression that everything that is said is being taken down in writing. Also, it is distracting to you. Writing and listening are both communication skills subject to their own particular ground rules. Neither can be done with maximum effectiveness if the interviewer is trying to do the other at the same time. The more note-taking that is done, the more it detracts from listening capacity. Take down a few key words if necessary, but focus most on listening to the applicant. If a written account of the interview is necessary, generate it immediately after the applicant leaves and the conversation is still fresh.

Use Appropriate Language

At all times deliver questions and comments in language appropriate to the apparent level of education, knowledge, and understanding of the prospective employee. If you are a chief psychologist interviewing an applicant for a psychology position, you will likely talk on a level at which you would ordinarily expect to communicate with another mental health professional. However, if you happen to be interviewing for the position of unit secretary and are talking with a person who has never worked in a correctional setting, you would alter your language accordingly. Always assume a reasonable degree of intelligence—you wish to avoid "talking down" to the person—but do not dazzle or confuse the applicant with unfamiliar terminology.

Do More Listening Than Talking

Throughout the interview be interested and attentive, never impatient or critical. Avoid talking too much about yourself or the institution. Remember, you are not selling yourself to the applicant—it is supposed to be the other way around. Also, you are not necessarily selling the institution or the job to the applicant. However, if the interview moves far enough in constructive directions, you may wish to answer the applicant's questions about the correctional work in general or the duties at this location in particular.

However, even on this score there are precautions to note. Specifics of certain features of employment like insurance, retirement, and other benefits with complex details should be left to the personnel department. This type of information should not be mentioned until the organization is ready to extend an offer of employment and the individual needs this information to aid in making the decision.

Indicate Some Type of Follow-up

Conclude the interview with a reasonable statement of what the applicant may expect to happen next. An interviewer generally cannot (and probably should not) make a definite statement at this time as to the applicant's prospects, but should suggest what may be expected and when it might occur.

For instance, the interviewer can always say something like: "We'll let you know our decision after we've finished all scheduled interviews, say within a week or 10 days," or "You should be getting a letter from us next week," or perhaps simply "We'll call you by next Friday." In any case, conclude the interview with some simple indication of impending follow-up. Never let the applicant go away with the feeling of "Don't call us, we'll call you."

■ Follow-Up

Follow-up is no problem when it is time to offer a job to an applicant. Extend the offer and the person will either accept or decline the job. In either case the interview process has been taken full circle.

However, even in instances when a job offer does not result from an interview, follow up and complete the interview cycle. This is a simple but deserved courtesy. Although the applicant was looking for a job, the organization was also looking for an employee, and this particular individual expended the time and effort to try to fill that need.

Appropriate follow-up takes very little time. Once a decision has been made to not extend a job offer, a short, polite letter to that effect is appropriate. Also, use this same approach to let individuals know that although a job offer will not be made at the present time the agency would like to keep their application on file for future consideration. Where appropriate, a letter of this type also can suggest that they may wish to apply at another institution within the agency. Depending on the agency's employment system, these letters might originate with the manager involved or they might go out from the personnel department.

In most organizations the personnel department will take care of applicant communications, but it is within the realm of reason for a manager to follow up with personnel to make sure this communication has occurred. In any case, and even for the most clearly unqualified of applicants, conclude the interview process with an answer. It is a courtesy due the applicant, and it serves to protect the agency's image as an employer in the community.

The employee selection interview can be a hazardous process filled with opportunities for miscalculation and misjudgment. It offers no guarantees that the organization will always locate the right person for the job. However, despite its flaws and pitfalls, the interview is the best available way of gaining information about prospective new employees. As such it will continue to remain one of the most important kinds of personal contact for the supervisor.

EXERCISES

Exercise 9-1: Preemployment Inquiries

Indicate whether each of the following questions is lawful (L) or unlawful (U) for you to ask an applicant in an interview situation:

1. Have you previously been employed under a different name?
2. What was your name before you married?
3. Where were you born?
4. Have you any condition that would prevent you from adequately performing in the position for which you are applying?
5. What was your title in your last employment, and who was your supervisor?
6. Do you have any handicaps?
7. Of what country do you hold citizenship?
8. Who is the relative you would designate to be notified in case of emergency?
9. What foreign languages do you read, write, or speak fluently?
10. What other countries have you visited?
11. How did you acquire your foreign language abilities?
12. At what schools did you receive your academic, vocational, or professional education?
13. What major credit cards do you hold?
14. Have you ever filed for bankruptcy?
15. Who would be willing to provide you with a professional reference for the position for which you are applying?
16. Are you a single parent?
17. Have you any commitments or responsibilities that may interfere with meeting work attendance requirements?
18. What relatives of yours, if any, are already employed here?
19. Did you receive an honorable discharge from the military?
20. Why do you want to work here?
21. Where did you live before moving to your present address?
22. Do you own or rent your home?
23. What experiences, skills, or other qualifications do you feel make you appropriate for the position under consideration?
24. Why did you leave your last employer?
25. What organizations, clubs, and societies do you belong to?

Answer Key
Key to Questions:
(L)awful: 1, 4, 5, 9, 10, 12, 15, 17, 18, 20, 23, 24
(U)nlawful: 2, 3, 6, 7, 8, 11, 13, 14, 16, 19, 21, 22, 25

Exercise 9-2: Would You Hire This Person?

This is a group exercise, with six or more members of the group filling the following roles:

Interviewer: Business Manager

You are the business manager for a small state correctional institution. You are interviewing applicants for a junior accountant vacancy. In your department you have two accountants, who are both qualified, reasonably effective individuals. You also have one junior accountant and two clerks. There is one open position for a junior accountant.

The personnel office is sending you the second and third job candidates to interview. The first you rejected as overqualified and looking for more money than the junior accountant position would warrant. It is your understanding that the stream of job applicants dried up after candidates two and three submitted their applications. You feel strongly about the need to fill this position in the near future and will have this foremost in mind when you interview candidates two and three. You do, however, have an alternative—you can hold the job open for a longer time and wait for the ideal candidate to appear.

After the initial review of applications you contacted some people you know to get an idea of the character of candidate number three. You learned that he has been with his present employer only 4 months but has not missed a day's work nor come in late. You did not check on applicant two because his application indicated he preferred you not contact his present employer.

Candidate Two

You are 42 years old and you have approximately 15 years of routine bookkeeping experience in a large commercial organization. You believe you are probably older than most persons who would be expected to apply for this position. You have done approximately the same level of work for about 10 years. You are still employed but you are not especially happy with your job. You believe you will probably go no higher in the organization, so you decided to look for something better.

Candidate Three

You are a 23-year-old high school graduate from another state. You did factory work for a year before taking 1 year of liberal arts courses at a community college. You then spent 3 years in the military service and after being discharged settled in this area, married, and took a laborer's job in a local cotton mill. You are not hesitant to reveal that you have been looking for a new job since the day you were hired at the mill. You generally agree that your background seems not to qualify you for accounting work but you have started a crash program of learning basic accounting on your own. You have done this primarily by acquiring books and practice sets and getting assistance from an acquaintance who is an accountant.

Instructions

- Divide all participants into three smaller groups of approximately equal size. Assign one of the three roles to each group. Each small group should designate a spokesperson to play the role assigned to the group.

- The business manager group should prepare its "interviewer" by developing a tentative line of questioning for the interviews. It should be this group's intention to learn as much as possible about each candidate (following acceptable lines of questioning) in a brief simulated interview.

- The groups representing candidates two and three should prepare their spokespersons by considering the kinds of questions they might expect to be asked. It is up to the "candidates" to respond to questioning, making reasonable assumptions as necessary, and "sell themselves" to the "interviewer."

- Devote approximately 10 minutes to group preparation, then allow about 5 to 7 minutes for each interview. Those participants not involved in the exercise as spokespersons should critique the interviews according to the interviewing guidelines provided in the chapter.

- Discuss the conduct of the interviews.

The One-to-One Relationship

10

Chapter Objectives

- Broadly define communication as *the transfer of meaning*.
- Emphasize the importance of establishing an effective one-to-one communicating relationship with each employee.
- Stress the essential two-way character of interpersonal communication.
- Highlight the common barriers to effective communication, and suggest how they can be avoided or overcome.
- Explore the true nature of "listening" and offer guidelines for improving listening capacity.
- Suggest guidelines for effective interpersonal communication in the supervisor-employee relationship.
- Stress the necessity for the supervisor to maintain an "open-door" attitude regarding employee communication.

I know that you believe you understand what you think I said, but I am not sure you realize that what you heard is not what I meant.

Anonymous

The difference between the almost-right word and the right word is really a large matter—it's the difference between the lightning bug and the lightning.

Mark Twain

Fortune 500 Company Executive or Prison Inmate?

It was early in the career of the new prison psychologist. He was confident, brimming with knowledge and the enthusiasm of someone who was finally doing something important. As part of his duties he was assigned to a classification team making initial assessments and assignments for incoming inmates at a large maximum-security institution. The chairman of the team was the institution's captain—a grizzled old veteran of the system, who also was a retired Marine lieutenant colonel.

In the course of describing his assessment of a particular inmate, the psychologist waxed eloquent. He concluded his summary by saying, "This inmate is a ruthless, ambitious, aggressive, and extremely intelligent individual who won't let anything stand in the way of getting what he wants."

The team chairman had been leaning back in his chair at the head of the table, seeming to barely listen. When these pearls of wisdom passed his ears, he leaned forward and said, "Son, you've just described to a 'T' the personality characteristics of the people who head up most of the Fortune 500 companies in this country. Now what did all that tell us about what kind of an inmate he's going to be?"

■ The Transfer of Meaning

Communication is critical in the correctional environment. On any given day the largest part of a correctional manager's job is likely to consist of interpersonal contact with other employees of the institution (and to a lesser extent to inmates), primarily those who report directly to the supervisor. This chapter recognizes that a supervisor's constant contact with employees is necessary to get things accomplished through them. It will concentrate on face-to-face communication specifically within the context of the supervisor-employee relationship.

A simple beginning for a complex subject might be to describe communication as "the transfer of meaning." In an organizational sense, communication involves the transmission of information and instructions from person to person in such a way as to accomplish mutual understanding on the part of both the sender and the receiver of every message. The goal of communication is always a complete and accurate transfer of meaning. This transfer must assure the person receiving the message understands the information in a manner identical to that of the person who originated the message.

The story related above is a classic in this sense. The psychologist correctly summarized all of the personality traits the inmate displayed. His meaning was clear to him and he thought this was useful data for others. But that communication failed to give the necessary essentials to those involved in managing the prison and making decisions about this offender. They needed to know the kind of inmate they were dealing with. So the intended transfer did not take place.

Regarding the supervisor-employee relationship, there is no single absolutely "correct" way of communicating with employees. Every employee is different from every other employee, and they are all different from the manager. What works well in a manager's relationship with employee "A" may not work at all with employee "B," and vice versa. It is safe to say that in relations with employees, there may be fully as many "correct" ways of communicating with employees as there are employees.

An effective relationship with each employee is not something that can simply be established and assumed to exist forever. No such relationship between two people is ever completely "established." Indeed the relationship must be constantly nurtured and conscientiously maintained if it is to serve appropriately the needs of the people involved and the organization.

Although interpersonal relations constitute a large part of the supervisor's job, it is not always easy to concentrate on people, and their problems and needs. There are hidden pressures that encourage managers to focus on things rather than people. After all, things cannot hurt, disappoint, frustrate, influence, disagree, or misunderstand. There is a measure of safety and security in things. However, the supervisor must learn to overcome the tendency to seek reduced vulnerability and look in the direction of the job's true demands.

The effective manager is called on to acquire and practice empathy—the capacity to put oneself in another's place and respond accordingly. Not many people are naturally attuned to the feelings and needs of those about them. But empathic sensitivity can be developed through conscious effort. Although most people tend to take communication capacity for granted, they are not really very good communicators by nature.

People can, however, learn to communicate effectively. To do so they first need to appreciate the difficulties inherent in human communication, and to recognize one critical tendency in ourselves. That is the tendency to believe that we are better communicators than we in fact really are.

People use words as primary tools of communication, in order to express meaning. Yet a tremendous gap exists between the thought or feeling experienced inside, and the words available for use to direct this toward information from another person. Likewise, a similarly large gap exists between the words taken in by the other person and the thought or feeling experienced as a result of those words. Consequently, it is possible only rarely (except in the most inconsequential situations) to transfer meaning accurately to another person in every precise dimension in which it was experienced.

It is absolutely impossible for the person "sending" the information to know what unusual connotations the recipient may attach to specific words, which will further impact the way the message is interpreted. Each individual is unique, due to countless factors and influences in each person's upbringing, background, education, and life experience. Because of these differences, people relate to the world and to specific linguistic cues in particular ways. Individual experiences are never identical, and that uniqueness affects the ability to communicate with each other. Whenever people communicate with others they are attempting to describe what has happened internally as a result of external experiences. As long as each individual can sense with only a single mind (his or her own) the process will remain imperfect. No person has control over the mind of another. Rather than directly transferring a thought or feeling, the best that can be done is put it into words and hope those words create a similar thought or feeling in someone else.

Interpersonal communication frequently fails because of the imperfections in the process. Too often communication fails simply because people assume the person they are talking with knows and understands the subject (after all, we know what we are talking about). Also, the process is fully as susceptible to failure when viewed from the other side. We assume we know what the other person means without taking steps to make certain this is so. The greatest number of communication failures probably occur in simple interchanges between people. You say something to another person and the two of you part, each with a meaning in mind. You believe you have been understood, and the other person believes he or she understands. Despite a supposed communication event, you have parted with completely different meanings in mind.

■ The Two-Way Street

Communication within the supervisor-employee relationship must be a two-way street that is heavily traveled in both directions. The supervisor's authority and the number of people with whom he or she interacts will limit the amount of communication flowing from employees up the chain of command. Therefore, it is generally up to the supervisor to go more than halfway in the communication process.

One-way communication—the simple delivery of orders and instructions in authoritarian fashion with no information flowing the other way—is no communication at all. Admittedly there are occasions when one-way communication seems to work effectively. Giving orders in a military situation or snapping out instructions in a correctional emergency are good examples of this type of communication. However, these one-way transactions work only because they have been preceded by a significant amount of two-way communication in the form of instruction, training, and practice.

There are pressures (sometimes of undeniable proportions) that encourage managers to communicate in a one-way fashion, that is, simply to deliver orders or instructions in a few words before moving on to other problems. The pressures of time are often upon us. It is always faster to deliver an instruction (even a seemingly well-thought-out one) than it is to engage in discussion or encourage a few simple words of feedback. One-way communication is clean and neat (at least it seems that way, until misunderstandings begin to surface) and seems to be accomplished without wasting time or words.

One-way communication often is the refuge of the insecure manager. It discourages feedback and suppresses discussion, and the manager is considerably less likely to be challenged and therefore cannot be found "wrong." One-way communication is also the province of an authoritarian manager. This is exemplified by the individual who behaves as though all meaning must flow from the "boss" downward and that nothing valid can be expected to flow in the opposite direction. This mode of communication will be familiar to those already in corrections; it is all too common. True two-way communication, on the other hand, takes time, often considerably more time than a corresponding one-way contact would have taken. It is a simple trade-off. More time is invested for the sake of achieving greater accuracy in the transfer of meaning.

Consider the originator of a message in a two-way communication situation. The "sender" of the message experiences a certain vulnerability by virtue of being open to feedback and thus open to question, discussion, and perhaps disagreement. This position of vulnerability is not acceptable to all people in their communicating relationships, but feedback is the all-important element in two-way communication. It is feedback that helps clarify what has been said and ultimately determines whether the message has been correctly received.

■ Barriers to Effective Communication

There are a number of common elements that present very obvious barriers to effective communication. Regrettably, most people are so habituated to their own communication style that they fail to see how interactions with others are impaired.

Semantics

Words, the most actively used tools of communication, are actually quite inadequate for the role required of them. People expect far too much of these small bits of language. A word is simply a symbol that is used to stand for something; it is not the thing itself. The only true meanings words have are the meanings given them through active use.

Words are inconsistent in their meanings; many of them possess a variety of "definitions." In fact, the 500 most frequently used words in the English language have a total of nearly 15,000 dictionary definitions. This suggests that the "average" word, if there is such a thing, can have 30 definitions, some quite similar to each other but some extremely different from all others.

Words are also likely to mean different things to different people. For example, how long or short a span of time does "prompt" mean? Some words can even mean different things to the same person at different times or under different circumstances. For example, "fix" may mean one thing when talking about repairing an automobile, and something else entirely if one's immediate interest is locating a reference point for navigation purposes or a drug deal.

There are words in the English language that can be described only as fuzzy in terms of meaning and usage. Take a good look at a few of the words that fill organizational communication—sometimes to the point of overflowing, such as the word "prompt." There also are other terms, such as "sufficient," "appropriate," "adequate," and other similarly vague terms. What meanings are going to be absorbed if the last sentence of a memorandum just received says, "You are requested to take prompt action, employing adequate measures to ensure appropriate response sufficient to the situation"?

One can sample many of the problems of semantics by reacting to a simple question long used in discussions on this subject. How tall is a tall man? Over six feet? Six feet six inches or taller? Five feet nine inches or taller? A person's answer probably depends in part on their own height, and ultimately the answer to the question depends entirely on who is doing the judging.

Emotions

Words often carry emotional overtones that vary from person to person and create additional barriers to interpersonal communication. Words that register neutrally or perhaps favorably on some people are likely to hit "sore spots" in other people. Aside from the obvious (profanity, obscenity, ethnic slurs, and downright insults) many words used without malicious intent will trigger negative emotional responses in listeners. Perhaps the word "stubborn" does not bother one person, but another individual resents being referred to by that word. Perhaps a supervisor places a favorable connotation on the term "eager beaver" and applies this to an employee. Contrary to the intent, the employee may resent the term as suggesting he or she is trying too hard or bordering on being annoying. (They would prefer to be called a "loyal and energetic employee.")

Positive emotions can be good to some extent in interpersonal communication. However, even positive emotions (joy or enthusiasm, for example) can cloud individual judgment and impair communication. That can occur when they are experienced to a degree that the feelings themselves become more important than the message being communicated.

Negative emotions (hurt, anger, and all their variations) definitely tend to impair effective communication. The correctional setting presents ample opportunities for these emotions to preload an otherwise neutral conversation. No matter how hard someone tries to keep language inoffensive, it is likely that language may occasionally touch someone else's sore spots and trigger an emotional reaction.

A line employee may have just been involved in a stressful situation involving name-calling by an inmate. The manager may unknowingly reap the emotional residue of that conversation when talking to that person a few minutes later. Consider the times you may have turned away from a conversation with the impression that you have somehow hurt, offended, or angered the other person. It is likely that you honestly did not have the slightest idea of what you might have said to cause that reaction. People do this unintentionally, and it is not likely that you can ever eliminate this from communication entirely. However, what you can do is be aware that this occurs with all people in interpersonal situations and, exercising empathy, try to understand the other person's position. Also, since you may hurt, offend, or anger someone unintentionally, always extend to the other person the benefit of the doubt. Remember that when someone touches one of your little sore spots and you react negatively, chances are it was completely accidental and no insult or harm was intended.

Generally, the higher the negative emotional level of one or both parties to an interpersonal exchange the less are the chances of meaningful communication. When emotion threatens to turn a discussion into an argument it is time to back away from the situation as diplomatically as possible and try again when tempers have abated.

Emotion in interpersonal exchange also tends to polarize views and drive participants toward opposite positions. Anger or its variations are also likely to be evident in the behavior of someone who has been put on the defensive. In such cases, the person often is using emotion unconsciously to cover a weakness or shield a position from attack.

It would be easy (but quite useless) to say we should not allow ourselves to be hurt or angry. People cannot help experiencing a feeling, and feelings themselves are neither right nor wrong—they are just there. However, be aware of the destructive potential of negative feelings.

In all dealings with employees and others, be reasonably careful of the words used, remain normally polite and friendly, and do not hesitate to be conciliatory if it appears necessary or helpful. Although managers occupy a so-called position of authority, they should never consider themselves too self-important to apologize when an error takes place in interpersonal communication. That thought bears repeating—a manager should be ready to apologize when he or she makes a mistake.

In all dealings with others, and especially in dealing with employees on critical issues and points of difference, avoid sweeping generalizations that are by their nature untrue. Two of the worst words that can be used are "always" and "never," as in "You're always late," or "You never submit it correctly." Rarely are those terms strictly true, but both are usually inflammatory.

Poor Listening

The following words of wisdom have appeared on many office walls: "How come it takes 2 years to learn to talk and 60 or 70 to learn to be quiet?" How much of the talking do you do in any interchange with an employee? This is a valid question because you cannot truly listen while you are talking.

Of the four verbal means of communication (writing, reading, speaking, and listening) effective listening requires the most communication time. Yet listening is the single verbal communication skill for which the least amount of solid, practical help is available. Although listening skills are not easy to describe or define, these skills can nevertheless be learned. Conscientious attention to the following suggestions and precautions can make a better listener.

Force yourself to concentrate on what the other person is saying. Do not turn off the speaker as uninteresting or fail to pay close attention to the subject because you "know it already." Some people speak in ways that fail to grab and hold the attention of others automatically. Conscious effort is necessary in those cases to know what is really being said. Knowing that focused attention may be required is often more than half the battle. The remainder becomes a matter of actually applying undivided attention. Also, keep in mind that even though a topic may not seem interesting or important to a listener, it usually is important to the person doing the speaking.

All but trained listeners are likely to rush their responses by reacting to only a portion of what is being said. The temptation is great. The listener hears a few words that trigger a response. The mind begins to race ahead and form the comment to deliver as soon as the other person pauses. Resist this temptation. Nothing more of what the speaker says is really heard. Sometimes the listener will interrupt the other person long before the thought is complete. Instead, deal with one complete thought or idea at a time, hearing the person out completely before offering a response.

Many people have a habit of listening only for "facts"—specific bits of information. Facts are useful, of course, but their usefulness is diminished if they are not considered within the context of the entire message. Listen to the whole message before making decisions or rendering judgments.

If it is at all possible do so, try not to interrupt the other person to offer a correction of something that may have been improperly stated. Certainly avoid interrupting to offer advice or to scold or otherwise criticize. This is especially important in face-to-face communication within the supervisor-employee relationship. Generally, interruptions are in order only if it is necessary to ask for expansion or clarification of something that has been said. Generally, they also may be used to inject related questions intended to encourage more relevant comments.

While listening, be aware of the possibility of emotional reactions within yourself. As discussed earlier, it is possible that the other person will unintentionally trigger an emotional reaction in you. Give the other person the benefit of the doubt and force yourself to concentrate on the message as being distinctly separate from its emotional overtones. You cannot help what you feel, but you can control what you do with those feelings. If you allow negative feelings to come to the foreground and dominate your reaction, your listening capacity will be sharply reduced.

■ Guidelines for Effective Interpersonal Communication

When you are doing the talking, you have the initiative, and at least nominal control over the medium. Use that advantage well. Before speaking up, put some effort into structuring the communication you are about to deliver.

First think out the what and why of the communication. Then, when you know what you are going to say and what you are trying to accomplish, you can consider how best to communicate it.

Consider your listener's needs, interests, and attitudes. Try to exercise empathy at all times, constantly judging what you are saying from your listener's point of view.

Deliver your comments in language properly suited to the level of knowledge, education, and experience of your listener. Never talk down to or over the head of the other person.

Except in the direst of emergencies (and even in emergencies, when you are not convinced that your listener is prepared to respond appropriately) follow up your communication immediately with a request for feedback. Ask to have the message played back to you in the listener's own words. This could be as simple as asking, "How would you describe what I've just asked you to do?" Ideally, the approach should not be one that suggests probing to find out if the other person understood—although that is a major purpose. Rather, try to convey the impression that this is a process of assuring that thoughts were communicated clearly and fully, which also is a major purpose of this step.

When you are doing the listening, there are still ways to enhance the flow of information and true communication.

Pay attention. Really listen—your undivided attention conquers many potential problems.

Always listen for meaning, striving constantly to determine what actually is being said and why it is being said.

Consider the whole person, searching out attitudes and feelings as well as meanings. Keep in mind that words and other signs (the nonverbal signals given out by the other person) cannot be fully separated from attitudes and feelings.

Be as patient as necessary to encourage the individual to try to communicate fully.

Be prepared to compromise as necessary to achieve agreement or understanding. Consider yielding completely on minor points or unimportant details. Compromise is not the dirty word it is regarded to be by some. Rather, reasonable compromise is often the most important step in establishing mutual understanding.

Return the message to the speaker in your own words. Use discussion to iron out differences until you both agree that the message sent and the message received are the same.

■ The Open-Door Attitude

Every manager has probably said at one time or another, "My door is always open." This is an easy statement to make, but it takes effort to assure that these words reflect more of an honest attitude than a timeworn platitude.

The manager's job is to help employees to get their work done, guiding and directing their overall efforts. In doing this, the manager must be prepared to answer questions, deal with problems, silence rumors, and put fears and suspicions to rest. Remember that employees do not work for a manager as much as they work with him or her. A large part of the manager's function is to "run interference" for employees so that they can accomplish their work as efficiently as possible. To do all of this the manager must be visible and readily available to employees and have an always growing one-to-one relationship with each employee. It is the manager's responsibility honestly to consider the employee as a whole person, not simply as just a producer.

To do so often requires the manager to be a person first and a manager second. Successful supervision requires human sensitivity. Without sensitivity, a supervisor has only rules, policies, and procedures, which are useful but which by themselves are grossly inadequate for the complete fulfillment of the supervisory role.

Effective one-to-one communication with each employee is the basis for mutual understanding between supervisor and employee. A healthy attitude for the supervisor assumes a Golden Rule approach to communication: Deal with others as you would wish to be dealt with yourself.

A one-to-one relationship with each employee is critical to the institution as a whole. The way each employee sees management—available, friendly, caring, and helpful, or perhaps the opposite—may be the way he or she comes to see the entire organization. Every member of management (and most particularly the first line manager) represents the organization to the employee. The impression the manager creates as a person will contribute, for good or ill, to the individual employee's impression of the organization as a whole.

In corrections, this has a secondary, but extremely important dimension. Inmates, in part, take their cues about the management of their prison from staff. If management communication gaffes generate a poor supervisor-employee climate (one characterized by mistrust or skepticism) it is likely that inmates will adopt that same attitude toward management. And anyone who has worked in corrections for any length of time will be able to describe situations (some of them critical) where trust and credibility were vital ingredients to finding a solution to that problem. A negotiator working with inmate representatives during a work stoppage or with a hostage-taker does not want to be hampered by adverse inmate perceptions about management.

EXERCISES

Exercise 10-1: What's in a Phrase?

Mort Harriman accepted the position of industrial procurement supervisor at a major state correctional institution that had a large metal factory. He was to be responsible for not only purchasing, but warehousing raw and finished goods and staging material flow to the factory floor.

In his mid-30s, Mort had considerable experience in related work in material management and procurement. Years before, following his graduation with a degree in general business, he was floor supervisor in the largest paint warehouse in a major metropolitan area. After 5 years in that capacity he moved on to become manager of purchasing for a sheet metal fabrication firm. Now, after 8 years, the economy had soured, and he had really felt lucky to have been hired at the nearby 1,000-bed state correctional facility, although at a lower wage. He had never before worked for a government agency.

Mort reported to Art Reynolds, the factory manager, and was impressed with him because he seemed to have his finger on everything on the factory floor. He seemed to be on top of the many details of running a factory inside a maximum security correctional institution. He was likewise impressed by the tempo, tone, and general enthusiasm of the factory. Despite the fact it was a significant change from the relaxed environment from which he had come, Mort was sure he would do well.

Mort came to work at the institution during an extremely busy period. Not only did he have the expected task of getting to know the workers in the warehouse and business office, but there was also the problem of relocating the factory's largest storeroom to a newly completed area. Also, it was the time of annual budget preparation and Mort's first week was the one in which each department's initial budget drafts were due.

On Monday, Mort's first day, he received an extensive orientation from Reynolds along with guidance concerning things he was expected to do. The last item covered was the matter of the budget. As Reynolds put it to Mort, "I realize you are new but you've walked right into the annual budget cycle. All of the other department heads are expected to have their first-cut budgets to me on Wednesday. You're going to be involved enough as it is, so let's say I'll expect to have your first rough cut as soon as you can do it but no later than noon Friday."

Later that day Mort began assembling preliminary numbers for the budget draft. He took it home with him and did a small amount of work on it.

On Tuesday some severe problems developed with the transfer of goods to the enlarged storage area and Mort found himself deeply involved. At the same time he was pursuing a series of face-to-face meetings with employees, in order to get to know them. By midweek Mort was thinking that never in several years on his past job had the work come at such an unexpected pace.

Shortly after noon on Friday Mort was seated at his desk handling a few items of correspondence when his phone rang. It was Reynolds, who asked, "Mort, where is that draft budget?"

Mort suddenly realized that he had not touched the budget since Monday evening. Recalling (or believing he recalled) Reynolds' words he said, somewhat defensively, "You told me to get it to you as soon as I could. I've been buried; I just haven't been able to get at it."

"I did say as soon as possible," Reynolds replied. "I could have used it Wednesday. I also said 'no later than Friday noon.' It's now Friday, past noon, and I don't have your draft."

Mort said, "I'm sorry I let it slip. I'll get it to you as soon as—" he stopped and caught himself for he had almost said, "as soon as I can." Instead he said, "I'll get it to you as quickly as I can put it together."

After a few seconds of silence on the other end of the line Reynolds said, "I'll be here until about 6:00 tonight. I expect your budget draft on my desk before I go home."

Mort cleared his desk and prepared to work on the budget, thinking somewhat glumly: "First week on the job and I'm already on the list." He could also not help thinking that his boss on his previous job would never have expected a new employee to get up and running so rapidly.

Instructions

Isolate the particular words or phrases that got Mort into trouble. Explore the likely reason why trouble resulted from a few seemingly innocent words.

What does this tell you about:

The context within which a message is delivered?

The apparent meanings of simple words?

Exercise 10-2: The Employee Who Is Never Wrong

"I know what I heard, and that's that," correctional officer Craig Masterson said in the no-nonsense tone that shift supervisor Harry Anderson had come to know so well.

"Captain Steele says otherwise, Craig," said Anderson. "He told me in no uncertain terms that the instructions he gave you were just the opposite of what you did."

"He's wrong," snapped Craig.

"He says that you were wrong, and he seemed quite sure about it." Harry paused thoughtfully before adding, "He took the trouble to explain the whole situation to me, and I have to say that I understood his instructions. At least I was able to give them back in my own words so he was satisfied that I understood."

Craig scowled, then shrugged and said, "Then he changed his story."

"You're suggesting that he lied to me?"

"I didn't say that. I'm just saying that he told me one thing and then apparently told you something else. Maybe he didn't understand what he was telling me. You know how he just kind of rattles off something quickly and runs away."

Harry sighed and said, "Craig, did you consider the possibility that you didn't understand? It isn't hard to misinterpret when everything happens so fast and—"

"I know what I heard," interrupted Craig. "When I know I'm wrong I'll say so. If I even think I may be wrong I'll say so. But in this case I know I'm right. It's not even remotely possible that I could have misinterpreted him."

Feeling that Craig had given him cause to say something that had been nagging at him for quite some time, Harry said, "It seems to me that you're never wrong, Craig."

Craig glared at his supervisor. "What do you mean by that?" he asked.

Harry took a deep breath and plunged in. "I've been watching you work various posts for 3 years, and in that time I've never known you to admit to being wrong about anything. This business with Captain Steele is just one more example. You always turn everything around so you come up clean. Is it so necessary that you be right about everything? Do you ever make a mistake?"

Craig's tone, already cool, became colder. "Like I said, I'll admit I'm wrong—but only when I am wrong. And I want to know the other times you're talking about, the times when I supposedly turned everything around."

Harry began, "Well, there was—" he stopped, shook his head, and said, "No, that was something else. In any case, you ought to know what I'm talking about. Think about it, you'll know what I'm saying. You've got an answer for everything, an answer that always places you in the right."

"You can't think of any specific incidents because there haven't been any," said Craig. He rose from his chair and continued, "You may be my supervisor, but I don't have to listen to this. Is there anything else you wanted to say about the Captain's problem?" he glared down at Harry.

Harry rose to his feet. "Well, it's just that this incident isn't really closed. Captain Steele told me to write it up for your performance record as a verbal warning."

"Well, you can bet I'll fight that," said Craig. "I won't sign a warning I don't deserve, and I won't say I'm wrong when I know I'm right."

When Craig left the office Harry began to regret having spoken to Craig as he did. He was convinced, however, that he had to try to get through to Craig about his apparent need to be "right" about everything.

Questions

1. When Harry "took the plunge" and went beyond the specific incident to talk about Craig's overall conduct, he made a mistake that is embodied in the statement "You always turn everything around so you come up clean." What was Harry's mistake?

2. How would you recommend attempting to determine the cause for the misunderstanding involving Craig and Captain Steele?

3. Taking Craig's departure from Harry's office at the end of the case description as your starting point, how could you propose to deal in the future with the employee who is "never wrong"?

Leadership: Style and Substance

11

Chapter Objectives

- Describe patterns of leadership, or leadership styles, ranging from rigid (autocratic) to open (participative).
- Review the implications of the classic "Theory X versus Theory Y" as it applies to leadership styles.
- Review opposing sets of assumptions about people that give rise to different leadership styles.
- Establish the necessity for sufficient flexibility in leadership to vary style according to circumstances.
- Identify the apparent primary defining characteristic of effective leadership.
- Establish the critical importance of the supervisor's visibility and availability to staff.
- Relate the employees' view of the supervisor to critical elements of leadership performance.

Real leaders are ordinary people with extraordinary determinations.

John Seaman Garns

Be willing to make decisions. That's the most important quality in a good leader. Don't fall victim to what I call the ready-aim-aim-aim-aim syndrome. You must be willing to fire.

T. Boone Pickens

Feared by Some, Respected by All: A True Leader

Some called him "Gentleman Jim." His career started out normally enough, as a correctional officer at a large penitentiary. But those above him quickly recognized that there was something different about this slim young man from the Midwest. He was quickly promoted through the custody ranks in the field. He eventually held a headquarters assignment supervising all security operations for the agency. The time came when he was selected to handle extremely sensitive special assignments for his agency and others (some with major international relations overtones). He returned to the field as a penitentiary warden, and ultimately became a regional director. When he retired, he carved out a career as an international consultant in corrections. He was feared by some, respected by all, and acknowledged universally as a leader in his chosen profession.

What qualities made the late James D. Henderson an acknowledged and successful leader in the field of corrections?

He set and maintained high standards for himself and those around him. He was an outstanding judge of character, promoting capable people and supporting them. He listened to others, formed opinions, made decisions, and stuck with them. He was an excellent problem-solver. He was loyal, and in turn expected loyalty. He was unfailingly respectful, polite, and professional in his dealings with people at all levels and in all walks of life. He disciplined when necessary, counseled when necessary, and chose each of those strategies wisely. He was adaptable; as corrections changed, he changed as well, but without abandoning his core principles. He was not afraid of hard work and believed in preparation for every contingency. He made a point of staying on top of developments in his profession while staying grounded in fundamentals. He had demonstrated personal bravery in the real world of prisons, and he never forgot his origins in the field. (When the question came up about possibly closing his regional office early because of an impending blizzard he said, "We'll close this office when they close A cell house at Leavenworth.") Because of his reputation for fair dealing, he even had the respect of inmates throughout the system, some of whom corresponded with him for years after their release.

■ Introducing Leadership

The above capsule career summary is a tribute to a real person—someone who typified what this chapter is about. His career followed no formula or pre-established pattern. The personal traits he brought to the table, blended with his experience and training, allowed him to gain recognition and respect within and outside of his agency.

His story is not unique. It is a variant on that of many correctional leaders. Indeed, most people have some potential for leadership, although in some people this potential may be limited. The essential difference between the leader and the non-leader is determined by the degree to which a person succeeds in learning about leadership and applying what has been learned.

Leadership is like many other human endeavors—talent helps, but it is not necessary to be extraordinarily talented to be successful. A particular manager may not be a "natural leader"—able to run a large organization or get hundreds of people to follow in some vast undertaking. But most managers stand at least an average chance of being able to furnish true leadership to the employees in a department or other work group.

There are vast differences in perceptions of what characterizes a leader and how a leader should behave. This discussion of leadership will begin with the consideration of style—the patterns of behavior projected by leaders as they work. And although it is tempting to talk about leadership on a grand scale (since most of the great leaders led armies, nations, churches, or corporations) this discussion is better served by limiting it to the context of the correctional manager's environment.

■ Patterns of Leadership

Leadership styles range along a continuum from purely authoritarian at one extreme to fully participative at the other. In the scale in **Figure 11-1**, from left to right, the following styles are delineated:

- Exploitative autocracy describes the harshest style of leadership. The exploitative autocrat not only wields absolute power over the people in the group but also uses the group primarily to serve personal interests. This type of leader literally exploits the followers.

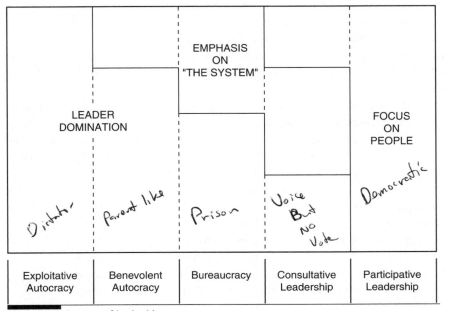

Figure 11-1 Patterns of leadership.

- Benevolent autocracy is when the leader wields absolute power, often with an iron hand, but is generally sincere in believing that the behavior of those in the group must be closely ordered and regulated for the good of the organization. References to autocratic leadership throughout the remainder of this chapter will refer generally to this particular leadership pattern.

- Bureaucratic leadership is a term that immediately raises visions of large federal and state government agencies. This applies especially where correctional organizations dealing with confinement are concerned, since most are organized in a quasi-military manner. As a leadership pattern it refers to maintaining a primary emphasis on rules and regulations. The bureaucratic leader "goes by the book," creating new rules and regulations as new situations arise. To a certain extent, the "book" itself often becomes more important than the purpose it is intended to serve.

- Consultative leadership is exhibited when the leader remains open to input from members of the group but, through either pronouncement, attitude, or practice, retains full decision-making authority. In many instances consultative leadership is appropriate (as even autocratic leadership of the benevolent kind is sometimes appropriate). But often, consultative leadership is practiced under a participative label. Some supervisors claim they are open to participation, however, in practice they are "open" only as long as the employees come up with the same decisions they would have made themselves.

- Full participative leadership exists when plans or decisions are made by all of the department's employees as a group or team. The supervisor is a key member of the group, providing advice, information, and assistance in any way possible. However, the supervisor has (in advance) made a decision to accept the outcome of the group process.

With the exception of exploitative autocracy (which serves the leader's goals rather than those of the organization), no particular leadership pattern can be labeled "wrong." What is right for one department may be wrong for another. What is right in one particular situation may be wrong under a different set of circumstances. And what was right at some time in the past may be wrong today.

Autocratic leadership patterns still exist in some organizations; Iraq under Sadaam Hussein comes to mind as an example of the exploitative variant. But in the Western world, time and changing social conditions have lessened the use of autocratic leadership in the workplace, particularly as employees have become generally better educated. Gone are the days when the average worker was illiterate and completely dependent on orders from above. This certainly has had a favorable impact on most organizations. But even today, despite these developments, many outmoded organizational assumptions affect the workplace. Autocratic and bureaucratic leadership remain in use when they could (and should) have given way to more consultative and participative management styles.

This shift in leadership patterns is an issue of note in corrections, as line security staff in many agencies are increasingly college-educated. Also, others have worked in noncorrectional organizations that use less-regimented forms of management. Employees appear to be less amenable to being supervised in the less participative styles. A notable number may seek employment elsewhere if they find their supervisors to be unenlightened in this regard.

Here it will be useful to briefly examine in a limited manner some of the assumptions on which so-called modern organizations are based. More complete consideration of the reasons why people work is provided in the discussion of motivation in Chapter 12.

■ Some Assumptions About People

Douglas McGregor, in his work "The Human Side of Enterprise," wrote of two opposing approaches to management: Theory X and Theory Y.[1] Theory X in its pure state is what has been referred to as autocratic leadership. Pure Theory Y is participative leadership. Each of these management theories is based on a number of assumptions, only the first of which (relating to management in general) is common to both. That common assumption, valid in any case, is that management remains responsible for organizing the elements of all productive activity. Its mission is bringing together the money, people, equipment, and supplies needed to accomplish the organization's goals. Beyond this assumption, however, the two theories proceed in opposite directions.

Theory X assumes people must be actively managed. They must be directed and motivated, and their actions must be controlled and their behavior modified to fit the needs of the organization. Without this active intervention by management, employees would be passive and even resistant to organizational needs. Therefore, staff must be persuaded, controlled, rewarded, or punished as necessary to accomplish the aims of the organization. That is because it is assumed that:

- The average person is by nature indolent, working as little as possible.
- The average person lacks ambition, shuns responsibility, and in general prefers to be led.
- The average person is inherently self-centered, resistant to change, and indifferent to the needs of the organization.

Theory Y, on the other hand, assumes that people are not naturally passive or resistant to organizational needs. If they appear to have become so, this condition is the result of experience in organizations. Under Theory Y management:

- Most people are motivated, have development potential, a willingness to assume responsibility, and a readiness to work toward organizational goals.
- It is management's responsibility to make it possible for people to recognize and develop these characteristics for themselves.
- The essential task of management is to arrange organizational conditions and methods of operation so people can best achieve their own goals by directing their efforts toward the goals of the organization.

■ Style and Circumstances

Autocratic leaders operate under Theory X assumptions. They choose to make all the decisions and hand them down as orders and instructions. Participative leaders generally ascribe to Theory Y assumptions and encourage employees to participate in joint decisions.

Some situations (institutional emergencies for instance) call for a highly directive Theory X approach. However, in most instances, there exists a choice of leadership styles available. These range from extremely closed to extremely open. The trick is to know which style to apply and when to apply it.

There may be some "Theory X people" in the department. They will likely be a minority—those few who actually prefer to be led and have their thinking done for them. However, there also may be a number of "Theory Y people" who are self-motivated and capable of significant self-direction. This is especially likely in departments employing large numbers of professionals, such as the medical or case management departments. Although the same rules (meaning personnel policies) apply uniformly to all employees, managers will deal differently with individuals in other ways. Some will be consulted and their participation sought. Others will simply be directed.

Avoid making assumptions about people. Know employees and try to understand each one as both a producer and a person. By working with people over a period of time (and especially by working at the business of getting to know them) a manager can learn a great deal about individual likes, dislikes, and capabilities. Learn about employees as individuals, and when necessary lead accordingly. If a manager is convinced that a certain employee genuinely prefers orders and instructions and this attitude is not inconsistent with job requirements, then use orders and instructions.

Although many employees of correctional organizations (particularly newer and younger staff) seem to expect some type of participative leadership, not everyone will. Maintain sufficient flexibility to accommodate the employee who wants or requires authoritarian supervision. It is just as unfair to expect people to become what they do not want to be as it is to allow a rigid structure to stifle those other employees who feel they have something more to contribute.

There is no single style of leadership that is appropriate to all people and situations at all times. Now, however, there is more reason than ever before to believe that consultative and participative leadership is most appropriate to modern organizations and today's educated workers.

■ Outmoded Views

In addition to harboring erroneous assumptions about people, many managers cling to outmoded notions of how a manager should behave. The correctional environment is particularly susceptible to fostering these stereotypical views. In all likelihood, this mind-set started with the historical role of the warden as a "mini-god" with virtually unbounded power within the institution. Even today, correctional staff see the warden in particular, and other top managers in general, as being "boss"—the essential giver of orders. The view is, "I'm the boss. I'm paid to make decisions and I'm responsible for the results of those decisions, so I (and I alone) will make those decisions."

Because the correctional culture places a high value on strength and assertiveness, some managers equate strong leadership with an autocratic style. They see participative leadership as "passing the buck" or "spreading the blame" and in general view participative leadership as a shirking of responsibilities. However, participative leadership is anything but abrogation of responsibility.

Recall from the material on delegation that although a manager can parcel out management authority and spread it among a number of employees, he or she remains responsible for the decisions and actions of those employees. The true participative leader—willingly remaining responsible for the decisions of the group—is displaying considerably more courage than the autocrat who simply decides and gives orders. To trust employees with a share of managerial authority while retaining full responsibility is a sign of strength, not a sign of weakness.

Another outmoded view of leadership is reflected in the belief that the leader should always know best. This view holds that employees look to the leader to "tell them how to do it." The true function of the leader is to help the employees find the best way to do it themselves. The leader does not take up a position at the rear of the pack and shove. Neither does the leader move in front of the crew and urge them to follow. Rather, the leader is somewhere in the pack—a facilitator, a remover of obstacles, and in general a catalytic agent that causes the entire group to move forward in the proper direction. The true leader is not the master of a department but rather its busiest, most responsible servant.

■ Leadership's Primary Characteristic

Many attempts have been made at creating a detailed list of qualities that characterize leaders. It is done all the time; great lists are generated that include standard noble but intangible characteristics. They include traits such as honesty, integrity, and initiative, as well as slightly more measurable criteria such as academic qualifications. People who write hiring requirements for managerial positions do it all the time, creating job specifications as full of noble characteristics as the Boy Scout Law. However, there is hardly a personal characteristic (education, experience, integrity, communication ability, energy, conscientiousness) that some supposedly successful leaders lack. However brief, no one list will exclusively delineate the personal characteristics of a leader.

Consider, then, is there anything that truly defines a leader? A leader is certainly not defined by organizational appointment, or simply conferring a management title. A title may describe a position but not a person. A great many so-called leadership positions are occupied by people who are anything but leaders in the true sense of the word.

The single factor that defines or characterizes a true leader is the acceptance by the followers. This means acceptance of the individual as a leader, not simply acknowledging that the position the individual occupies requires obedience.

Many managers are not especially respected, are perhaps even ridiculed or joked about when not present, but are nevertheless obeyed. Obedience will often exist (perhaps grudgingly) because of the authority of the position itself. But willing obedience will be extended consistently only by those employees who have accepted the supervisor's leadership.

This positive, persuasive style of leadership is effective. As Alan Greenspan, chairman of the Federal Reserve Board, once said, "You can lead an organization through persuasion or formal edict. I have never found the arbitrary use of authority to control an organization either effective or, for that matter, personally interesting. If colleagues cannot be persuaded of the correctness of a decision, it is probably worthwhile to rethink it."

Acceptance by one's followers cannot be mandated; it must be earned. Without this acceptance a supervisor is a manager in title only and a leader not at all.

■ Word Play: Leadership Versus "Management"

Leadership has been attracting a great deal of attention and acquiring newer shades of meaning in the Total Quality Management (TQM) movement. Empowerment is "in" and delegation attracts little concern (although empowerment is no more than proper delegation with a slick new finish on it, as noted in Chapter 6). Correspondingly, leadership is "in" at present and is likely to remain that way for a while. Many concerned with the quality movement speak of the need for "not management, but leadership."

With this highly positive connotation placed on the word *leadership*, one is left with an eroded picture of management. It conveys the impression that somehow mere management does not measure up to leadership. However, management and leadership remain two words that are generally synonymous and freely interchangeable in many uses.

As suggested earlier, someone can be a manager (or supervisor, director, coordinator, administrator, chief executive officer, or whatever) in title without being a true leader. However, this in no way renders management as a whole anything less than, nor anything significantly different from, leadership. There is poor management and there is good management; there is poor leadership and there is good leadership. One can be a manager in title without being a leader. But in terms of function, one cannot manage without leading and cannot lead without managing. Good leadership and good management go together, as do poor leadership and poor management.

Management is often described as both art and science. The same might be said about leadership, but it is usually considered as more of an art. Leadership is more likely to inspire thoughts of the human element, whereas management conjures up images of "techniques" and "tools." Whether speaking of management or leadership, however, one is dealing with the process of accomplishing goals through the efforts of people. Only so much of this process can be quantified and reduced to rules and techniques. The remainder will come from the heart and the gut. It can be developed from within, but it can never be instilled from without. Whether leadership or management, it is this unquantifiable, frequently elusive "soft" side that puts the word "good" or another adjective in front of management or leadership. And leadership and management cannot be justly compared without the use of qualifying adjectives.

■ Can You Lead "By the Book"?

Now a disclaimer for this chapter and indeed for the entire book is needed. The previous section noted the so-called "art" of leadership and the perceived "techniques and tools" of management. Throughout this book, the reader will find references to, and reliance on, a variety of approaches to management—principles, practices, and philosophies that can assist in understanding how to be a better manager. Yet none of these should be taken as constituting the final and exclusive word on management.

Indeed, accept the contention that good leadership (or good management) is both art and science, and one must conclude that leading or managing totally "by the book" (or by any one of these principles, practices or philosophies) is not possible. Management is a highly individualized enterprise, based on the people involved, the type of organization, the organization's culture, the labor management environment in which it functions, and many other important factors. Despite this, one of the errors commonly committed by struggling leaders is overreliance on what might be referred to as cookie-cutter management.

The cookie-cutter approach is what occurs when one attempts to apply specifically named management techniques (or "kinds" of management often cynically described as "flavors of the month"). Think of what occurs when you roll out a batch of cookie dough and apply a cookie cutter. You obtain a desired shape, but you also get leftover material. Even if you roll out the dough and cut another time or two, you are still left with a remaining lump of dough that fails to fit the required shape. Whenever effort is applied to produce a desired shape from some form of input, some material is squeezed into the desired shape. Like the dough left outside of the cookie cutter, the extra material that falls away during the shaping process is ignored, for all practical purposes lost to consideration. This is what occurs when managers attempt to apply "formula" management.

It is reasonable for managers to seek order in their work. However, people tend at times to look beyond simple order and seek formula approaches, recipes for doing this or that task or handling particular kinds of problems. People look for cookie cutters, unconsciously willing to settle for the neat boundaries created by the instrument and equally willing to ignore what falls outside. However, management's problems cannot be consistently and adequately addressed by processes that by their very nature attempt to force issues into certain configurations.

Cookie-cutter management is especially prevalent in management literature. Any number of management authors have named their own approaches to management using labels that each fervently hopes will catch on and become the next "flavor of the month." This sells books, attracts speaking engagements, and enhances an author's value as a consultant.

Consider, for example, the "excellence" movement inspired by *In Search of Excellence* by Peters and Waterman.[2] While this extremely enlightening book did not itself espouse a cookie-cutter approach to management, the concepts it advanced were taken up by others who did exactly that—created dozens of cookie-cutter approaches as they attempted to formularize and proceduralize these concepts into "excellence programs." Much the same happened years earlier with management by objectives (MBO). The entire MBO concept grew from a single chapter of an excellent book written by Peter F. Drucker in the 1950s.[3] As presented by Drucker, MBO was largely honesty and common sense; it was not the periodic exercise with forms and notebooks that it became for so many managers. Drucker himself even cautioned against allowing such a process to become a paper mill, yet many of those who picked up on his work and ran with it turned it into one of the biggest wastes of time and paper with which managers have ever had to contend.

Managers have likewise been offered cookie-cutter approaches in the forms of quality circles and many of the permutations of the principles of TQM. There also have been many less notable approaches that have come and gone as successive authors have attempted to originate the next "flavor of the month," to be the one to launch the next MBO or TQM.

The good news about cookie cutters is that they each contain concepts of potential value to managers at all levels. The bad news is that none of the cookie cutters includes everything that a manager might need to know because by its very nature the cookie cutter always leaves some material outside. Also, none of the cookie cutters can instill in the individual the qualities and characteristics of a successful leader. Common sense, honesty, integrity, insight, pride, enthusiasm, and the belief in every employee's value and potential contribution cannot be proceduralized. These qualities are among the elements that fuel the art of leadership. If they are not there in sufficient quantity to inspire people to follow willingly, the best that can ever be obtained from the latest cookie-cutter approach is a brief burst of success—perhaps real, perhaps only perceived—to be washed away in a returning tide of cynicism. Managers cannot lead or manage "by the book." The book provides only the science, but the art springs from the capabilities and actions of the individual manager.

■ An Employee's View

Not all employees in a department will view their supervisor in the same light or develop the same impression of the supervisor as a leader. Each employee will experience their supervisor's leadership style in bits and pieces that add up to a total impression.

While employee opinions can be important in a functional sense, there is little to be gained from worrying about the opinions of employees. Indeed, it pays to remember that a manager's self-image rarely coincides with the way they are viewed by employees. What the manager sees as strengths may not be seen as strong points by employees. Conversely, what employees see as strong points may not even have occurred to the manager as significant traits.

There are several aspects of supervisory performance that are likely to influence employees' assessment of their supervisor. Subordinates may not use the same terms used to describe these aspects of performance. Neither may they use labels such as "autocratic," "authoritarian," or "participative." However, the view employees form of their supervisor is usually related to the way the supervisor comes across in regard to some or all of the following:

- Does the supervisor communicate openly, sharing all necessary or helpful information with the employee group? Openness to communication is associated more with consultative or participative leadership, and a closed communication posture is associated more with autocratic leadership. Granted, there are occasions in correctional settings when sensitive information must be closely held, but those occasions usually are rare.
- Does the supervisor display awareness of people's problems and needs? The participative leader tends more toward awareness of individual problems while the autocratic leader often appears unaware or even uncaring.
- Does the supervisor display trust and confidence in employees? The autocratic leader displays little trust or confidence, relying mostly on close supervision. The true participative leader is able to extend trust and confidence.
- What means are used to motivate employees? The autocratic leader frequently relies on fear and punishment to move people forward. The participative leader motivates through involvement and reward whenever possible.
- Does the supervisor provide support to employees? The employees of the autocratic leader often find they stand alone when things go wrong and they need the supervisor's backing. Support for employees and their decisions and actions is a hallmark of the participative leader.
- Does the supervisor request input on job problems? The autocratic leader tends to go it alone, but the consultative or participative leader is generally open to input from the work group.

■ The Visible Supervisor

Supervisors experience many pressures that encourage them to "face upward" in the organization toward higher management. After all, the supervisor's praise, reward, and recognition come from this upward direction. It is natural for supervisors who seek career growth to look upward. They recognize advancement is often facilitated by the extent to which they are organizationally visible outside of their own areas of responsibility. Opportunities to serve as acting associate warden or to participate in cross-training in other correctional departments can be legitimate developmental tools used in this light.

In facing upward, however, the supervisor runs the risk of losing sight of the rank-and-file members of his or her own department. In the long run the employees and their day-to-day performance have the greatest effect on the performance of the supervisor. But it is all too easy for supervisors to be so busy facing upward. They are meeting with their own supervisor, serving on committees, attending outside functions, and such. In doing so they lose touch with the people who can truly make or break supervisors by how they perform.

To provide true leadership to the work group, the supervisor must be (and be seen as) an integral part of the work unit. To get things done effectively through employees, supervisors should:

- Be visible and available, spending most of the time where really needed
- Show concern for the employees' problems
- Maintain a true open-door attitude so that employees can always access the supervisor when necessary
- Rely on immediate feedback to let all employees know exactly where they stand

A great deal of true leadership is provided by visible example, and the more visible the manager is to employees, the greater the chance of gaining their acceptance.

Indeed, personal visibility is thought by some to be a key to successful management—correctional or otherwise. That is the heart of the MBWA (Management by Walking Around) school of thought—one that is highly regarded in correctional work.

One of the most respected correctional administrators in the United States, the late James D. Henderson, spoke to this issue, saying,

> . . . *personal visibility builds staff confidence in their leaders. Instead of visualizing some front office paper-pushers, line staff can relate to managers they see regularly, have an opportunity to talk to, and who listen to their concerns and problems. I'm talking about personal contacts with staff in every area of the institution. I have toured institutions where neither staff nor inmates knew the warden who was walking along with me, providing a clear indication of the problems existing in that facility. In another location, the unit log in segregation indicated that the warden visited there for a total of three minutes. That's simply not enough time in such a critical area of the institution. In yet another case involving a serious escape, supervisory personnel had avoided visiting death row for weeks at a time.* [4]

■ True Leadership

As noted, in the last analysis the only factor that truly defines a leader is the acceptance of leadership by the followers. It has been further suggested that a style that leans toward participative leadership is more likely to be found acceptable by the majority of today's workers than one that leans toward autocratic behavior. However, no single pattern of leadership behavior is appropriate to all situations at all times. A manager may lean toward participative leadership most of the time. However, there still may be times when the situation calls for autocratic behavior—the manager needs to be able to make a decision, issue an order, and expect results. Crisis intervention is a reality in correctional work.

A good manager must be able to rapidly shift gears to this mode, even if it is not his or her normal pattern. Also, there may be times when bureaucratic behavior (strict interpretation and application of the rules and regulations) may be required. True leadership is flexible; it responds to both individual and organizational needs and is shaped to fit the needs of the moment.

EXERCISES

Exercise 11-1: The Buck Stops Here

As a chief correctional supervisor in a small minimum-security facility, you report to the warden.

The plate on the warden's desk saying "The Buck Stops Here" reasonably describes her approach to the job. She never avoids a decision or a problem, even of the most controversial or unpleasant sort, and for this you respect her.

However, the warden makes her decisions in a vacuum with no input from other department heads involved in the issue at hand. She is apparently conscientious in her attempts to come up with the best solution each time. But when she transmits the instructions for carrying out her decisions she does so without giving you or anyone else the opportunity to provide the perspective of the person who has to translate the order into action.

Questions

1. When you receive an order from the warden which you know is inappropriate (and assume you know so because you are much closer to the problem) how can you make yourself heard without deliberately rejecting her style of leadership?

2. Are there ways to effectively increase the flow of information to a supervisor so that the leadership pattern is strengthened by having better data?

3. How (and here is the ticklish part with many supervisors) might you bring this subject up in a way that shows your legitimate concerns?

Exercise 11-2: A View of You as a Leader

This exercise may be difficult for you—not difficult to accomplish, but difficult to accept what you learn from it. However, it can be helpful in suggesting areas in which your employees' view of your leadership style differs from your view.

The six questions appearing below are taken from the six points discussed in the chapter section titled "An Employee's View." Each is provided with its own scale, extending from 0 (fully autocratic) to 10 (completely participative).

1. Do I communicate fully and openly?

   0   1   2   3   4   5   6   7   8   9   10
   (Not at all)   (Completely)

2. Am I aware of people's problems and needs?

   0   1   2   3   4   5   6   7   8   9   10
   (Unaware)   (Fully aware)

3. Do I display trust and confidence?

   0   1   2   3   4   5   6   7   8   9   10
   (Not at all)   (Fully)

4. Do I motivate using fear/punishment or reward/appreciation?

   0   1   2   3   4   5   6   7   8   9   10
   (Punishment)   (Reward)

5. Do I furnish backing and support in a pinch?

0   1   2   3   4   5   6   7   8   9   10

(Never)   (Always)

6. Am I open to employees' input on problems?

0   1   2   3   4   5   6   7   8   9   10

(Rarely if ever)   (Usually)

Prepare a simple handout sheet including the questions and the scales as they appear here and a few lines of instructions. You need only instruct employees on how to use the scales, explaining that you are seeking (related to a management development exercise, if you wish) an employee view of the department's leader. Stress that completing the form is optional and they should not use their names. Provide a drop-off point so you will be unable to determine who did or did not complete a form.

Before you receive any completed forms, rate yourself on the same six questions using the same form.

After you receive the completed forms, compare the employee ratings with your own, both individually and by taking an average rating offered by all employees on each question.

You and your employees may not see your strong or weak points in the same light. If you are unhappy with a particular response, remember that a poor rating does not mean you are necessarily that way. But it can mean that you are viewed that way because it is "how you are coming across" to people. The results may suggest aspects of your management/leadership style that could use your attention.

ENDNOTES

1. McGregor, Douglas M. "The Human Side of Enterprise," *Management Review* 46, no. 11 (November 1957): 22–28, 88–92.
2. Peters, T. and R. H. Waterman Jr., *In Search of Excellence.* New York: Warner Books, 1982.
3. Drucker, P. F. *The Practice of Management.* New York: Harper & Row, 1954.
4. Henderson, James D. and Richard L. Phillips, "Developing a Safe, Humane Institution Through the Basics of Corrections." *Federal Correctional Journal*, vol. 1, no. 1: 15–19.

12

Organizational Communication: Looking Up, Down, and Laterally

Chapter Objectives

- Compare and contrast the characteristics of upward communication and downward communication in the organizational setting, with special attention to the barriers to upward communication.
- Identify the specific reasons why information flows downward far more readily than upward in the organizational setting.
- Examine the supervisor's central role in organizational communication.
- Provide suggestions for strengthening communications with other organizational elements including immediate superiors.
- Suggest ways of dealing with the communications network of the informal organization, that is, "the grapevine."
- Reinforce the importance of the supervisor's visibility to the department's employees.
- Examine the implications of the supervisor's organizational posture, that is, "upward facing" versus "downward facing."

What we got here is . . . failure . . . to communicate. . . .

the Captain in the movie *Cool Hand Luke*

I wish I had done a better job of communicating with my people. If people understand the why, they will work for it. I never got that across.

Roger Smith, former chairman of General Motors

FROM THE INSIDE   "There's Gonna Be Trouble"

The institution was in tense ferment.

Inmates were telling their work supervisors that they had better not come to work tomorrow, because there was going to be trouble. Cell house officers were getting the cold shoulder from inmates who usually would talk to them freely. This information was reported up the chain of command. Snitch notes were coming to the special investigative supervisor, lieutenants, and captain. These reports also were filtered up. Then the inmate sources dried up and all that was left was inferential data—inmates stocking up on commissary items, not talking to staff any more, and mailing personal pictures out of the institution. The warden and associate wardens considered all of this information and more. All of it suggested that something was going to go down.

Staff rumbles began. Some said that there would be an inmate work stoppage the next day, some said a riot. There was talk that the female employees would not be allowed to go into the secure portion of the institution tomorrow. Some were saying they thought that radical inmates were intent on taking over the prison as some kind of political statement (it was the 1970s and such things were thought possible, if not likely).

Meetings were held. Staff were advised of contingency plans. Emergency squads were formed and held at the ready. Extra personnel were on duty in key locations. Backup emergency response gear was moved into position. All this had to be done on short notice, informing staff of the precautions being taken without unduly alarming them.

But there was alarm, and there were tensions as each shift passed. Supervisors continued to talk to line personnel and the few inmates who would speak to them. They worked to gather additional information about inmate activities, and at the same time to reassure staff that the administration was prepared to respond quickly.

The next day there was a disturbance about noon. Tragically, in the first few minutes of the riot, an officer was stabbed to death in one of the big cell houses. Fires were set in the industrial building and two cell houses. The dining room went up in riot. Four officers were dragged into the laundry building where they were held hostage for about 12 hours. The fires were suppressed, response teams restored control, and a negotiation team freed the hostages shortly after midnight that night.

■ What Goes Down May Not Come Up

The introductory quote is prescient, not only because it comes from a correctional setting, but because of the way it identifies a key element in correctional management. When he used those remarkably expressive words and tones, the Captain in the 1970s movie *Cool Hand Luke* was talking about his problems in getting a recalcitrant inmate to "get your mind right." But he spoke as well for the problems of supervising anyone (and keeping necessary information flowing) in the correctional environment.

The lead quotes are about one-way communication problems. But it is safe to assume that in any organization, communication should, and in fact must, move both upward and downward through the structure. The vignette about the riot tells a more complicated story. It is that of an administration trying to get information to flow up and down about a budding problem, preparing to respond, and assuring staff that things were under control.

As this story shows, and an informed management view of organizational communication will recognize, information does not flow both upward and downward with equal ease. Downward communication is facilitated largely by management's control of its own actions and by its control of most of the means of communication in the work setting. In contrast, a great deal of upward communication remains dependent on stimulation, encouragement, and the creation of a climate conducive to communication. The essential differences between upward communication and downward communication can be highlighted by looking at a number of factors that inhibit upward communication. *Barriers*

Organizational Concerns

Physical distance between supervisor and employee (as in the security department where staff are on posts throughout and even outside the correctional setting) can inhibit upward communication. Simply put, the more time a manager spends physically separated from the location where most of the work unit's employees are located, the tougher the communication is. This was highlighted in the discussion on "span of control" in Chapter 5, where it was pointed out that supervisors with employees who have considerable on-the-job mobility or are scattered over an extended physical area need to take steps to keep in touch. The more time spent physically removed from subordinate employees, the more likely a supervisor is to miss something that might have otherwise been communicated up the chain of command.

The number of levels in the organization structure can also inhibit upward communication. If a piece of information originating with a single food service employee truly deserves to reach the warden, the chances of it being properly and accurately communicated upward diminish with each managerial level it must go through. Each level that a message must pass through presents another set of opportunities for the message to be misinterpreted, sidetracked, or stopped entirely.

The relative complexity of a given problem or situation can also hamper upward communication. Some employees are often unable to define complex problems fully, especially if these problems appear to involve jobs or departments other than their own. Also, some employees lack sufficient command of communication skills to enable them to translate their thoughts and observations into concise, understandable messages, so they simply do not bother to try.

Problems Involving Managers

The attitude exhibited by a supervisor or manager also can have a great deal to do with how well information flows upward. If a supervisor's manner and attitude should seem to say "no news is good news" or "don't tell me anything I don't want to hear," employees are likely to be discouraged from speaking up. Also, if the supervisor appears to behave defensively (perhaps seeming to regard opinions, problems, or requests as personal jabs) employees are likewise discouraged from speaking up.

Some managers frequently exhibit apparent resistance to becoming involved with the personal problems of employees. This resistance or reluctance (coming no doubt from the supervisor's understandable uneasiness with hearing things that might be considered "confessional" in nature) is understandable. But it serves as a wall that unfortunately keeps out desired feedback as well as unwanted information. Recall the notion of the employee as a "whole person," and learn to accept the high likelihood that many supposedly personal problems have their work-related sides.

The manager's available time can be a factor inhibiting the upward flow of information as well. Effective listening is a time-consuming process, and under the pressure of many high-priority tasks it is easy to find yourself dealing with some items superficially or not at all.

Probably one of the greatest inhibitors of upward communication lies in actions of the recent past: management's failure to respond to some earlier communication. A question, problem, or observation coming from an employee remains one-way communication (and thus not true communication at all) unless an effort is made to "close the loop" by providing the required feedback. This is not to say that employees should always expect to receive the responses they desire. Rather, they need simply to receive something indicating that their messages have been considered and that management is being responsive. Feedback (after proper investigation, consideration, or consultation as appropriate) may be simple: "Thanks very much for pointing it out; it's being taken care of"; "I'm sorry, but it can't be done (for such and such a reason)." However, employees who receive no feedback are likely to regard their concerns as "swallowed up by the system" and will be discouraged from communicating at all.

Problems Involving Employees

Employees will invariably see downward communication as occurring more freely and frequently than upward communication could possibly take place. Tradition, authority, and prestige all favor the management hierarchy and downward communication over upward information flow. After all, you feel free to call on your employees at just about any time, just as your manager is likely to feel free to call on you at any time. However, rarely will nonsupervisory employees feel that same degree of freedom in their ability to call on the supervisor.

Also, most of the mechanisms and means of organizational communication are in the hands of management and favor the downward flow of communication. These include bulletin boards, public address systems, employee newsletters and other printed matter, and duplicating services. The individual employee with something to communicate must usually do so either by writing it out or relating it orally to another person. Some things never get communicated upward because some employees are unwilling to communicate upward. Some people simply will not relay a problem or concern to the supervisor because it has a personal dimension that they do not wish to reveal. Some may also hesitate to point out certain problems for fear of being blamed for causing them. Similarly, some will say nothing they might consider to be self-incriminating or self-deprecating in any way.

A final but sometimes insurmountable barrier to upward communication is found in emotion and prejudice on the part of a few employees. To a limited number of nonsupervisory employees, management (even though it may be enlightened, humane, and people-centered) is to be regarded as exploitative and untrustworthy simply because it is management. Frequently this attitude extends so far as to regard the new supervisor moving up from the ranks as considered to have abandoned the "good guys" and joined the "bad guys." Rarely will an employee harboring such a view of management discuss any serious concerns with a supervisor—unless the supervisor has conscientiously worked to earn the employee's trust and confidence.

The Manager's Role in Organizational Communication

Have you ever found yourself, perhaps in anger or frustration, saying, "The trouble with this place is there's no communication"? If the truth could be determined, most supervisors have said this (or something very much like it) more than just a few times. However, the next time you feel inclined to cry "no communication," you might do well to consider that a great deal of the supervisor's role in organizational communication depends on you and what you do and say. Before dwelling on the shortcomings that "they" exhibit ("they" being the often-blamed but never specifically identified villains of "they won't let me do it," "they didn't tell me," for example) it might prove more productive to work on your own communications practices. You cannot change someone else's habits and practices, but you can encourage them to change these for themselves and you can best do this by changing your behavior.

Look at the simplified diagram of **Figure 12-1**. The supervisor is in the middle with lines of communication running to other people in the organization. The strictly formal lines of communication are those numbered 1, 2, 5, and 6. These depict the direct reporting of relationships that exist between line management and employees, as well as between line management and upper management. Lines 3 and 4 suggest a large number of less rigid, but still formal, communicating relationships. These relationships are still formal because although a given supervisor neither manages nor reports to any of the people who work in or manage other institution functions or departments, communication is nevertheless required of many of them.

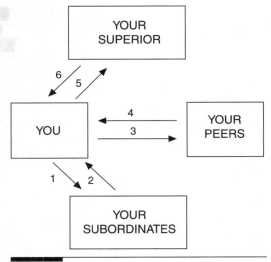

Figure 12-1 Your communication channels.

First appreciate that communication along three of the six lines shown (those outgoing lines, numbered 1, 3, and 5) is completely in the manager's hands. The sender of a message—the originator of a communication—has full control over all information being disseminated. The manager controls what is sent, when it is sent, why it is sent, how it is sent, to whom it is sent, and where it is sent. And a manager can always send something. Even when an intended communication is dependent on first receiving something from another party, the manager can at least say "I don't know yet," "I'm still waiting," or "I'll call you as soon as I hear."

An example from one of the author's daughter's not-too-well-managed workplaces is applicable. As this first was being written, she and two other employees had been selected for special training, but its start had been delayed for more than a month. Her immediate supervisor (who may or may not have known more than she communicated to the three line employees), in response to repeated questions on when the training would start, would only tell them, "The training will not start before November 11th." These employees could read a calendar and see that November 11th was a Tuesday—an odd day to choose to start a training program. They also were very acutely aware that November 10th was the day their 90-day probationary period ended. Repeatedly posing the question had resulted in the same, almost rote, answer. From this information, it could not escape all three employees that there was at least a possibility that the company had delayed the training until one or more of them had been "let go." Cryptic, incomplete, and vague information in this instance did more to frustrate the employees than inform them. Far better that the supervisor had told them she did not know exactly when it was going to take place, and the date had nothing to do with their probationary status. The final outcome of the situation was that the training did start on November 11th, but the morale of the three affected employees had been adversely impacted by the inability of the organization to clearly and effectively communicate with them.

The point of this discussion is twofold. First, outgoing channels of communication are completely under the manager's control. Second, the best way to get information moving along incoming lines of communication is to assure that the outgoing lines are open and operating.

Some of a manager's control in the second instance is due to the authority of the position. These people report upward in the organization, so most of them are expected to provide certain information at certain times. An additional measure of control will stem from the manager's degree of success in requesting and receiving feedback from employees. This is where assignment of completion targets and deadlines and periodic reports come into the picture, along with effective use of follow-up. In short, manage in a way that assures a great deal of work-related communication; a manager's style tells employees what is needed and expected.

However, there remains sometimes a considerable amount of information a supervisor would like to know (individual complaints, claims of unfair treatment, and dissension among employees) that cannot be mandated by structure or approach. Managers will receive this added information only by earning the trust and confidence of employees and by showing through their actions that they are willing to communicate openly, honestly, and in confidence.

A manager has considerably less control over the information arriving by way of line 4, the channel running from other organizational elements. There is no supervisory relationship (either up or down) between the manager and any of these people. It is possible that some personal or organizational factors may convey some implied authority with some of these people, but no formal authority will exist with many of them. Therefore, the communicating relationship with these people has to be based on cooperation. When communicating with others in this category, display a willingness to communicate. Think about who else truly needs this information, who should be aware of it, and to whom it should go as a matter of information or simple courtesy.

Again, the best way to get information moving in is first to see that it is freely moving out. Do so even if it is necessary to shake off the old "50-50 ethic." That is the suggestion that in communication (as well as in other endeavors) it is only "fair" for each party to go halfway. Some people will respond to conscientious efforts in kind; others, however, will respond inadequately or not at all. Nevertheless, be prepared consistently to go more than halfway more than half of the time. If this sounds unfair, then look at it from a selfish point of view: you are the person benefiting most from these extra communications efforts.

The one channel of communication that has not yet been discussed is line 6—from your superior to you. This area presents more problems than the others. It is on this channel that a manager stands the least chance of exerting some appreciable measure of control. Obviously, there is positional authority to help in this relationship; in fact, the reverse is true. Developing a sound communicating relationship with your supervisor, then, deserves special attention.

■ Nuts and Bolts

There are a number of fundamental issues that this kind of communication analysis must necessarily address.

Getting the Next-Level Supervisor to Communicate

If your immediate supervisor is a conscientious practitioner of the art of effective communication, you may be able to skim through these few paragraphs and not worry at all about what they contain. On the other hand, you may have sufficient problems with this particular relationship to make these few words worth considering.

Once again, how you are communicated with is to some extent a reflection of how you communicate. The following are a few ways in which you can "tune up" your communications along line 5 in an attempt to stimulate more effective communication along line 6.

Be Selective in What You Communicate

Do not expect your supervisor to do your job for you and take care of your problems. When something comes up and your initial impulse is to turn to your supervisor for information, advice, or assistance, before doing so make sure it is something you cannot take care of yourself. One of your legitimate functions as an employee is to be a "problem filter" for your supervisor, screening out those matters that should be resolved at a lower level.

Do Your Homework

It has been stated several times in this book that the existence of problems is one of the major reasons for having managers. No manager really needs more problems than are already present. Problems make themselves felt in sufficient numbers and at all organizational levels; what your supervisor needs are solutions. Even though many problems you encounter might be beyond the range of your decision-making authority, you can do more toward solving them than simply passing them one block up the organizational chart.

Remember that you are just one of several people reporting to the same superior, and what may look like a trickle of trouble to you may strike the supervisor as a flood of grief. Rather, when you must pass a problem up the line, you should do so having analyzed it, assessed its implications, prepared perhaps two or three alternative solutions, and possibly recommended the answer that looks best to you. In short, instead of saying, "Here's a problem. What do you want me to do?" you should be saying: "Here's a problem; here's why it's a problem; here are one, two (or however many) possible solutions; here's the answer I think is best, and here's why."

Structure Your Communications

This is a variation of the previous point in which you were advised to do the spadework necessary to make decision-making easier for your superior. When you need information, especially small bits of advice or minor decisions, put your questions in writing in such a way that they can be answered in one or two words.

To cite an example, an education supervisor was having difficulty obtaining a few minutes of discussion time with the associate warden about a program change she wanted to make. Unable to get time with her supervisor when she needed to, she reexamined the situation, expressing the problem in the form of three concise questions. After typing the questions on a single sheet of paper, she left them with the associate warden's secretary. The next morning the answers were on the education supervisor's desk (see **Exhibit 12-1**).

Make Yourself Available

Some supervisors are thoughtful enough to say something like: "The best time to get me is first thing in the morning before the telephone starts to ring," or "I'm likely to be free between 4:00 and 4:30." Even without such assistance, however, you may often find you are in a position to know your manager's comings and goings and develop a sense for the better times to try for a brief audience. Although a conscientious manager will try to be available to someone with a problem, this conscientiousness may not be particularly visible to you. As noted earlier, you may be but one of several supervisors reporting to this manager and the activity you see may be only the "tip of the iceberg." So make yourself available to your supervisor, and in doing so strive to consume as little time as possible. This latter point concerning time is raised not because the supervisor's

Mr. Parsons:

1. Can I go ahead with rescheduling the evening GED program as we discussed?

    Yes

2. Will I have at least a half an hour with the warden to discuss this after the next staff meeting?

    Yes - schedule with his secretary

3. Will you be available to talk with him at that time?

    No - I'll be at another meeting then   B.P.
    5/17/04

    Paul J. 5/16/04

Exhibit 12-1 A structured communication.

time is valuable (although it certainly may be) but because your time is valuable. Most of your time belongs to your department, not to your supervisor.

■ The Grapevine

What is known as "the grapevine" may be more accurately described as the organization's informal communications network. Prisons are famous for grapevines, because there are two separate but interlocking networks in operation—staff and inmate—in a highly pressurized setting. In fact, because of the closed environment that correctional facilities present, grapevine issues are particularly important, and will be discussed at greater length than one ordinarily might think necessary.

Every organization has a formal structure, a network of reporting relationships describable by the well-known organization chart. The formal lines of communication in a correctional agency or institution follow many of the relationships suggested on the organization chart. However, managers also have, as everyone in the organization has, a number of informal channels of communication. Informal relationships with friends, acquaintances, and others in the work organization also lead to a flow of communication. Further, many of these relationships at least partially describe the informal organization—an implied structure that exists based on numerous related effects of respect, acknowledgment, deference, or prestige accorded to various individuals primarily because of personality impact. Parallel structures exist in both the employee and inmate arenas.

You have seen the informal organization at work when a certain two or three nonsupervisory employees happen to stand out from the group, perhaps even speaking for others, although they have no official standing. This situation can also exist when a single supervisor is regarded by the workgroup as "senior" over a number of others at the same level because of some particular trait or combination of traits. In short, interpersonal relationships and people's regard for one another describe the informal organization, at best a phantom structure that is always shifting and realigning.

People will talk. The grapevine is not required by management, and it is certainly not controlled by management. It runs merrily back and forth across departmental lines and rapidly changes its course. It also crosses between staff and inmates, a fact that often is lost on some people. The grapevine is dynamic but unreliable. It carries a great deal of information and misinformation. And it is here to stay.

Be aware of the grapevine. Tune in; listen to what it is carrying and learn from it. Supervisors are likely to be isolated from some of the bits and pieces the grapevine carries. At the very least, they will learn of some things long after they have been known to staff for some time. How much a manager hears is frequently dependent on how well they relate with employees and peers.

When tuned in to the grapevine, a manager is going to hear a few things that are simply not correct. When learning about something that is disturbing or seems clearly inappropriate, check it out if possible. Supervisors are responsible for setting the facts of the story right whenever the opportunity presents itself. But be sure to have the story straight so as to not simply heap more speculation onto a growing rumor.

The grapevine sometimes possesses the distinct advantages of speed and depth of penetration. Some bits of news can travel through the institution at an astonishing rate, carried not only by staff, but inmates as well. It often reaches people who would never think to read a bulletin board or look at an employee newsletter. The grapevine can carry the good as well as the bad. And since it will always be a factor, it is advantageous to feed it some real facts whenever possible so it will have something useful to carry.

Finally, consider that it sometimes is not the truth or falsehood of a rumor that is as important as the fact that the story is out there on the grapevine at all. Some information on the informal communication network telegraphs employee attitudes and beliefs, even if the specific data is not wholly accurate. An astute manager will try to discern when this kind of dynamic is at work, and can use it to anticipate developments within the organization.

■ Which Way Do You Face?

Employees are likely to infer a great deal about their supervisor's overall attitude by how effectively the supervisor communicates. It is also likely that many such inferences will be influenced by management visibility and availability. This element is exemplified by how much employees see of their supervisor in and around the department, and how readily they can get a few minutes of managerial time when they need it. Again, having the "Management by Walking Around" ethic can assist in improving a supervisor's overall ability to communicate.

However, in most organizations there are many pressures that cause managers to "face upward" and in general be most visible and available to a manager's superiors and other members of higher management. After all, notice from above leads to pay increases, promotions, and other rewards. There are some traps in this reasoning, however. Not all higher managers are necessarily impressed by your ready availability to them. In fact, an effective top manager may begin to see what they believe is too much of a subordinate manager, and will begin to wonder who is running the department.

The point is that it is desirable for the supervisor to spend most of the time "facing downward" toward the ranks of the employees in the department. It is there that the real action, the real challenges, and the true opportunities are found.

Supervisors consciously develop and cultivate a variety of communicating relationships in the organization, but the most important relationships are those established with employees. Regardless of the number and capability of the managers in any correctional organization, it remains largely the nonmanagerial employees who do the hands-on work of running the institution. Supervisors are there to assure that the portion of the correctional facility for which they are responsible is managed in the best possible way. A manager's primary attention belongs to the people who do the work. In the last analysis, it remains line employees who, through their job performance, can determine whether a manager succeeds or fails.

EXERCISES

Exercise 12-1: The Crunch

You and four other department heads are at a meeting with your immediate supervisor, the associate warden to whom the five of you report. Also present are two other department heads and the warden's executive assistant. The subject of the meeting is sensitive.

Just minutes into the meeting your supervisor makes a statement that you know to be incorrect. You attempt to intervene, but she asks you to hold your comments. She seems to be focusing almost entirely on the other department heads and the executive assistant.

Your boss proceeds to build an argument on her incorrect statement and you can sense that she is verbally "painting herself into a corner." Since you have already been silenced once you are hesitant to speak up again, and although you are sure of your information you have no way of "proving" anything without making a trip to your office and rummaging through some files. Within the confines of the conference room it would simply be your word against hers, and she is the boss.

What should you do? In deciding on a possible course of action, consider the implications of:

- Keeping quiet and allowing her to proceed in apparent error.
- Intruding, forcefully if necessary, until your information is heard by the group.

Exercise 12-2: The Unrequested Information

One morning, about 15 minutes before your work day officially starts, you are enjoying a solitary cup of coffee when you are joined by Mr. Hatfield, one of your employees. Hatfield proceeds to advise you ("in strictest confidence; please don't say that I told you") that another department head, Mrs. Thomas, has been talking about you. She has been telling others (both line and managerial personnel) supposedly why you were passed over for promotion last month, and that the reasons were not very complimentary. Hatfield proclaims that he doesn't ordinarily carry stories but felt that you "had a right to know, for the good of the institution."

It is well known that you and Thomas have a history of not seeing eye to eye on many things. This has been evidenced by what you thought were completely professional disagreements over budget items and a number of policy issues.

Should you:

- Thank Mr. Hatfield and ask him to report anything else he might hear?
- Acknowledge his concern "for the good of the department" but tell him to bring you no further such stories?
- Thank him, ask him to say nothing to anyone else, and decide for yourself to keep an eye on Mrs. Thomas?

Explain why you chose the particular answer you selected. You may modify, qualify, or further explain your answer as you believe may be necessary.

13

Motivation: Intangible Forces Working For and Against Management

Chapter Objectives

- Establish a perspective on "what employees want" from the organizations they work for and what does or does not contribute to satisfaction in work.

- Examine a number of reasons why people commonly leave their jobs.

- Review some classic theories of human motivation and the basic forces at work in motivation and suggest the varying influences of different forces on employees in today's organizations.

- Examine the value of material rewards as motivators and consider the relative strength of money in shaping an individual's perspective on employment.

- Identify and discuss the true motivators and the apparent potential dissatisfiers in employment.

- Describe the supervisor's role in creating the environment in which employees will become self-motivated.

The only way to motivate an employee is to give him challenging work in which he can assume responsibility.

Frederick R. Herzberg

I don't like work—no man does—but I like what is in work—the chance to find yourself. Your own reality—for yourself, not for others—what no other man can ever know.

Joseph Conrad

Good Benefits Can't Keep Them All

The pay was good. The workplace was clean and well maintained. A healthy uniform allowance was provided. Benefits were more than adequate. There was virtually 100 percent job security. All this was in a tight job market, where such things were hard to find.

People who had been promised certain job assignments were not given them. Work schedules were changed on short notice. People were temporarily moved into jobs for which they had not been properly trained. Mandatory overtime was the norm. Managerial personnel were not honest with employees about changes in the agency.

Despite all the material positives of the job—pay, benefits, the physical conditions of the workplace—people were leaving in droves. Poor management was driving them off faster than the personnel office could replace them.

■ Satisfaction in Work

Motivation is the initiative or drive causing a person to direct behavior toward satisfaction of some personal need. Each person has needs, and many of these needs are the reasons why they work. Depending on how well needs are met through employment, employees may be more or less satisfied with their role in the organization.

This section will begin with the premise that, in the long run, the satisfied employee will more likely be a better producer than the employee who is generally dissatisfied. A satisfied employee is usually more enthusiastic, more willing to work, and more of a self-starter.

Unfortunately, traditional organizations inadvertently do a great deal to assure that a fair amount of dissatisfaction will exist in the workplace. That is the point of the introductory vignette for this chapter—an actual situation that took place in a major law enforcement agency. Work often has been structured, subdivided, and systematized to the extent that human factors beyond mere job performance are ignored. When this happens, the inherent challenge of work as a normal human activity is diluted or dissipated. Correctional work tends to display these characteristics, despite the fact that individual action and interaction with inmates is a key element in many correctional jobs.

One of the authors once had as a colleague a physician who was formerly the team doctor for a professional football team. When asked once about how a particular type of knee injury (from which the author was recovering) affected pro football players, his answer started out with, "Well, football players are a lot like people...." Well, prison employees are a lot like people—they respond to work in terms of personal needs and desires. Just as in other workplaces, correctional managers need to seek ways to allow employees to satisfy their basic needs through their work.

The Real Reasons People Quit

A 1969 report issued by the Administrative Management Society presented the results of an extensive survey intended to determine the reasons most frequently given by people quitting jobs.[1] The seven reasons identified were:

1. **Lack of recognition.** The number one reason for people leaving jobs involved lack of recognition, the feeling that what they did was not fully appreciated, and that their contributions were not acknowledged.

2. **Lack of advancement.** The second most cited reason for making a job change was the desire to go to an organization with more opportunities for advancement, promotion, or professional growth.

3. **Money.** Monetary reasons were generally twofold, with a split between people seeking more income and people seeking change because they believed they were paid unfairly relative to others.

4. **Too many supervisors.** Although likely to include a number of people who legitimately worked for more than one supervisor, this reason largely involved hazy lines of authority under which workers were unclear as to who they were actually answerable to a great deal of the time.

5. **Personality conflicts.** This general reason for quitting included problems with coworkers as well as conflicts with immediate supervisors and other members of management.

6. **Underqualification.** People who quit jobs for this reason were those not fully qualified for their positions to begin with. Placed in such jobs through poor selection practices and perhaps weak personal decisions, such persons resigned generally to escape the feeling of "being in over their heads."

7. **Overqualification.** This is the opposite of reason number six, involving people who were more qualified or capable than required by the job. Overqualified people generally cited lack of challenge and lack of interest in the work.

These survey results suggest a picture of what people feel they need from their work. However, it is not a clear picture. Many people who leave jobs for any of the above reasons would probably not be able to say clearly if they did so because they were seeking something or trying to escape something. However, the seven reasons reflect a number of basic human needs. People have material needs and thus the need for money. People also have needs relative to their recognition as individuals and needs that relate to their regard for themselves.

Demands on the Organization

People want the organizations they work for to supply them with a number of things. These do not fall in any particular order. What is important to one person may matter very little to another. Generally, however, the following list encompasses most of what employees expect of their employers:

- Capable leadership that can be respected and admired.
- Decent working conditions—surroundings that promote safety and physical well-being. This is particularly important in the correctional setting.
- Acceptance as a member of a group.
- Recognition as an individual or partner, not simply as a servant of the "system."
- Fair treatment relative to that received by others.
- A reasonable degree of job security. This would be considered a plus for government employment, although as private sector correctional operations become more widespread, this is less of a factor in those facilities.
- Knowledge of the results of individual efforts.
- Knowledge of the organization's policies, rules, and regulations.
- Recognition for special effort or good performance.
- Respect for the individual's religious, moral, and political beliefs.
- Assurance that all others are doing their share of the work.
- Fair monetary compensation.

■ Motivating Forces: The Basic Needs

Before continuing, it is worth highlighting a small word casually thrown about since the beginning of the chapter. The word is "needs." It is necessary, in considering human motivation, to take as broad a view of this term as possible. If someone is pursuing a promotion or other reward, it might be tempting to say that the person does not really need this (considering a need as something essential) but rather may simply want it. However, to do so is not referring to needs in such a way as to force a definition of absolute essentials. Neither does it separate them from other things we could call wants, desires, wishes, or aims. Rather, these are the things people pursue simply because they represent fulfillment. In this sense they are indeed "needs" because individuals see them as essential to personal fulfillment.

In his well-known "need hierarchy," A. H. Maslow described the basic human needs as follows:[2]

- Physiological needs. These are the most fundamental needs—things required to sustain life, such as food and shelter.
- Safety needs. These include the need to feel reasonably free from harm from others and reasonably free from economic deprivation (call this "job security").
- Love needs. These include the need to be liked by others and to be accepted as part of a group (be it a work group, family, or social group). Needs at this level involve a sense of belonging.
- Esteem needs. At this level in the hierarchy people experience needs for recognition, for approval, for assurance that what they are doing is appreciated.
- Self-actualization. According to Maslow, the need for self-actualization represents "a pressure toward unity of personality, toward spontaneous expressiveness—toward being creative, toward being good, and a lot else."

Maslow states that people proceed through the need hierarchy from the most fundamental needs toward the highest-order needs. Once a need is satisfied, another arises to take its place. Thus, we experience needs of increasingly higher order. If an individual is in need of food, clothing, and shelter, then the basic physiological and safety needs are motivating behavior. In the workplace, when employees experience unmet physiological and safety needs, then factors like job satisfaction and interesting work experiences will not mean a great deal. At that level an individual employee is most interested in generating an income with which to acquire the basics of life. However, once these lower-order needs have been satisfied to a reasonable extent, they begin to experience the love needs, look for acceptance, and attempt to take a place in various groups. Thus, an individual progresses through the hierarchy until ultimately motivated by the need for self-actualization.

The legitimacy of the need hierarchy has been rather firmly established. However, there are vast differences among people as to individual needs. Thus there are corresponding differences in what is required to satisfy those needs.

For instance, one person's need for assurance of reasonable job security may be filled by the knowledge that the job will last at least another 3 months. Another person may feel uneasy unless assured the job will last until retirement. Also, love needs and esteem needs may be quite powerful in an individual who requires constant reassurance of worth and capability. Another person may experience much lesser needs at this level simply because of the presence of a higher degree of confidence and more sense of self-worth. Regardless of the needs actually experienced by any given individual, the matter of progression through the need hierarchy holds true. As a need is reasonably satisfied, another, higher-order need arises to take its place.

People have vastly different reasons for working. As it was in the United States years ago, today in some of the underdeveloped countries people concentrate most of their energies on simply remaining fed, clothed, and sheltered at a minimal level. They rarely go beyond the satisfaction of basic physiological and safety needs. In a modern industrial society, however, these lower-order needs are satisfied for most people most of the time. Employees then experience higher-order needs and proceed to seek satisfaction. Even the person who feels continually driven by the same goal (for instance, money) will be doing so for changing reasons. Once reasonable economic security is no longer a concern, money may be seen as the means of securing leisure time, social acceptability, status, prestige, perhaps even power and influence. These are all expressions of higher-order needs.

■ What Makes Them Perform?

In a 1966 survey conducted by the United States Chamber of Commerce, first-line supervisors in twenty-four organizations were asked to rate ten so-called morale factors. They were asked to place them in the order in which they believed these factors would be important to their employees.[3] In short, they were asked to rank these factors as motivating forces. A second phase of the survey then required all the employees of the same supervisors to rank the same ten factors in order of importance to them as individual workers.

The supervisors estimated that the ten factors would appeal to their employees in order of importance as follows:

1. Good wages
2. Job security
3. Opportunity for promotion and growth
4. Good working conditions
5. Interesting work
6. Organizational loyalty to employees
7. Tactful disciplining
8. Full appreciation of work done
9. Understanding of personal problems
10. Being included in things

The employees placed the same morale factors in the following order of importance:

1. Full appreciation of work done
2. Being included in things
3. Understanding of personal problems
4. Job security
5. Good wages
6. Interesting work
7. Opportunity for promotion and growth
8. Organizational loyalty to workers
9. Good working conditions
10. Tactful disciplining

Note that the order of importance expressed by the employees places primary emphasis on satisfaction of higher-order needs, and focuses less on those usually associated with purely materialistic factors. Certain economic motives (such as wages and job security) were considered important; they do in fact appear in the upper-middle portion of the employees' list. However, to these people who were employed and had normal expectations of remaining employed, the economically related factors did not appeal to them as primary expectations of their work.

In addition to the difference in position of the economic motives on the two lists, note also that the three factors rated highest by the employees were rated lowest by the supervisors. This raises some obvious questions about the importance of higher-order needs and job satisfaction and about the relative value of money as a motivator.

You will note some apparent inconsistencies between the results of this survey and "The Real Reasons People Quit" listed earlier in this chapter. One of the more glaring inconsistencies involves the employee's view of the opportunity for advancement or for "promotion and growth." In this survey, the opportunity for promotion and growth earned only seventh place in the employees' order of importance. However, lack of advancement opportunity was the second most cited reason for people leaving their jobs. This suggests that some particular forces might be dissatisfiers if they are weak or absent, but might not be particularly strong motivators if they are present. In other words, if advancement opportunity is there, it may not cause people to work harder to fulfill certain needs. However, if this kind of opportunity is absent it may well cause feelings of dissatisfaction that could lead a person to quit.

Then, too, the strength of such a factor will vary according to individual needs. The person with strong desires for advancement may leave if opportunity is not present. But the person in whom this need is not as strong may stay and be happy regardless of the lack of advancement opportunities.

■ Money as a Motivator

Most people would probably agree that money is an important reason for working. However, it has frequently been shown that money does not necessarily motivate people to work more effectively. Yet many managers continue to regard money as the principal key to motivation.

Earlier it was suggested that the factors leading to job satisfaction are not necessarily the same factors that can lead to dissatisfaction. It is possible to separate the factors that have a bearing on job satisfaction or dissatisfaction into two types: (1) those inherent in the job itself and (2) those external to the job and thus belonging to the environment within which the job is performed.

The true motivators, it appears, are those factors inherent in the job and realized through the workers' own efforts at accomplishment. The true motivators thus include such things as achievement, the opportunity for achievement, the opportunity to assume responsibility, the actual performance of meaningful work, and the opportunity to learn, develop, and grow. All these are parts of the job itself. Their presence in a given job does not guarantee that the worker will necessarily take advantage of the opportunity to pursue them. However, if they are not there, the worker who might otherwise be motivated cannot put them there.

Among the external factors (those occurring in the job environment) are organization policy, salary, fringe benefits, character of supervision, and working conditions. When these factors do not reach an individual employee's level of expectation, they can lead to dissatisfaction. However, even when environmental factors are well satisfied, they generally provide no sustained motivation. The true motivators grow on and from themselves, but the environmental factors need to be continually increased or reinforced to provide any lasting benefit.

Money is important, and many people would not be against receiving more money for doing what they now do. However, money's main functions are primarily to help avoid pain or discomfort. To put it bluntly, having an adequate salary helps avoid the feeling of economic deprivation. It should be noted that since essentially all public correctional systems operate within a civil service structure that classifies jobs in a rational way, this pay/job equity factor may not be a major element for staff in those organizations. (Although disparity with other law enforcement agencies may be a factor—the widespread problem of jail deputies being paid less than road deputies in sheriff's departments is well known.) Whether that is the case in private corrections is not as clear. But since private correctional agencies often compete with other law enforcement agencies for staff, it is likely that some pay/benefits balance exists here as well.

Money and other external factors that relate to money have one desirable characteristic that the true motivating factors do not possess. They can be measured on an objective scale. How does one begin to measure such intangibles as achievement and the feeling of worth resulting from the doing of meaningful work? How do managers apportion such "rewards" among the workforce? The fact is they cannot. Money, however, can be measured, so it is constantly measured and used as a "motivator."

■ Motivation and the First-Line Supervisor

The age-old carrot-and-stick approach, alternating reward and punishment for certain behavior, simply refuses to work once people have reached an adequate subsistence level and are motivated primarily by higher-order needs. The supervisor could do well to remember that economic rewards are but a portion of the total "reward" the employee works for.

Leadership and employee motivation are directly related to each other. The quality of supervision will generally have a significant bearing on the willingness of employees to perform. Leadership that employees can respect and admire is far more likely to produce positive performance than leadership that employees see as harsh or arbitrary.

One of the keys to motivating employees is the supervisor's level of self-motivation. Managers set the example for employees. If the supervisor is genuinely interested, stimulated, energetic, and caring, this will show and many (although regrettably not all) employees will respond in kind.

Employees are unique individuals and their "motives" actually represent personality characteristics that each brings to work. Some may be ambitious and desire money and leadership positions. Some may be fulfilled largely by helping others. Some may satisfy their needs through overcoming obstacles to accomplish difficult tasks. These drives remain much the same within each person, so what a manager is actually attempting to influence in employees is "aroused motivation." The goal is to awaken in them the drive to seek fulfillment of certain needs.

Generally, managers provide aroused motivation by:

- Valuing employees as individuals and treating each as such
- Providing challenge in the workplace whenever possible
- Increasing or varying job responsibilities when possible
- Helping employees to grow in such ways as to benefit both them and the organization.

It is not always possible to provide a great deal of challenge in certain jobs or provide certain workers with increased responsibility. But it is always possible to bring one or two of the prominent motivating forces into play in any situation. A simple "you did a good job" can be a powerful motivator, and a "thank you" can be valued compensation for someone's efforts. Ultimately most managers discover that they cannot "motivate" an individual as such. Rather, it is possible only to create the climate within which the person will become self-motivated.

EXERCISES

Exercise 13-1: "What's This Place Coming To?"

Like most other states, this one was experiencing a major increase in the number of convicted felons sentenced and awaiting transfer to its facilities. The strategy used for several years—allowing shorter-term cases to "back up" in county jails—was no longer viable. Sheriffs were exerting a great deal of pressure on the legislature to do something to give them more space for their misdemeanant cases by removing the felons.

With state prisons seriously crowded, there didn't seem to be any way out of the crisis but to contract beds with a private corrections firm. The newspapers in the major communities throughout the state carried occasional stories about the efforts of the Department of Corrections to acquire more beds from the private sector. More than one of these articles had suggested that youth offenders and lower-security adult inmates would be sent to those private correctional facilities and the institutions housing them would be renovated for confining more dangerous, higher-security cases.

A number of employees at the state's low-security youth reformatory (especially those who had worked there a long time) were concerned about the future. One of these was Mark Patterson, who had been supervisor of education for the last 14 years. He ran a very large school and vocational training complex to meet the needs of the young people confined at his facility.

Patterson had expressed his concern more than once about these reported plans to various members of administration, including the warden. He received no answers from his immediate supervisor (the associate warden) who seemed to know little more about the future than Patterson. From the other members of the administration he received only references to published documents describing how the legislature was considering this plan, and what some of the implications might be. These references only raised more questions in his mind because they suggested the reformatory would be a prime candidate for upgrading to a medium-security adult facility.

One morning before work Patterson encountered a neighbor at the local gas station. They talked a few minutes before the neighbor said, "Oh, that thing about the reformatory in the paper today—what is it going to do to your job?"

Mark was puzzled, but rather than display his lack of specific knowledge he said, "Oh, I don't really know yet. We'll have to wait and see." After the neighbor paid for his gas, Mark picked up a copy of the local paper.

The lead story was that the Department of Corrections had made some far-reaching recommendations to its legislative oversight committee for redistribution of bed space in various locations. Among the many adjustments, all of the youth offenders at the reformatory would be sent to a contract facility and the institution would indeed be converted to an adult institution. Ominously, the article said that there would be some staff adjustments, since adult offenders needed different kinds of programs than the youth cases.

When Patterson went to work that day he noticed solemn faces wherever he went, and particularly in his department. Before the morning was over he discovered that several of his best teachers were already talking about applying for jobs at the juvenile training school, which was not mentioned in the article as being involved in the plan. Several others wondered out loud if they could move to the private facility and continue doing what they were doing. Without exception, they voiced misgivings and outright concerns over working with a higher-security adult offender.

Patterson met briefly with the associate warden, who indicated he had no prior knowledge of the change. He further suggested that Mark was under consideration for an upcoming opening as executive assistant to the warden, but even this news had little effect on Mark's disposition.

Questions

1. What are some of the possible short-range effects on the morale, performance, and individual effectiveness of the education personnel and staff throughout the reformatory?

2. What "morale factors" (refer to the section, "What Makes Them Perform?") are most affected by the impending changes at the reformatory?

3. Keeping in mind that the employees' need to be "included in things" must be balanced against the institution's inability to know precisely what is happening outside of the correctional facility itself, how might this matter have been approached so as to minimize negative reactions among the employees?

Exercise 13-2: The Promotion

With considerable advance notice, the very large parole office's records supervisor retired. Although the job was advertised under normal civil service procedures, within the four walls of the building it was assumed that you (the assistant supervisor) would be appointed to the supervisor's job. However, a month after your supervisor's departure the department was still running without a full-fledged supervisor. Day-to-day operations had been left in your hands as "acting" supervisor, but the deputy chief parole officer had begun to make some of the administrative decisions affecting records operations.

After another month had passed you learned "through the grapevine" that the agency had interviewed several candidates from outside for the supervisor's position. Nobody had been hired, however.

During the next several weeks you tried several times to discuss your uncertain status with the deputy. Each time you tried you were put off; once you were told simply to "keep doing what you're doing."

Four months after the supervisor's departure, you were promoted to records supervisor. The first instruction you received from the deputy was to abolish the position of assistant supervisor.

Questions

1. What can you say about the likely state of your ability to motivate yourself in your "new" position?

2. What can you say about your level of confidence in the relative stability of your position, and how might this affect your performance?

3. At the time you assume the supervisor's position officially, what is likely to be the motivational state of your staff? Why?

ENDNOTES

1. Fournies, Ferdinand. "The Real Reason People Quit." *Administrative Management,* October 1969: 45–46.
2. Maslow, A.H. "A Theory of Human Motivation." *Psychological Review,* 50 (1943): 370–396.
3. Chamber of Commerce of the United States, Washington Review, 1966.

Performance Appraisal: Cornerstone of Employee Development

14

Chapter Objectives

- Review the primary objectives of performance appraisal and establish appraisal as an essential management process.
- Identify and review common approaches to employee performance appraisal.
- Assess common appraisal problems and suggest why many appraisal programs fail.
- Stress the need for reasonably objective appraisal as opposed to personality-based appraisal processes.
- Outline the requirements of an effective performance appraisal system.
- Highlight the requirements or characteristics necessary to make the organization's performance appraisal system as legally defensible as possible.
- Introduce standard-based appraisal as a desirable long-range consideration in improving the organization's evaluation process.
- Introduce the concept of "constructive performance appraisal."
- Describe how supervisors can improve and upgrade an existing appraisal system.

The privilege of encouragement is one that may be exercised by every executive and supervisor and it should be cultivated, not so much as a working tool to be employed objectively, but as an act of deserved kindness and intelligent leadership.

Anonymous

A Secret Resolution

This is a chapter about performance evaluation. But sometimes important events don't ever appear in a performance evaluation. This is a story about one of those times.

The unit manager worked closely, but often adversarially, with the supervisor of education—let's call her Barbara Morgan. He (a relatively young white man) and she (an older African-American woman) tangled openly about many issues—agency policy, internal facility politics, inmate program approvals, and so on. But they also had just enough in common to understand where the other was coming from, and each respected the strengths and knowledge the other brought to their areas of responsibility. Despite their differences, an interestingly respectful relationship actually developed between these two department heads over the years.

One of the unit manager's counselors, Rob Hamblin, was bordering on what one might call a "redneck." While he was a bright fellow who did seem to be amenable to supervision, he clearly did not have anything that resembled respect for the education supervisor. Nothing was expressed openly, but these two were never going to bond.

One day, as the unit manager was touring a housing area, he was paged by the supervisor of education. Sounding distressed, she asked him to meet her in her office right away. He did so, and heard an amazing story.

Ms. Morgan related that she and Mr. Hamblin had been involved in an argument. She admitted that she had said some very provocative things to him as the conversation ended. Then, she said, as he was leaving Hamblin swatted her on the buttocks with the palm of his hand. She (no shy, retiring person herself), in turn, punched him in the mouth. This happened in the education area of the institution, but from her account no inmates witnessed the episode.

The education supervisor was in tears by that point in the story. She conceded that she had provoked Hamblin. She was upset that she had been hit. She was upset that she had hit Hamblin. She didn't want to file charges. She didn't want the associate warden to know about the incident (because she suspected he was looking for an excuse to fire her). She simply didn't know what to do, and she asked the unit manager for help.

The manager told her to stay put in the office and not talk to anyone, and then returned to his own office. No sooner than he arrived, Hamblin came on the unit and to the office. He related a very similar story, and the

It is easier to discover a deficiency in individuals, in states, and in Providence, than to see their real import and value.

Georg Wilhelm Friedrich Hegel

FROM THE
INSIDE

same dilemma. He had no idea why he had done such a stupid thing. He had been provoked, but he had no right to do what he did. He could see his career (and perhaps years of freedom if assault charges were pressed) in the balance. He didn't want the repercussions of this event to get out of hand. He had done something really wrong, and wanted help. He, too, was in tears.

The unit manager told him that he thought there might be a way to prevent this from resulting in formal action. He suggested that Hamblin immediately go to Morgan and apologize, and ask for forgiveness. It sounds corny in a day and age when litigation springs vigorously from every imagined insult and slight. But there was a time when people still did those things, and they sometimes worked out all right.

So the unit manager called Ms. Morgan, and he and Hamblin went to her office. Hamblin did everything short of getting on his knees and begging forgiveness. Morgan, in turn, expressed her regret for her part in provoking Hamblin during the argument and for hitting Hamblin. Morgan made it clear that she didn't want the original assault to result in any action against Hamblin. Hamblin swore he had never done such a thing before and never would again. They didn't kiss and make up, but they handled the conversation as adults and professionals.

Both appeared to the unit manager to have learned an important lesson, so it was agreed that the incident and the resolution would remain confidential among them. The two former antagonists continued to work together until Hamblin was transferred some time later.

So far as the three people involved knew, no one else in the institution ever knew of the incident. The associate warden never learned that one of his department heads had been involved in provoking a fracas with a co-worker, so it never factored into any evaluation of her performance. Because Morgan was clear on this point, the unit manager didn't use this incident in any evaluation of Hamblin's performance. And the associate warden never learned how skillfully the situation had been defused, so the unit manager's evaluation never reflected his actions either.

The Manager's Darkest Hour?

Why refer to performance appraisal as "the supervisor's darkest hour"? The evaluation of employee performance through the application of some formal appraisal system is part of almost every supervisor's job. Yet as common as the requirement for performance appraisal may be, supervisors often dread that process and all that goes with it. Nevertheless, in all modern organizations (corrections included) employee appraisal remains a basic responsibility of all persons who direct the work of others. Indeed, in many public correctional agencies the origin and structure of the performance evaluation process is closely tied to the collective bargaining process, and thus is an even more critical activity for supervisors to carry out properly and effectively.

Of course performance appraisal operates both inside and outside of whatever formal system the correctional agency may have. The opening vignette is one instance of how employee misbehavior can be handled outside the formal system. But inside the system, a supervisor who formally reprimands an employee for a breach of policy is, in effect, appraising performance. Likewise, the supervisor who informally compliments an employee for a task well done or who criticizes an employee for committing an error is also appraising performance. Thus any instance of criticism or praise, whether offered within or outside the context of a formal performance appraisal system, constitutes employee evaluation.

Pursued within the context of a formal, mandated system (as it is in most correctional agencies) performance appraisal requires a great deal of the supervisor's time and attention. Most supervisors have had occasion to discover that there is often not enough time to do everything they must do, so appraisal is left to compete with many other activities for the available time. Since appraisal requires information that the supervisor must accumulate over an extended period of time, difficult elements of the appraisal process are always competing for space on the supervisor's list of priorities. This happens with respect to performance appraisal because it appears to have no direct impact on the accomplishment of the day-to-day work in the department. Therefore, it often gravitates to the lower part of the supervisor's priority list until attention is required.

Some appraisal systems call for the evaluation of all employees at the same time, ordinarily once each year; interim reviews may occur quarterly or every 6 months. Under the all-at-once approach, the supervisor often views the performance appraisal task as overwhelming. Other essential tasks may suffer because appraisals must be done under impending time limits. Because of the pressures created by ignoring other work, the supervisor may fail to do justice to the appraisals. If the supervisor has a large number of appraisals to do, perhaps the ones undertaken first receive the most care while the latter appraisals receive diminishing time and attention. Sometimes there is a requirement to do a number of appraisals within a specified number of days or weeks. Often this requirement is imposed at what is an inappropriate time (and rarely is there an appropriate time for a task that intrudes so deeply into the daily routine). Consequently, many supervisors come to regard appraisal as, at best, a necessary evil or, at worst, an unnecessary and resented intrusion.

Fortunately, not all appraisal systems call for the evaluation of all employees at the same time. A significant number of systems call for the appraisal of employees on their employment anniversary dates. Although this approach is to be preferred over the all-at-once approach, it too can have its problems. The supervisor who must evaluate every employee at once may be extremely busy for a few weeks, but he or she knows that once the appraisals are finished the process will, for all practical purposes, go away for the greatest part of the year. However, when appraisals are done on employees' anniversary dates, the manager still has the same number of appraisals to do but they are staggered throughout the year. Under this approach, appraisals are never "caught up" and the process hangs over the supervisor as a nagging task that, through its constant presence, places subtle but steady pressure on the supervisor.

Not all of the pressures associated with performance appraisal are as subtle as those created by the anniversary-date approach. There are also direct pressures that come from higher management (usually by way of the personnel department) aimed at getting appraisals accomplished according to some schedule. The personnel department must invariably remind evaluators of due dates and must otherwise keep tabs on the process. And thus there is another reason why many supervisors come to view appraisal in a decidedly unfavorable light as "the personnel department's system" or "just more of personnel's paperwork."

A great deal of discomfort about performance appraisal also arises from the perceived uncertainties inherent in the process. Supervisors may be well aware that they are expected to document their judgments about their employees. This judgmental activity is something that many supervisors are uncomfortable doing; potential confrontation is seldom something to look forward to. They are also well aware that if they are to be truly conscientious in this process, they may have to discuss unfavorable judgments with employees. Thus, in addition to reacting negatively to what they see as a requirement imposed upon them from above, supervisors also often tend to react negatively to what they see as a highly subjective process in which their opinions and judgments may ultimately be challenged and thus must be defended.

In short, in many organizations the supervisors tend to view performance appraisal primarily as a requirement of "the system" rather than as a key element of the essential supervisor-employee relationship.

It is important to note again that, as any experienced administrator knows, there inevitably are variations in any nice, tidy personnel management system. First of all, as a practical matter, not every aspect of an employee's performance is going to be documented. Things happen that never come to the attention of a supervisor. There also are times when an event or action should not be recorded or used against an employee for sound management reasons. An otherwise good employee does something that is totally out of character and may deserve a "pass." (The opening anecdote presents an interesting illustration of the issue of what actually does and does not go into an evaluation, as well as how personal judgment enters into many on-the-fly management decisions.) A court settlement may preclude an incident or pattern of performance from being included in an evaluation.

In the above-cited instance, one can argue that the unit manager was a peer, not a superior, and did not have the right to do what he did. But that flies in the face of reality—many things happen in an institution that could have implications for higher management, but which are resolved among peers. In fact, these take place regularly as part of a healthy delegation environment. They also can take place when peers use each other as resources, as occurred in this instance.

One also can argue that this was a criminal act (assault or perhaps sexual harassment) and it should not have been concealed. But again realistically, many low-level criminal acts are never reported, and in that instance the woman involved very clearly did want the episode to come to further light. This was due to her expressed fear that the incident would precipitate her termination, based on her history of adversarial relations with her supervisor. The unit manager made a judgment that the incident was sufficiently out of character for both individuals to warrant the informal resolution he brokered.

And finally, in so doing, the unit manager created another example of a supervisor not knowing something about the performance of a subordinate. In this case the rather skillful defusing of a potentially serious problem for the institution and the individuals involved remained undisclosed. Something that under other circumstances might have demonstrated the unit manager's positive managerial talent was never brought to light.

■ Why Appraise at All?

As discussed in Chapter 13, most employees are not uniquely motivated by the visible rewards that exist as part of the organizational setting. Salary, fringe benefits, and working conditions are not the only things that employees work for.

Recall the position of "full appreciation of work done" as a potentially powerful motivator of employee performance. People who are doing good work need to know they are doing good work, and they need to know that what they do is appreciated. This knowledge and appreciation are essential parts of the "psychic income" that every employee needs to receive in some measure in addition to the real income associated with the position.

In addition to knowing they are doing well and that their work is appreciated, employees need to know when they are not doing particularly well and what they can do to correct their behavior. Criticism itself, even so-called constructive criticism, does not bring about long-lasting behavioral change. It is one thing to criticize. It is something else entirely to criticize and be able to supply alternatives for behavioral change and improvement.

Performance appraisal is needed because all employees deserve to know where they stand in the eyes of the supervisor and the organization. Beyond this, however, employees need to know where they stand so as to be able to do something positive about future performance.

■ The Objectives of Appraisal

The primary objectives of a performance appraisal system should be:
- To encourage improved performance in the job each employee presently holds.
- To provide growth opportunity for those employees who wish to pursue possibilities for promotion.
- To provide the organization with people qualified for promotion to more responsible positions

In many instances, the true objectives of a performance appraisal are not well served. An appalling number of appraisal systems are oriented functionally almost entirely toward criticism and fault finding. Certainly these systems were not intended to be used in this fashion, but their weaknesses (primarily their focus on the past) have brought about their general misuse. Rather than simply looking at the past and stopping there, an effective appraisal system should seek to utilize the past only as a starting point from which to move into the future. When the appraisal interview becomes history and the form finds a home in the personnel file, the employee should be able to reasonably answer these two questions:

1. How am I doing in the eyes of my supervisor (and thus in the eyes of the organization)?
2. What are my future possibilities?

This brief review of the performance appraisal process will describe the common approaches to employee evaluation, consider some reasons why appraisal programs frequently fail, comment on the need for performance appraisal, and consider ways of more fully utilizing performance appraisal as an effective management technique.

Essential Elements of an Effective Appraisal System

Every correctional agency's appraisal system is different and so the following discussion necessarily is somewhat general. However, for any performance appraisal system to have a realistic chance of being effective, it must meet a number of conditions. If it does not, it will not produce the expected results. However, even if the system does meet all the conditions, success is not necessarily guaranteed. To be fully effective, a system requires thorough, conscientious application by managers who believe in the value of performance appraisal. Careless or indifferent application can kill even the best systems or turn them into mere paper exercises.

The essential elements of an effective performance appraisal system are:

- **System Objectives**—Overall, the system must serve the true objectives of performance appraisal as previously described.
- **Appropriateness of Criteria**—System criteria (characteristics upon which the employees are evaluated) must be as closely related as possible to the kinds of work being evaluated. A single approach never fits all of the jobs to be evaluated within a single organization, especially a correctional facility, with its many and varied occupations.
- **Performance Standards**—Evaluations should reflect specific performance standards that are established based on the employee's job description. These standards should be set in terms of objective measurements. That is, the supervisor should be able, for a significant number of evaluation criteria, to come up with a numerical measure of results that may be compared with an established standard.
- **Employee Knowledge of Criteria**—Well in advance of being evaluated, employees must know the criteria on which they will be evaluated. Employees must be fully aware of the job description tasks as they are presently known to the manager, and they should be fully aware of all applicable job standards. The usual method for establishing this base is by providing each employee with a copy of the standards at the beginning of the rating period.

- **Management Education**—Managers should be thoroughly oriented in the use of the system and thoroughly trained organization-wide in the consistent application of the process.
- **Ongoing Use**—Once it is completed, an evaluation should serve as a live, working record to be used as a starting point for monitoring progress. This is especially important for unfavorable evaluations, which should be sufficiently complete as to spell out specific steps and time frames to be involved in correction and improvement.
- **Appraisal Interview**—The appraisal interview should be meaningful. It should not be avoided because the manager is uncomfortable nor should it be treated once-over-lightly and disposed of quickly simply because the manager feels the pressures of time. The appraisal interview should be a true two-way exchange and receive the manager's full attention for whatever time is required.
- **Self-Contained Record**—Once it is placed in the employee's personnel file, a completed performance appraisal should stand on its own. Cross-reference to an appraisal manual, evaluation key, or a list of explanations should not be necessary in determining what any particular rating means.
- **System Administration**—The system must be administered appropriately and effectively. All scheduled review dates must be observed. Managers must receive appraisal forms and reminders sufficiently in advance of the appraisal due date. They must receive interim reminders as necessary to assure that appraisals are not allowed to run late. In short, someone needs to pay constant attention to the process of keeping the system moving. The fact that this kind of auxiliary support system seems to be necessary in almost every correctional organization (and others as well) is in itself a testimony to the problem most supervisors have with this portion of their jobs. For any other aspect of a supervisor's job, this level of "hand-holding" or external monitoring would be considered evidence of a performance problem on the part of the supervisor. And yet, it is an almost universal phenomenon.

■ Traditional Appraisal Methods

Over the years a number of appraisal systems have evolved, some depending on a greater or lesser amount of structure than others. Each agency has a specific appraisal method that managers are required to use. However, this chapter will try to show generally how a variety of systems work. In addition, it will show that there have been continuing efforts to make appraisal systems more objective, more reliable, and less dependent on the unsupported judgment of the people doing the evaluating. The major approaches to performance appraisal are discussed next.

Rating Scales

Ratings scales, the oldest and most widely used appraisal procedures, are of two general types.

Continuous Scales

In this system, in referring to a particular evaluation characteristic, the evaluator places a mark somewhere along a continuous scale (see **Figure 14-1**). There is usually a numerical scale involved, so the evaluator is actually assigning a certain number of "points" to the individual for that particular characteristic. Generally, the evaluator is aware of some position on the scale that constitutes "average" or "satisfactory" performance.

Discrete Scales

In this system, each characteristic is associated with a number of descriptions covering the possible range of employee performance. The evaluator simply checks the box, or perhaps the column, accompanying the most appropriate description (see **Figure 14-2**).

Rating scale methods are easy to understand and easy to use, at least in a superficial manner. They permit numerical tabulation of scores in terms of measures of average tendency, skewness (the tendency of a group of employees to cluster on either side of a so-called average), and dispersion.

Rating scales also are relatively easy to construct, and they permit ready comparison of scores among employees. However, rating scales have several severe disadvantages. Do total scores of 78 for Barbara and 83 for Carlos really mean anything significant?

These systems are also subject to assumptions of the ability of a high score on one characteristic to compensate for a low score on another. For instance, if an employee scores low relative to quantity of work produced, can this really be counterbalanced by high scores for attendance, attitude, and job knowledge?

Ratings frequently tend to cluster on the high side when scales are used. Supervisors may tend to rate their employees high because they want them to

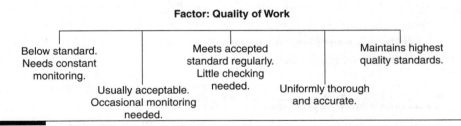

Factor: Quality of Work

Below standard. Needs constant monitoring.

Usually acceptable. Occasional monitoring needed.

Meets accepted standard regularly. Little checking needed.

Uniformly thorough and accurate.

Maintains highest quality standards.

Figure 14-1 One characteristic from a continuous-type rating scale.

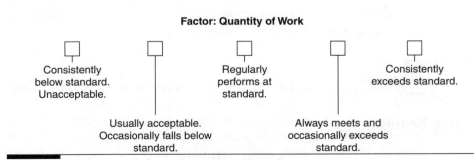

Factor: Quantity of Work

Consistently below standard. Unacceptable.

Usually acceptable. Occasionally falls below standard.

Regularly performs at standard.

Always meets and occasionally exceeds standard.

Consistently exceeds standard.

Figure 14-2 One characteristic from a discrete-type rating scale.

receive promotional consideration and their fair share of pay raises. They also want to feel good about themselves; they often take this course because it is easier to praise than it is to leave oneself open to the appearance of being critical. Also, different supervisors tend to rate differently. Some consider *average* as precisely that—average acceptable work, nothing to be ashamed of. However, other supervisors seem to think of *average* as something of a dirty word and thus tend to rate most employees on the high side of the scale.

Employee Comparison

Employee comparison methods were developed to overcome certain disadvantages of the rating scale approaches. Employee comparison may involve the ranking method or the forced distribution method.

Ranking

This method forces the supervisor to rate all employees on an overall basis according to their job performance and value to the institution. One approach is simply to look at subordinates and decide initially who is the best and who is the poorest performer and then to pick the second and next-to-last persons in rank order by applying the same judgment to the remaining employees. This is simple enough to accomplish, but the process is highly judgmental and strongly influenced by personality factors. Also, some employee must end up as "low person on the totem pole," and this may not be a fair assessment overall. It also is somewhat difficult to apply to a large department. The chief of security in a major penitentiary with several hundred correctional officers is going to find it difficult to use this method, even if the strategy is to apply it at the shift level rather than across the entire department.

Forced Distribution

This method prevents the supervisor from clustering all employees in any particular part of the scale. It requires the evaluator to distribute the ratings in a pattern conforming with a normal frequency distribution. The supervisor must place, for instance, 10 percent of the employees in the top category, 20 percent in the next higher category, 40 percent in the middle bracket, and so on (see **Figure 14-3**). The objective of this technique is to spread out the evaluations. However, while it is true that the general population may be distributed according to a normal curve, in an organization, managers are dealing with a select group of persons. If employees have been properly trained and probationary periods correctly used to eliminate the genuine misfits, then the true distribution of abilities and performance in the work group should be decidedly skewed. That is, the group's "average" should be better than the general average assumed by the so-called normal distribution (see **Figure 14-4**).

Checklists

Yet another appraisal technique involves the use of checklists to enumerate specific performance traits. The weighted checklist is the most common.

This system consists of a number of statements that describe various modes and levels of behavior for a particular job or category of jobs. Every statement has a weight or scale value associated with it. When rating an employee the su-

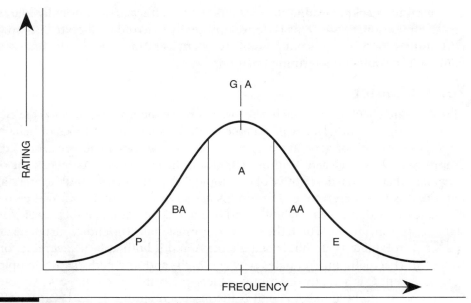

Figure 14-3 Employee comparison: forced distribution using the "normal curve."

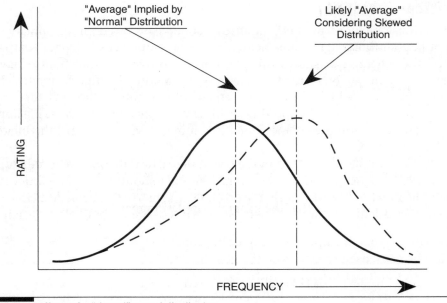

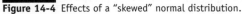

Figure 14-4 Effects of a "skewed" normal distribution.

pervisor checks those statements that most closely describe the behavior exhibited by the individual. The completed rating sheet is then scored by averaging the weights of all the descriptive statements checked by the rater. This is much like the rating scale approach except for the application of the weights. Some evaluation characteristics are worth more or less than others.

Often in checklist evaluation systems the weights are intentionally kept secret from the supervisor. This is done supposedly to avoid deliberate bias on the part of the supervisor; it is not possible to "slant" a rating to make the final score come out in some predetermined manner.

Forced Choice

Like the pure checklist approach, the forced-choice method requires the development of a significant number of statements describing various types of behavior for a particular job or general category of jobs. These statements are arranged in groups of four or five each. Within each group the evaluator must check the one statement that is most descriptive of the employee's performance and the one statement that is least descriptive. The groups are so designed that each will contain two statements that appear favorable and two that appear unfavorable. A set of five statements from among which the supervisor must make the choice just described is shown in **Exhibit 14-1**. While statements A and B both appear favorable, only statement B actually differentiates between high- and low-performance employees. Statement C is actually descriptive of low-performance employees. Although E also appears to be unfavorable, it is inconsequential in this set because of the presence of C. Statement D is neutral. Once again, the actual value or weight of the statements is kept secret from the supervisor.

Critical Incident

The critical incident method requires a supervisor to adopt the practice of recording all those significant incidents in each employee's behavior that indicate either effective or successful action or ineffective action or poor behavior. The use of a notebook or card system provides reminders of performance characteristics under which various incidents can be recorded. For instance, if an employee saved the day by spotting an inmate scaling a fence and taking appropriate action to stop the escape attempt, you might want to record the incident under "initiative."

There is a potential hazard in the use of the critical incident method. Supervisors are busy people, and often everything that should be recorded does not reach the notebook, particularly the positive or constructive episodes. However, negative incidents, because of their "seriousness," are more likely to reach the

---

Circle the letter for the statement that is most descriptive of the employee's performance and the letter for the statement that is least descriptive of the employee's performance:

Most	Least	
A	A	Makes mistakes only infrequently
B	B	Is respected by fellow employees
C	C	Fails to follow instructions completely
D	D	Feels own job is more important than other jobs
E	E	Does not exhibit self-reliance when expressing own views

Exhibit 14-1 Illustrative group of statements from a forced-choice appraisal.

pages of the book than are many occasions of positive performance. Also, this approach can lead to overly close supervision, with the employees feeling that the supervisor is watching over their shoulders and that everything they do will be written down in the "little black book" or on their "5 x 8 card."

Field Review

Under the field review appraisal method the supervisor has no forms to fill out. Rather, the supervisor is interviewed by a representative of the personnel department who asks questions about the performance of each employee. The interviewer writes up the results of the interview in narrative form and reviews them with the supervisor for suggestions, modifications, and approval. No rating forms or factors or degrees or weights are involved; rather, simple overall ratings are obtained.

The field review approach relieves the supervisor of paperwork. It also assures a greater likelihood that supervisors will give adequate and timely attention to appraisals because the personnel department largely controls the process. However, the process takes the valuable time of two management representatives (the supervisor and the personnel interviewer). It also requires the presence of far more personnel department manpower than most institutions feel they can afford.

Free-Form Essay

This method simply requires the supervisor to write down impressions about the employee in essay fashion. If desired by the organization, comments can be grouped under headings such as: job performance, job knowledge, and goals for future consideration. To do a creditable job under this method, the supervisor must devote considerable time and thought to the evaluation. On the plus side, this process encourages the supervisor to become more observant and analytical. On the other hand, the free-form essay approach generally demands more time than the average supervisor is willing or able to spend. Also, appraisals generated by this method are often more reflective of the skill and effort of the writer than of the true performance of the employees.

Group Appraisal

Under this approach an employee is evaluated at the same time by the immediate supervisor plus three or four other supervisors who have knowledge of that employee's work performance. This method may be particularly adaptable to correctional shift work, where staff are likely to be assigned to multiple posts and work for multiple supervisors over the course of a rating period.

The virtue of this method is its thoroughness. It is also possible for multiple evaluators to modify or cancel out bias displayed by the immediate supervisor. However, the drawbacks of this approach are such that it is rarely used. It is extremely time-consuming, tying up perhaps four or five members of management to evaluate a single employee. Moreover, it often is inapplicable because there may be few managers beyond the immediate supervisor who are sufficiently familiar with the employee's performance.

■ Common Appraisal Problems

A common problem encountered in performance evaluation is the "halo effect." This refers to the tendency of an evaluator to allow the rating assigned to one or more characteristics, or recent events, to influence excessively the rating on other performance characteristics. The rating scale methods are particularly susceptible to the halo effect.

For instance, if a manager declared an employee to be excellent in terms of "initiative" and "dependability," he or she also might be inclined to rate high relative to "judgment" and "adaptability." As a practical reality, it is extremely difficult to completely separate consideration of one performance factor from the others. That is because many performance characteristics actually include shades of others, and there is no guaranteed way of eliminating their interactions—this can contribute to the halo effect.

As noted earlier as a common problem in most rating systems, many supervisors tend to be liberal in their evaluations, that is, to give their employees consistently high ratings. Most approaches to rating are partially based on the assumption that the majority of the work force will be "average" performers. However, many people (supervisors included) do not like to be considered "only average."

Central tendency or clustering is another problem, one that some of the rating methods just described have attempted to overcome. Some supervisors are reluctant to evaluate people in terms of the outer ends of the scale. To many supervisors it is "safest" to evaluate all employees consistently. This often leads to a situation in which "everyone is average." This is contrary to the likelihood that in a work group of any considerable size there are, in fact, performers who are both better and worse than the so-called average. Some performance evaluation systems rely on an informal method of structuring the number of evaluations in each category. That can help avoid both central tendency ratings and the prior problem of overly liberal evaluation systems.

Interpersonal relationships also can pose a considerable problem in performance evaluation. The supervisor cannot help but be influenced, even if only unconsciously, by personal likes and dislikes. Often a significant part of an evaluation will be based on how well the supervisor likes the employee rather than how well the employee actually performs. This is somewhat different than the halo effect, which refers to a short-term impact of performance on evaluation, rather than a longer-term influence based on personal relationship.

■ Why Appraisal Programs Often Fail

Many performance appraisal programs fail outright, or at least partially fail to do the jobs they were intended to do. A number of the reasons for appraisal program failure stem from weaknesses in the systems as already described. Some, however, result from deeper-seated reasons.

Although most appraisal programs are supposed to be based on observable behavior and performance, many often fail because they actually require the

1. Quality of work	6. Initiative	7. Dependability
2. Volume of work	4. Job knowledge	8. Attitude
3. Effectiveness	5. Adaptability	9. Attendance

Exhibit 14-2 Listing of rating characteristics from an actual appraisal form.

supervisor to render personality judgments. Consider the difficulty of truly evaluating a number of employees relative to each other as they relate to their work in terms of "characteristics" that are often actually personality characteristics. There are many appraisal systems in which the performance characteristics defy objective assessment (see **Exhibit 14-2**). How can a supervisor truly rate someone on a characteristic such as "adaptability"? The problems are compounded by problems of semantics—what in fact are "initiative," "judgment," and so on?

Another reason for failure is that supervisors are unqualified to judge personality characteristics. Very few people are qualified to do this in an accurate way, yet supervisors are put into a position of having to do so time and time again.

As part of the pressure toward making judgments about personality traits, many systems fail to allow for distinguishing between the cause and results of behavior. An employee who comes across as irritable and constantly argues with others is likely to get marked down on "attitude." However, the results of the behavior (the clashes with other people) are the only real evidence the supervisor has to go on. To say that these interpersonal clashes result from a poor attitude is to try to assign a cause to the behavior. This assignment of cause to result is (in almost all cases) an unqualified leap for the supervisor. It is inappropriate to call an employee who appears unwilling to adapt to a supervisor's ideas "stubborn," but most appraisal systems constantly require second-guessing cause in this fashion.

Programs can also fail because of the uncomfortable position of the person doing the evaluating. It is an extremely serious matter to probe the personality of an employee in a fashion that results in a permanent record in the employee's personnel file. The supervisor is put into a position of power over the employee's potential for promotion, pay increases, and favorable references in the future. Many supervisors dislike being put into this position and compensate by keeping all their evaluations high or at least harmless.

Some systems fail mechanically, owing to poor system administration. Perhaps appraisal forms and notices do not come out on time, are not followed up, do not get discussed with employees, or do not get completed at all. The mechanics of any appraisal system must be such that a system is "kept moving." Because of natural resistance to the uncomfortable task of appraisal, a system can die of its own weight unless it is continually nudged along.

Poor follow-up on appraisals can weaken a program, if not cause the program to fail entirely. This suggests, as mentioned at the beginning of the chapter, that an appraisal should not be an evaluation of the past to be filed away and forgotten. Since it should ideally be a guide to future action, a performance appraisal should be reflected in an active document that is used in the employee-supervisor relationship during the months to come.

That is a strong argument for a properly structured critical incident system. Calling in the employee in on a scheduled basis to review all performance entries in the notebook or on the index card is a perfect opportunity for the supervisor to scan prior entries and follow up on both positive and negative episodes. This can aid both supervisor and employee in picking up on both functional and dysfunctional trends. In many systems, unfortunately, the only time the last performance appraisal is pulled out of the file is when the next appraisal is due.

■ Legal Aspects of Performance Appraisal

Performance appraisal now carries with it many potential legal traps. An increasing number of wrongful termination lawsuits or individual labor-management disputes are an outgrowth of inadequate performance appraisal procedures. The stage is set for a wrongful discharge complaint when an employee is let go for any performance-related reason and yet his or her evaluations on file show "good" or "standard" or "satisfactory" performance.

Performance appraisal information is also playing an increasingly active role in complaints filed under the Age Discrimination in Employment Act (ADEA). Such actions most commonly involve complaints concerning promotions, retirements or layoffs, and discharges.[1] In all of these kinds of actions there are questions of employee performance. Regardless of what the defendant organizations say about the performance of the individuals who complain they were discriminated against, the courts generally rely on the documentation of performance found in the personnel file—most commonly the written performance appraisals.

In these situations, the biggest trap in performance appraisal, referred to earlier in this chapter, comes back to haunt many appraisers. That trap is the lenient appraisal—stating on the record that the individual's work is acceptable when in truth it is unsatisfactory in some respect.

Drawing on how performance appraisal has fared in the legal system through a significant number of cases, it is possible to make some reasonable conclusions as to the characteristics of a legally defensible system. No matter what correctional system a manager works for, an effective, legally defensible appraisal system would include the following elements:

1. The system is based on the job, with the appraisal criteria arising from an analysis of the legitimate requirements of the position. This is the embodiment of the oft-repeated admonition to focus on the job itself and not the person who does the job.

2. Performance is assessed using objective criteria as much as possible, given the unique requirements of the job. The reasons behind the assessments must amount to more than simply the unsupported subjective assessment of the appraiser. These criteria must be known to the employee in advance of the rating period.

3. The appraisers have been trained in the use of the system and possess written instructions on how the appraisal is to be completed. This establishes that any appraiser is as reasonably capable of evaluating performance as any other appraiser using the same system, and that the system can be expected to be applied as intended.

4. The results of each appraisal are reviewed and discussed with the employee. This is a major concern. Documentation of performance problems and efforts to correct them are necessary if an employee fails to improve and must be let go. But it is also necessary to prove that the employee knew about the difficulties. The legally defensible appraisal system can be used to demonstrate that the employee knew of the problems and was given the opportunity to correct them.[2]

Standard-Based Appraisal: A Long-Range Target

This section discusses a performance appraisal system that, in the view of the authors, represents a highly functional, realistically structured method of evaluating and documenting actual employee performance. Because it is not in common use in corrections, this discussion might strike some as overly long. But this method dovetails so well with many of the fundamental management principles espoused throughout this book that the following material should be taken (if for no other reason) as an example of how to practically apply them.

The difficulty often encountered in the setting of performance standards and the sheer volume of work sometimes involved in this process should not be taken lightly. Going to the standard-based appraisal on an organization-wide basis requires that performance standards be developed for the majority of the tasks in every active job category in the organization. More often than not, this first requires that the organization's job descriptions be updated and revised so that most of the individual task descriptions lend themselves to the development of standards. It also means that the organization must adopt (which usually means install from scratch) some form of work measurement system. However, with sufficient effort, it is usually possible, for any specific position, to come up with several measures that are applicable to various aspects of each employee's work.

Effective job standards ordinarily reflect a concern for four major dimensions of an employee's work: (1) quantity, (2) quality, (3) use of time, and (4) use of financial resources (cost). These four areas are, of course, interrelated—time is money, quantity is related to time, and so on—but it is nevertheless possible to focus performance standards in each of these four key areas. An employee performance standard may take any of the following general forms:

1. Quantity

    Number of reports, inmates, or other items processed

    Number or percentage of occurrences of specific types of acts

2. Quality

    Number of errors, repeats, or rejections (as in reviewing clerical output)

    Employee turnover rate (as could apply in evaluating a supervisor)

3. Time

Deadlines missed, turnaround time (actual as compared with desired)

Acceptable work accomplished within a time unit

4. Cost

Amount of cost savings or cost per inmate (as in food service operations)

Cost per item (as in production in correctional industries)

Budget variance, both bottom line and line item (again applying to the appraisal of supervisors, to include things like overtime management).

Consider an example of a performance standard that relates to quality of work. Suppose an acceptable error rate for a particular activity has been determined to be four mistakes per month (each necessitating, of course, rework or repeat). Suppose further that the normal anticipated range of error has been determined to be three to five mistakes per month. The performance standard for the particular activity is set at three to five errors per month, so the individual who generated three, four, or five errors in a month for this activity may be said to have met the standard.

Consider another example, one of the very few that may be applicable to all employees in an organization. Assume that an organization that allows its employees up to 12.0 paid sick days a year has generated a long-run average of 7.2 sick days used per employee per year. Further, suppose that this organization, wishing to provide a modest incentive to reduce sick-time usage, establishes (perhaps arbitrarily but certainly realistically) that 6 days per year would be an acceptable level of sick time usage. Again, allowing some flexibility about the desired performance target, it might be decided that 5 to 7 sick days represents standard performance relative to sick time usage, less than 5 days' sick time implies that the standard has been exceeded, and more than 7 days' sick time indicates that the standard has not been met. This may be a valid standard to use. However, one would not be surprised if an employee union would resist this proposition, saying that the amount granted by statute is the standard to which employees should be held, rather than some lower number.

It will not always be possible to attach a measurable standard of performance to every last task on an employee's job description. There is no way of establishing that the average performance for a correctional officer will involve finding an average of 5 make-shift knives for every 100 cells searched, for instance. For most employees, however, it will in fact be possible to come up with performance standards that measure most of the major tasks an employee performs.

While one should be careful not to make supervision a counting exercise (and certainly should not involve artificially devised and functionally unnecessary tracking systems), methods of this type can be useful in some job categories. Such a standard might be the number of cells searched for a correctional officer (since normal procedure would be to log all cell searches). In a records office it could entail the number of sentence computations performed in a week, again something that is normally tracked and easily checked.

It should be noted that the preceding simple examples used only three outcomes in comparing performance with a standard. The employee either failed to meet the standard, met the standard, or exceeded the standard. Little else is needed, least of all the judgmental variations suggested by the rating-scale approach and several of the appraisal ratings appearing in Exhibit 14-2. Rarely are more than three simple outcomes required, and often it is possible to rely on only two outcomes. Many employees have job responsibilities, the fulfillment of which can be assessed simply by noting whether they did or did not get accomplished. In some cases there are only two outcomes—*yes* or *no*. Did the officer search at least five cells on each shift and were those searches logged in the unit log? *Yes* or *no* tells the story.

The essential intent in using performance standards in appraisal is to quantify the outcomes of performance as much as possible and thus reduce the necessity for managerial judgment to the lowest possible level. This can be accomplished only when appraisal is not based on general personality or performance characteristics but is based rather on specific job responsibilities and standards of performance for the fulfillment of those responsibilities.

■ Constructive Appraisal

There is another performance appraisal method available in addition to those already discussed—the constructive appraisal approach. This is not a specific system but rather a broad approach that may be used in place of a more formal system under certain circumstances or perhaps used as part of or in support of an existing system. However, before an approach such as this is used in any formal way as a supplement to an official evaluation system, the supervisor must be sure that there are no labor-management, statutory, or regulatory constraints to doing so. Where there are such bars to broad-scale implementation of a new system of this type, it may be possible to do so on a pilot basis.

The constructive approach to performance appraisal involves those employees who are capable of setting goals for themselves and thus helping determine the basis on which they will be evaluated. This approach is workable in most institutions and is particularly applicable to managerial, professional, and technical personnel. It is also applicable to most other employees who have a reasonable degree of self-determination or control over their immediate work environment and the order and manner of performance of their tasks. For instance, many records staff, business office and finance employees, and even many secretaries and clerks can be appropriately evaluated using the approach.

The four elements of constructive appraisal are discussed next.

Job Description Analysis

To set the stage for future activity, the employee is asked to analyze the job description independently and indicate where it should be expanded, contracted, or altered in any way. While the employee is doing this, the supervisor should also be reviewing the job description. They meet to discuss all elements of the job and work on correcting or modifying the job description until they jointly agree on job content and on the relative emphasis each part of the job should receive. This step allows the supervisor and the employee to come to an agreement on what the employee should be doing. As a by-product of this process, the department is assured of having job descriptions that are as current as they can reasonably be made.

Employee Performance Objectives

Working independently, the employee develops a few simple targets or objectives covering some key aspects of the job. Within reason, these objectives should have the following elements. They should be manageable, that is, attainable through the employee's own effort and not dependent in any way on forces beyond the employee's control. They should be realistic, that is, attainable within a reasonable period of time. And they should be challenging, representing at least a modest improvement over past performance.

These objectives should be expressed in a few simple statements that cover several aspects of the employee's job but do not necessarily try to cover everything. They should simply constitute a modest plan of improvement that the employee will pursue while still concentrating primarily on day-to-day activities.

The objectives suggested by the employee should be as specific as possible. Ideally an objective will embody statements of what, how much, and when. An objective formulated by a secretary may look like: "To reduce typographical errors coming (what) by at least 50 percent (how much) before the end of the year (when)." Another example could be a caseworker developing an objective that states, "To submit all classification studies on new inmates to the chief of classification within 30 days of the inmates' commitment to the institution."

Negotiate Objectives

Once the employee has developed the objectives, these should be discussed with the supervisor. The supervisor should make no effort to impose any personal objectives on the employee, while also assuring they remain consistent with the overall objectives of the department and the organization. Within those boundaries, it is the supervisor's job to help the employee keep the objectives realistic, manageable, and challenging.

Once the supervisor and employee have agreed on a program of objectives, this program should be committed to writing, perhaps in a separate memorandum. In some systems it may be possible to incorporate them into the employee appraisal form.

Discuss Results

Periodically the supervisor and employee should meet to discuss where the employee is in terms of progress toward the objectives. This will not necessarily occur at the time of the next scheduled performance appraisal. Some objectives may be realizable within 2 or 3 months and some may take 2 or 3 years to attain. This approach usually requires the supervisor and employee to get together on the subject of employee performance more often than a rigid performance appraisal program would require.

Certainly the employee will not always achieve the objectives set. Some objectives will be exceeded; some will barely be approached. However, the most important facet of this appraisal approach is the total process of involving the employee in setting objectives, working to attain them, and working with the supervisor to analyze the differences between planned performance and actual results. Again, it is evident that the entire structure of this method lends itself more to smaller departments rather than large ones.

Under the constructive appraisal approach, the employee knows well in advance the basis for evaluation. Having participated in establishing the objectives, the employee is better able to understand the goal to be attained. Also, having participated in establishing the objectives, the employee is usually more willing to work for improvement.

This appraisal approach also encourages the employee and the supervisor to come to complete agreement on the content of the employee's job. This is no small accomplishment in many departments. That is because without such agreement, it often appears that a job exists in three distinctly different forms: (1) the supervisor envisions the job in one fashion, (2) the employee sees it another way, and (3) it exists "on paper" in perhaps a somewhat different form.

Most importantly, constructive appraisal takes place entirely within the context of the supervisor-employee relationship and can serve only to strengthen that relationship. More opportunities for growth and job satisfaction are created. Moreover, appraisal's most valuable function is being performed—in that it is helping to create the climate necessary to support open communication between supervisor and employee.

The more often constructive appraisal contacts occur between employee and supervisor, the stronger the supervisor-employee relationship becomes. When the ideal employee-supervisor relationship exists, performance appraisal becomes a mere formality, because under these conditions both supervisor and employee know fully where the employee stands at any given time.

■ The Appraisal Interview

A great deal of what was said about the mechanics of interviewing in Chapter 9 is applicable to the formal performance appraisal interview. The interview should occur on time, that is, when it was scheduled to occur or when the employee was told it would occur. The setting should be one that allows for privacy, freedom from interruptions, and free and open discussion. The appraisal interview should focus on joint problem solving.

■ Living with an Existing System

More often than not the particular approach to performance appraisal used in the organization is decided elsewhere. A manager may be using an appraisal method that was in place before arrival in the agency. It may be necessary to use an appraisal system that was designed and implemented by others. In such cases it was done in a personnel administration section, sometimes at the state level rather than within the department of corrections itself.

Many performance appraisal systems make use of some variation of the rating scale method. If that is the case in your agency, accept what has been said about the pitfalls of checklists (especially those that require a manager to render personality judgments). Be aware that you are required to operate in risky territory. Try to make the best use possible of the system you are required to use, and whenever possible to have solid behavioral or performance-based data to back personal judgments.

Use the system because you must, but make every effort to use it wisely and fairly. Awareness of the pitfalls of standard appraisal approaches, such as the halo effect, central tendency, the making of unqualified judgments, and others, is a large part of the protection you need to keep you from falling into these traps. Regardless of the system used, make every effort to emphasize performance or production (that is, results) rather than personality traits.

Even within the confines of an existing system it is possible to get away from scales and boxes, at least partly, by going beyond checklist requirements. With supervisory and professional employees it may be possible to use constructive appraisal along with the checklist method. These employees may not be covered by a specific managerial appraisal system or be subject to a collective bargaining agreement that specifies a particular appraisal system. Even with lower-grade-level employees or those doing repetitive manual or clerical jobs, it may be possible to open the appraisal process to some degree of input.

Consider asking some of your employees one or both of the following questions:

1. What do you do now that you believe you could do better?

2. How would you change your job if you could?

You may get nothing in response to these questions, but at least you will have made the effort to let someone know you are interested. However, you may get far more than you could ever hope for in terms of positive suggestions. Some of the best suggestions for improving performance in these more repetitive tasks come from the people who do the jobs day in and day out. After all, regardless of how the supervisor may view the job from the outside, when it comes to the inside details there is no one who knows more about the job than the person who does it.

If you are permitted to do so, consider also the use of self-appraisal along with the existing performance appraisal system. Give the employee a blank copy of the appraisal form, and ask for a self-rating to be brought to the appraisal interview (so you do not see it prior to generating your own evaluation of the employee).

There are, of course, some weaknesses in self-appraisal. Some employees use the opportunity to "ego trip," giving themselves high marks in many areas. However, in practice most employees tend to be more critical of themselves than are their supervisors. In any case, the process will help provide some insight into an employee's self-concept.

Self-appraisal also gives the manager insight into employee strengths and weaknesses, both as observed by supervisor and perceived by employee. The process can also suggest where manager and employee might best concentrate their efforts at joint problem solving.

Take the case where the rating system in use calls for assessment of the employee on twelve characteristics. Suppose the manager and the employee are reasonably close together in their independent assessment of nine or ten of these characteristics. Then in all likelihood the two or three characteristics on which there is significant variance constitute the most potentially productive starting ground for joint problem solving.

The agency's performance appraisal system and how a manager uses it may relate directly to the general philosophy of leadership in the organization. If permitted the latitude by statute or policy, autocratic leadership will usually perpetuate the use of a highly structured, rigid appraisal system. This will leave little room for constructive appraisal or employee input in any form. Conversely, a more participative leadership style may encourage appraisal methods that are correspondingly more open to employee input and involvement. Regardless of the organization's formal appraisal system, however, within the manager's own department there usually is some freedom to back up the system with some constructive and more individualized steps, and turn appraisal into a more growth-producing process.

■ A Simple Objective

Performance appraisal systems are intended to improve performance in the job the employee now holds and to develop the employee for possible promotion. However, consider an employee who performs (and perhaps has been performing for years) a simple job in quite acceptable fashion, leaving little room for improvement and no opportunity for growth or advancement. Even in this limited set of circumstances performance appraisal has a simple but still extremely important objective—to encourage the employee to continue delivering the same acceptable performance.

Thus, performance appraisal is essential to employees at all levels. Proper appraisal stimulates improvement, encourages growth and advancement, and conveys appreciation of individual effort.

EXERCISES

Exercise 14-1: It's Review Time Again

"Well, Jack, I'm sure you know why you're here—it's performance appraisal time again. I want you to know that I've seen a lot of good work coming from you these past 12 months—well, 14 months really, since we're a little off schedule as usual. I appreciate it, and I'm sure the warden appreciates it, too. There's always room for improvement, of course, but let me hit the good stuff first.

"Your output has been great, and I'm especially satisfied with the way you tackled the special classification project when we got that big transfer of inmates from the reformatory. You showed plenty of good judgment in the decisions you made and in your recommendations.

"There are a couple of things that bother me, however. But I know I can speak straight from the shoulder. Your aggressiveness is still something of a problem. I can think of two, maybe three times when I've had to work real hard with the captain and lieutenants to get them calmed down after you were interviewing inmates and didn't get the inmates back to the unit in time for count. It wasn't so much that you forgot to call in the out counts as it was the way you got so defensive when the shift supervisor called you about it. I'm sure you'll agree that tact and diplomacy aren't your strong suits. I point this out because your lack of sensitivity to working smoothly with other people isn't going to do you any good if you're thinking about moving up someday.

"And another thing . . ."

Instructions

Analyze and critique the foregoing "opener" of Jack's annual appraisal interview. Be especially sensitive to:

- Apparent "system" weaknesses
- The rendering of personality judgments
- The treatment of the causes and results of behavior

Exercise 14-2: Ms. Parker's Appraisal

This is another role play, with roles assigned in accord with the following context:

Mr. Haskins, an ambitious, relatively young physician's assistant, was recently promoted to administrator of a mid-sized correctional hospital. Success oriented, he recognized the importance of running an efficient medical operation in the prison.

Haskins inherited a subordinate, Ms. Parker, head nurse of the secure medical/surgical ward. Having worked at the facility for 7 years and proven herself to be quite capable, Ms. Parker required minimal direct supervision. She was considered a competent manager in her own right. Staff assigned to her section were loyal, stable, and highly motivated. They worked well together and seemed to emphasize quality care.

Ms. Parker and her staff were acutely aware of Mr. Haskins' promotion. Within a few days there were frequent references to Haskins as "the new kid on the block" and "the snoopervisor." Ms. Parker quickly came to resent the flurry of questions, criticisms, and suggestions that seemed to come out in every discussion she had with Mr. Haskins.

In formally appraising Ms. Parker, Haskins made some critical comments. While not complaining specifically about Ms. Parker's work, Haskins rendered some harsh judgments concerning "negative attitude." Ms. Parker was referred to as "resistant," "sarcastic," and "irritating." Their concepts of supervision clearly differed markedly, and there was reason to conclude that a basic clash of personalities was making itself evident in the performance appraisal itself.

The above information reveals the potential for turning the appraisal interview from a learning situation into a conflict. Your task is to consider the possibility of avoiding the personality clash and viewing the meeting in a way that would facilitate a positive approach to appraisal.

Mr. Haskins: You honestly believe that Ms. Parker is automatically resistant to change regardless of the nature of a particular change. She truly strikes you as irritable and generally sarcastic. You realize, however, that some of her behavior may result from a combination of factors, including your "new ideas" and your relative youth.

Ms. Parker: You have been distressed by the steady stream of communications from Mr. Haskins, including many items you consider unnecessary, nitpicking, and change merely for the sake of change. You have worked in this institution for years. You feel you know a great deal about quality inmate-patient care and that you have a great deal to contribute. You would welcome a setting that gives you the opportunity to present what you know in such a way that it would be considered fairly.

Mr. Haskins and Ms. Parker: Both assume they have already had one tentative appraisal contact during which many of the aforementioned hard feelings came out. They are ready to try again.

The role-play participants are to conduct an appraisal meeting intended to establish a desirable new beginning for the relationship between Mr. Haskins and Ms. Parker. It is suggested that the participants consider the positive approach to appraisal and attempt to reach agreement on two or more sample performance targets. Participants may use tasks from their own areas of responsibility in the appraisal discussion. They may even move the setting from nursing to some other functional area to accommodate participants' backgrounds, as long as the two primary role players agree in advance on what is or is not included.

The remaining members of the group should take either side of the appraisal in a discussion. (It works best if group members "choose sides" before starting and do not change during the exercise.) Afterward, all participants should critique the results of the role play in general discussion.

ENDNOTES

1. Schuster, Michael H. and Christopher S. Miller, "Performance Appraisal and the Age Discrimination in Employment Act." *Personnel Administrator,* 29, no. 3 (March 1984): 48.

2. McConnell, Charles R. *The Health Care Manager's Guide to Performance Appraisal.* Maryland: Aspen Publishers, Inc., 1993, p. 207.

Criticism and Discipline: Guts, Tact, and Justice

15

Chapter Objectives

- Establish the need for "rules and regulations" in the operation of any organization.
- Examine the nature of criticism and offer some self-improvement guidelines for taking criticism and subsequently criticizing others when necessary.
- Describe the characteristics of properly delivered criticism.
- Describe the nature of appropriate discipline and introduce the need for progressive discipline in the organizational setting.
- Establish the primary purpose of disciplinary action.
- Distinguish between problems of conduct and problems of performance and describe how each should be addressed.
- Provide guidelines for the use of fair and effective disciplinary action.
- Establish the importance of proper documentation of disciplinary action.

Indifference is probably the severest criticism that can be applied to anything.

Ann Schade

I never gave them hell. I just tell the truth and they think it's hell.

Harry S. Truman

FROM THE INSIDE

A Lesson in Gender Issues

The chief of classification at the penitentiary supervised a clerical pool of five women. As allowed by procedure, from time to time they went into the inmate contact areas of the facility, but for the most part the five of them worked in a secure room off the main corridor of the institution. One of the five was an attractive young woman in her early 20s—we'll call her Sara.

The chief spent little time in the pool area itself, but did notice that often when he was there one of the caseworkers, Jack, recently divorced, seemed to be hanging around Sara's desk a lot. This situation seemed to the chief to be aggravated by Sara's tendency to wear scoop-neck dresses and blouses that sometimes could have used an extra button. Sara was taking up a lot of Jack's time and attention, and it wasn't always work related.

But that wasn't the only problem. Word started getting around that when Sara made the occasional trip into inmate contact areas that she was drawing unusual attention. She was good looking enough as it was, but her outfits (perfectly acceptable in free-world work settings) were making waves in the population.

The chief called Sara into his office for a chat. He tactfully told her that she should consider that she was working in what was essentially an all-male setting. He suggested that she consider choosing clothing that was a bit more modest, and asked if that would be a problem. She said that it wouldn't be, and thanked him for the advice. The ripples in the population slowed down, and Jack spent more time with inmates and less time peering over Sara's desk. (The chief had had a talk with Jack about his time priorities as well.)

In today's world, the chief might have gotten into trouble handling this situation as he did. To avoid problems, he probably would have called on a female supervisor from some other department to talk to Sara. He would be right to do so today, because times have changed. But in those days, there was less sensitivity to gender-related issues, and this approach worked and was not offensive to Sara.

The latter assertion can be made confidently because as it turned out, Sara showed career promise beyond her capabilities as a secretary. She moved up in the organization, went into personnel management, and later into training. Years later, she and the chief (also in another job) were assigned together again, and she would come and chat with him from time to time.

Early in those discussions she related her recollection of the above story. She said that she had appreciated it then, because she was young and naive, and had not thought about her impact on the men around her. She felt that it had been done in such a diplomatic and appropriate manner that she regularly used the episode as an illustration in her training programs. She held it out as an example of how to deal with sensitive gender-related issues, and to correct employee behavior without being overbearing or counterproductive.

■ The Need for Rules

Rules do not exist solely to benefit the organization (correctional or otherwise) in some abstract way, nor are they traps set by management to snare the unwary employee. Neither are rules, as the old saying might suggest, "made to be broken." They exist for good reasons. And in some respects, there is no workplace more driven by rules and structure than the correctional institution.

Rules outline a general pattern of behavior that employees are expected to observe and practice. Ordinarily created with the rights and needs of the majority in mind, rules exist to protect employees in a number of ways. They are guidelines for individual and group behavior and action, and as such they represent the organization's expectations of its employees. And in corrections, first and foremost, rules exist to safeguard the institution and its staff and inmates.

Rules should be reasonable both as to their stringency and as to their number. Because correctional agencies organize and control the lives of individuals who, by definition, are not particularly good at following rules and laws, the regulation of their conduct by institutional rules is a particularly important issue. Today's legal environment certainly requires that the structure of correctional life be spelled out clearly—for staff and inmates.

This need to clearly and minutely spell out the rules and regulations for inmates can carry over into staff issues as well. When organizations give way to the temptation to write a rule or regulation to cover every last contingency, the result is an unworkable bureaucratic structure. Many correctional organizations seem to operate in a virtual policy maze in which it is sometimes impossible for staff to do anything without running afoul of some regulation.

In addition to being reasonable in quantity and strictness, rules should attempt to serve the common good without infringing on individual rights and freedoms. Since nothing remains constant for long, rules should be regularly examined for their applicability to present circumstances and conscientiously updated as real needs change.

For instance, it would be helpful for the "rules of the organization" (perhaps called the employee handbook or the personnel policy and procedure manual) to be thoroughly reviewed for possible changes at least once each year. This review can easily fit into the annual policy review that most correctional organizations conduct for all inmate and administrative management policies.

Just as most prisons provide inmates with a copy of the rules and regulations that apply to them, the employee rules and regulation should be made known to every staff member. The organization should assure that every employee has had the opportunity to become thoroughly familiar with the rules. It may seem unnecessary to mention this, but it has been proven that many employees are not familiar with their own organization's rules. Even in systems that require each new employee to sign a statement indicating receipt and review of the employee handbook, many people will sign without reading the rules.

Rules and their enforcement are but a part of a legitimate concern about criticism and discipline. Many forms of behavior, often involving marginal, questionable, or otherwise hazy aspects of job performance, are deserving of criticism. There are few rules for the supervisor to use in criticizing an employee except rules of the broadest possible kind, or rules of reason or judgment. Even specific rules defining employee conduct (those that can result in disciplinary action if they are not followed) are not always as clear-cut as managers would like to think they are.

The "From the Inside" vignette represents this type of issue. There was no written rule about female staff dress. A problem emerged—one that could have had serious personal safety implications. The supervisor tactfully addressed it in a nonthreatening way, and the employee responded. No formal action was needed and the problem was solved. As a subtext, note that the supervisor also talked to the male employee involved about his workplace conduct. Again, this was not done in a disciplinary context, but one of constructive observation and suggestion. In both cases, this low-key approach worked.

■ Criticism

Anyone who has been a supervisor for any length of time will have found it necessary to directly criticize someone because of conduct or work performance. Also, those with experience in supervisory positions will have found they are sometimes targets, rather than sources, of criticism. The first-line supervisor is in the position of having to take criticism from above for "everything that's wrong" with worker conduct and performance. First-line supervisors also have to take criticism from below for "everything that's wrong" with the organization and its policies and practices.

Think about the last time you were criticized. Try to remember how you felt, how fair or unfair the criticism seemed to be, and how the critic's attitude and words actually struck you. Then, keeping in mind how you believe the criticism should actually have been handled, try to put yourself in the position of the critic. You are likely to find that criticism is no more pleasant to give than it is to receive.

Also, think about the tone of the criticism. Was it conveyed in a rational way, or in the heat of the moment after a mistake or a poorly handled situation came to light? The tone and manner in which criticism is conveyed is critical to how it is received. It also has a strong bearing on how effective it is in bringing about the desired correction in performance.

When you are criticized, learn from it. Learn what you should do and what you should avoid in criticizing others by carefully reviewing how you have been criticized under certain circumstances and how you reacted to that criticism. Learning to take criticism is the first step toward learning how to criticize others appropriately.

Taking It

When you are criticized, whether by an employee, your supervisor or another member of management, or someone from outside the institution, remember the following rules for accepting criticism.

Keep Your Temper

Make up your mind that no matter what is said you are going to remain calm. Do not jump to conclusions based on the possibly harsh, uncomplimentary, or unfair remarks you may be hearing. If you allow your critic to "get to you," you will experience emotional reactions that will serve only to diminish your true listening capacity and hamper your ability to deal effectively with what you are hearing. Whatever you do, stay calm.

Listen Completely

When criticized, a person's natural temptation is to start developing a defensive stance while the critic is still speaking. To strive for an open mind and the ability to listen completely, try imagining yourself in the position of a neutral third party. Act as though you were hearing about someone else, not about yourself.

Consider the Source

Frequently, this cliché is suggested as a way to shrug off anything coming your way. However, this expression often retains a degree of validity. Ask yourself the following questions:

- What are the person's credentials? Is this person qualified to criticize in this instance?
- What are this person's motives? Are there any vested interests involved?
- Are the comments valid, or do I just happen to be a convenient target?

Whenever you are criticized, especially by your supervisor or another member of management, try to determine whether the person is genuinely trying to help you improve.

Evaluate the Criticism

Try to judge whether your critic had all the facts available and whether, quite simply, the criticism made sense. Search for positive suggestions in the criticism, and try to determine whether you will benefit by following your critic's advice.

Keep It in Perspective

Some people (and it does not pay for a supervisor to be among them) do not take criticism well. They can be absolutely crushed by a few harsh or critical words. Realize, however, that criticism, no matter how harsh or ill directed, is not the end of the world. More often than not things quickly go back to "business as usual" and interpersonal relationships drift toward their original state. Criticism, unless carried to destructive extremes, is simply words strung together. As Mark Twain said, "If criticism had any real power to destroy, the skunk would have been extinct long ago."

Follow Up

Having evaluated criticism directed at you, you can do with it what you wish, depending on how deserved you believe it was. If you can honestly accept what you have heard, then take steps to change your behavior accordingly. However, if you feel you honestly cannot accept the criticism, that it is perhaps misdirected, undeserved, or unduly harsh, you have still gotten something from it. At the very least, you have learned a bit more about the person who did the criticizing.

Giving It

Supervisors are often reluctant to criticize employees. This reluctance may be rationalized as part of an unwillingness to take the chance of hurting others' feelings. However, it is likely that the reluctance to criticize is just as readily attributable to the supervisor's natural resistance to performing an unpleasant task.

Certain kinds of behavior may deserve criticism and may perhaps even be deserving of disciplinary action—criticism plus warning or punishment. The reasons for criticism are generally incidents of misconduct or poor performance or the manifestation of a poor attitude.

Misconduct hinges largely around the existence and application of rules; its presence is often relatively clear cut. Poor performance, on the other hand, often relates to assessments of quality and other judgments that are at least partly subjective. So the line separating what is deserved from what is undeserved is likely to be indistinct. Problems of employee attitude, especially if not immediately reflected in misconduct or poor performance, are the most difficult to deal with and call for the greatest care on the supervisor's part.

Criticism is necessary, but only if it is truly constructive. Indeed, few would deny the value of receiving "constructive criticism" when it is due. However, criticism is still criticism, and regardless of the "sugar coating" afforded by the word *constructive* it is still likely to have a bit of a sting to it.

First and foremost, criticism should always include guides for correction. Simply telling people what they have done wrong without suggesting how they might do it correctly amounts to no more than placing blame. Criticism should always focus on the problem—on the results of a person's behavior and not on the person who created the problem. Beyond simply suggesting corrective measures, criticism should also allow the supervisor and employee to work together to develop a new approach jointly. Criticism, properly applied, can be used to strengthen the joint problem-solving aspect of the supervisor-employee relationship.

Criticism should also be:

- **Timely**—It should be delivered as soon as possible after the behavior occurs or the results are discovered. "Saving up" criticism for some special upcoming contact (for instance, a performance appraisal interview) is ineffective and quite possibly destructive.
- **Private**—Never criticize an employee in the presence of other people. The words that pass between the two of you are no one else's business.

Even when you walk in on a situation in which a number of people are present, take whatever steps are necessary to stop what is going on and then meet with the problem employee in private before delivering individual criticism.

- **Rational**—Never criticize in anger. Take the risk of blunting the impact of criticism by allowing time to cool down and assess the situation calmly before criticizing. Often when anger has dissipated a person will be able to see dimensions of the problem that were previously hidden by an agitated emotional state.

Consider some of the more common causes of incidents, errors, and omissions leading to criticism of employees: lack of adequate job knowledge; poor understanding of management's expectations; inability to perform as expected; so-called attitude problems. Recognize that each of these causes embodies some reflection of the supervisor's responsibility for the knowledge, understanding, capability, and attitude of the employee. It would be beneficial, then, for the supervisor to begin considering each situation apparently deserving of criticism by asking: Have I fulfilled all of my responsibilities to this employee?

■ Discipline

A bit of research into the origin of words will reveal that *discipline* comes from the same root as the word *disciple* and as such actually means "to teach so as to mold." Originally, teaching was the key to discipline—to shaping or molding the disciple. However, most people have come to think of discipline, and thus disciplinary action, as punishment.

The true objective of discipline should not be punishment. It should be correction. Thus, one of the primary requirements of disciplinary action is that the employee be afforded the opportunity to correct the behavior that prompted the action.

In all but the few legitimate instances in which immediate termination is appropriate, corrective action must be progressive. That is, it must follow through a number of steps in which subsequent similar infractions are addressed with increasing severity. Indeed, most government agencies have embodied progressive discipline into their personnel management regulations. Some even have established tables of sanctions that include recommended action to be taken for first, second, and third infractions.

At each step in the progressive process two pieces of information must be made clear to the employee: what the person must do to correct the problem, and what may follow if the problem is not corrected. Throughout the process it is essential to view the sequence of occurrences exactly as it would be later viewed (and often is viewed) by outside entities such as a labor-management review panel or even a judge. Envision the question, "Was the employee given every opportunity to correct the offending behavior?" Be ready to answer it in excruciating detail.

In applying corrective action it is also necessary to make a distinction between problems involving conduct and those involving performance. Although traditional progressive disciplinary processes may work well for conduct issues, such processes are not appropriate for performance problems. The employee who has broken no work rules and violated no policies is not appropriately handled with a progressive disciplinary process designed to deal with work performance that falls short of standard or expectation.

The supervisor always needs to ask: Is the employee not capable of performing as expected under present circumstances? If the employee cannot, for one reason or another, perform as expected, it is not a discipline problem. To treat it as one is to burden the process with a layer of negativity that serves only to impede correction.

Conduct Problems: Traditional Progressive Discipline

Counseling

The first step taken to address a specific kind of errant behavior should be counseling. One to one, supervisor to employee, the employee should be told what was done that was wrong and why it was wrong. The supervisor should explain what the organization's rules are concerning this behavior (with specific use of handbooks, policy manuals, and other written references), and what the possible consequences of this kind of behavior are. The discussion should end with delineating the period of time within which correction is expected. All of this needs to be accomplished without reference to any kind of "warning." It is simply an important, job-related discussion between supervisor and employee. And of critical importance, any such counseling session should be thoroughly documented by the supervisor in notes retained in departmental files.

Oral Warning

Repeated errant behavior following counseling should be addressed using the more formal early stages of the progressive discipline process. This next stage in most systems is the oral warning.

The oral warning should be documented by the supervisor, preferably on a form created for that purpose by the agency. If it is documented, is this not actually a written warning? It certainly may seem so, but the difference between a written and oral warning lies in what goes into the employee's personnel file. The record of an oral warning should be retained in department files. It should go into the official personnel files only as part of a subsequent warning for the same kind of behavior.

If it is truly to be an "oral" warning, why document it at all? Because the oral warning is a step in the progressive disciplinary process, and when an employment relationship breaks down and legal problems result, it can become necessary to provide evidence that each step in the process was followed.

Written Warning

The written warning follows in turn as necessary, with this documentation automatically included in the employee's personnel file. An employee whose improper behavior has not been corrected following counseling, oral warning, and written warning is in a position in which failure to change will lead to loss of income, via suspension and perhaps loss of employment. By this stage the supervisor and employee have been together on the subject of the employee's behavior problem at least three and perhaps four, five, or six times. It is time for the supervisor to bring other organizational resources into the process.

Before Suspension

At this step, each agency will have different procedures to follow. However, if no specific steps are required prior to suspending an employee, then the supervisor may at this point consider referring the employee to one of three available sources of further assistance: the employee assistance program (if the agency or institution has one), a medical resource for a physical examination, or the personnel department.

If, in any of their numerous contacts, the employee has given the supervisor reason to believe he or she may be experiencing mental health problems of any kind, a referral to the employee assistance program may be in order. If a physical problem related to the employment situation may be involved, then a referral for a physical examination may disclose reasons for the poor performance. If the problem appears to be unrelated to these two factors, the referral should be to personnel or perhaps to an employee relations specialist or employee ombudsman if the agency has such a position.

This referral puts the employee in contact with someone who can possibly point the way toward resolution of some underlying problem. Also, a knowledgeable person other than the supervisor is brought into the process, and this new participant may be able to get through to the employee where the supervisor could not. Finally, this step gives the employee one more distinct opportunity to correct the problem behavior.

Suspension and Discharge

If the referral step is unsuccessful, suspension without pay and eventual discharge may follow, as necessary. However, a well-functioning referral program for employee behavior problems will significantly reduce the use of the clearly punitive steps of suspension and discharge.

When it appears there is cause to discharge an employee, the manager should take the case to the personnel department for thorough review before taking action. In today's legal environment, virtually all government agencies require human resources or administrative review and concurrence in such cases. In some cases, legal counsel may also be involved. A review of this type would be prudent in the private sector as well. This review should be aimed at determining whether all bases have been covered from a statutory and regulatory perspective, and whether the record clearly demonstrates that the employee was given the opportunity to correct the inappropriate behavior. Because of the time required, this review serves another extremely important function. It assures that no employee is ever fired on the spot or otherwise terminated in the anger of the moment.

This is not to say that occasions arise when an employee should not be allowed to remain on the job. Some severe infractions must of course be dealt with as they occur. An employee caught bringing drugs into the institution should not be allowed back in. A staff member found stealing government property, embezzling funds, or committing some other outright criminal act should not be allowed to be in the workplace until the matter is adjudicated. However, an immediate firing is never the answer. The offending employee should instead be removed from his or her post, sent home, and placed on home duty, administrative leave, or indefinite suspension pending resolution of the matter.

Not all infractions will require the application of all the foregoing steps. A mild problem, such as tardiness (within a few minutes of starting time) may, if it becomes chronic, eventually require all of the steps previously described. A more serious infraction (such as a tower officer caught sleeping on post) may call for a written warning or suspension on the first violation and discharge on the second violation. The organization's personnel department ordinarily provides guidance for determining the severity of disciplinary action for specific infractions.

Performance Problems: "Warnings" Not Applicable

Substandard performance can be simply described as the production of unsatisfactory results that prevent an employee from attaining or maintaining the job standard. The job standard may be further defined as the acceptable level of output achieved by employees in the same or similar job classification, or the standard level of results. These should be defined in advance by the supervisor and clearly identify the quantity and quality of output expected.

It is inappropriate to apply a progressive disciplinary process to an employee who is exhibiting substandard work. The substandard performer has not "broken the rules." To group this person with the supposed rule breakers is to lend a negative aspect to a process that must be as positively oriented as possible.

For the Newly Identified Substandard Performer

The supervisor's primary objective should not be to get rid of the substandard performer, as is frequently the case, but rather to show the substandard performer how to perform acceptably. The supervisor should:

- Review the job standard with the employee, ensuring that the standard is known and understood and, ideally, that the employee accepts this as a reasonable expectation.
- Counsel the employee, developing an action plan specifying what must be done to attain an acceptable level of performance and a timetable for doing so. This step should be documented thoroughly (as any counseling session should be documented) with a copy of the complete plan supplied to the employee.
- Conscientiously monitor the employee's progress against the improvement plan, providing assistance as necessary.
- Remove apparent obstacles to the employee's success when possible and make reasonable accommodations to enhance employee performance.
- If this process does not correct the problem within the agreed-upon time period, it should be repeated in essentially identical fashion, including the creation of a mutually acceptable plan of correction. This plan may include necessary modifications from the earlier plan, based on what was encountered in the process, as well as a new agreed-upon target date.

The Important "Plus"

The second time the employee is taken through the corrective counseling and instruction, the process should include the same referral process as for a progressive discipline case. The advantages of doing this are that a new person who can possibly help is brought into the process. In addition, the employee is given the chance to address a possible underlying problem that may be causing the apparent performance problem.

How Many Times?

Depending on how much, if any, improvement is noted from one time to the next, the foregoing process might be applied three or four times. Finally, when every reasonable opportunity to improve has been extended multiple times without lasting change occurring, a last-ditch deadline should be established. When this deadline arrives, employment should end unless all conditions of the plan of improvement have been met. Of course, a wise supervisor who has reached this stage has already been in touch with the personnel department to advise them that a problem situation is developing, and that adverse action may be the outcome.

Dismissal, Not Discharge

When it is necessary to release an employee for reasons of substandard performance, the action should be identified as a dismissal for failure to meet job standards, and not as a discharge. The distinction is important. A discharge for cause is a "firing," and a person so terminated has lost employment through inappropriate conduct. In most parts of the country that individual is not eligible for unemployment compensation. However, an individual who is dismissed for inability to meet the standards of the job is not considered wholly responsible for the termination. Consequently, the action is regarded more as a layoff than a discharge and the person so terminated is usually eligible for unemployment compensation.

■ Guidelines for Fair and Effective Discipline

Be Reasonable

At all times strive to keep the severity of disciplinary action consistent with the infraction. Many agencies have established specific guidelines for applying discipline in specific categories of offenses. Others maintain data on the sanctions applied for given charges brought against an employee; try to access that information. In the unlikely event there is no external guiding structure, remember the particular action associated with similar infractions in the past. Try to avoid any unjustified variation in the scope of the punishment involved.

Avoid Making Examples

Every employee deserves fair, consistent treatment relative to all other employees. Resist the occasional temptation to make an example of an individual for the sake of discouraging the same kind of behavior by others. Making examples serves only to create fear and resentment, and it destroys the effectiveness of discipline. The rationale also probably is indefensible in an administrative hearing or in court, should an employee challenge the action.

Follow the Rules

The rules in the employee handbook will not cover every situation, but they will apply to many occasions when disciplinary action is warranted. Use the rules and use them consistently. What applies to one employee should apply to all others for the same offense. Use written warnings when the system calls for them. Do not simply "slide over" an incident deserving a warning without actually issuing the warning. Omissions of this type can return to later haunt the slipshod manager.

Respect Privacy

Do not meddle in the life of an employee outside of the institution. What an employee does during off hours is ordinarily none of a supervisor's business. The only time a supervisor can be concerned with an employee's private life is when there is reason to believe the person is engaged in something that can harm job performance or negatively affect the reputation of the agency.

Even then the manager must be cautious to avoid violating someone's privacy. It is vital to ensure there is a "nexus," or logical connection, between the questionable off-duty conduct and the employee's role and performance in the workplace. This is often easy to establish for law enforcement-type positions, such as the case of a correctional employee or police officer using illegal drugs off duty. But management cannot assume automatically that every objectionable or unconventional personal act is logically connected with the workplace in a way that would justify some form of personnel action.

Avoid Favoritism

Under no circumstances should a manager favor any employees over others. Undoubtedly there will be some employees who are personally more likeable than others; people are different and those reactions are natural. However, make every effort to assure that disciplinary actions dispensed are consistent among all employees, regardless of personal feelings toward various people.

Act Only on Clear Evidence

Take disciplinary action only when absolutely certain a truly "guilty party" is involved. Hearsay or secondhand evidence is totally inadequate and undeserving of more than passing attention. It is far better to run the risk of allowing someone who deserves punishment to slip through rather than to discipline an innocent person unjustly. Many agencies require some form of investigation and multilayered review to ensure that disciplinary actions are fully supported in fact.

Avoid Dwelling on History

When an incident is over and disciplinary action has been taken, forget it. Do not bring the incident up again. Do not continually remind the employee of what happened. Unless the nature of the misconduct itself requires supervisory follow-up, the only extent to which a manager can justifiably deal with history in most cases of disciplinary action is through the necessary accumulation of written warnings. This practice, of course, must be spelled out in agency rules and regulations.

The Inevitable Documentation

The need to provide documentation has been mentioned repeatedly, even to the extent of calling for documentation of so-called "oral" warnings. It is often too easy to delay the documentation and perhaps forget about it entirely, mainly because documentation is usually not an immediate concern. Most of the time documentation is never used again, whether immediately or later. But when it is needed to defend a personnel action, it is "worth its weight in gold."

After most discipline-related actions take place, the involved parties go forward to other concerns and never look back. Occasionally, however, an employment action is challenged, either internally through a grievance or appeal process or externally through a labor-management proceeding or a lawsuit. And when an action is challenged, all of the documentation related to it in any way is brought to the foreground.

Documents related to performance or conduct problems are used to reasonably establish whether something did or did not occur. They show whether an employee was or was not spoken with about a certain problem. They provide a starting point for establishing the factual basis for the action taken. They can demonstrate whether the employee agreed or disagreed with a certain course of action. They can be used to establish whether the employee was given the opportunity to improve or correct. Human memories fade or become "selective" regarding certain kinds of information, and documentation is essential to either support or refute certain contentions. The importance of the documentation of employment actions cannot be stressed too strongly. In the words of one investigator for an employee advocacy agency, "If it's not properly documented, all signed and dated, we assume it never happened."

■ Guts, Tact, and Justice

Criticizing employees and parceling out disciplinary action in a way that is productive, not destructive, takes courage and tact. In most instances it is normal for the supervisor to feel uneasy. One can go so far as to suggest that a manager should never want to reach the point of becoming completely comfortable with criticism and discipline. Doing so might suggest a callousness that runs contrary to the character of an effective supervisor.

An employee who has done something deserving of criticism or disciplinary action nevertheless deserves full consideration as an individual. The manager involved in the process must continually be aware of feelings—his or her own as well as others'. However, regardless of individual feelings, what the manager says and does must be out of consideration for the needs of the institution, its inmates, and its employees.

The way a supervisor goes about this task is important in a number of ways. First, the employee involved can be improved or destroyed as a valuable resource for the organization. The time and resources expended in training and supervising that person can be swept away very quickly by ill-advised or tactless attempts at discipline. If the employee was salvageable, inappropriately applied discipline can ruin any chances of retaining that employee.

Second, the organization itself can suffer if an ineffectively disciplined employee goes back out in the workplace and begins to sow seeds of discontent. It also is not unknown for staff to engage in actions that directly sabotage operations. This can extend to cooperating with inmates to undermine management credibility. In the extreme, some truly bitter employees have been known to cooperate with inmates to introduce contraband and otherwise compromise security—all because they were bitter about what they felt was a poorly handled disciplinary action.

Finally, there is the impact that discipline can have on other staff. Earlier, it was noted that a supervisor should never use discipline to "set an example" for other staff. However, the subtly different reality of the situation is that other employees will be aware of the discipline meted out to their fellow workers. They will not only calibrate their own actions somewhat as a result ("Wow—if they did that to Bill they'd tag me for doing that too!"). But they also will develop a sense of the overall fairness with which management approaches these issues. That is a major element in staff morale and effectively managing the workforce over the long run.

EXERCISES

Exercise 15-1: Did She Have It Coming?

"That was a stupid thing to do," said Ron Walker, food service administrator at a large, medium-security correctional facility.

"What do you mean?" asked the recently promoted assistant food service administrator, Susan Aldred, flushing noticeably at the words.

"You fiddled around making those bulk vegetable purchases so long that you stalled us right into a price hike. This late in the budget year there's no way we'll recover that much in other purchases. Thanks to you we'll go about $4,500 over budget for the year."

"So I made a mistake," Aldred retorted.

"Mistake? More like a colossal blunder. Forty-five hundred bucks! I don't know whatever convinced you that you know the vegetable market. The way prices have been going you should know you've got to get in and cut a contract fast." Walker shook his head and repeated, "Forty-five hundred!"

Aldred stood and glared down at Walker. "So I slipped, and I know it. In the 2 months I've been in this job I've saved twice that in other areas—how come I don't hear about that?"

"Because that's your job," Walker replied curtly.

"Well maybe I need a new one," Aldred said, and stormed out of the office.

Questions

1. Do you believe Aldred "had it coming" or that some criticism was deserved?
2. What essential element is missing from Walker's criticism of Aldred?
3. How might this situation have been approached to minimize the chances of an emotional interchange?

Exercise 15-2: A Good Employee, But . . .

Lieutenant Rich Patterson was uncomfortable about a personnel action he was considering. He decided to discuss it with Marcia Saenz, another lieutenant. He began with, "I have no idea how I should deal with Pat Nelson. I just don't recall ever facing one like this before."

Marcia asked, "What's the problem?"

"Excessive absenteeism," Richard answered. "Pat has used up all of her sick time virtually overnight, and most of her sick days have been before or after scheduled days off. You know what that looks like!"

"What's unusual about that? Unfortunately, we've got several people who use their sick time as fast as they earn it. And most of them get 'sick' on very convenient days."

"What's unusual is the fact that it's Pat Nelson. She's been a pretty good officer for more than 7 years, but all of a sudden she starts what looks a lot like sick leave abuse. She's used up all of her sick leave in 7 months. And recently, she was out for 3 days without even calling in."

Marcia said, "You can terminate her for that."

"I know," said Rich.

"Especially when you take her other absences into account. You've warned her about them?"

After a moment's silence Rich said, "No, not in writing. Just once, face to face."

"Any record of it? Fill out a disciplinary dialogue form for her to sign?"

"No," said Rich. "I really hated to. I know I should have taken some kind of action by now, but I can't seem to make myself do it."

Marcia asked, "Why not?"

"Because she's always been such a good employee. She's always shown up on time and done what she's told to do, she handles inmates well, and until this came up she was one of our best officers. She's still that way, except for this thing."

Rich shrugged and continued, "I guess what I'm really hung up on is, how do I discipline someone who is usually a good employee, and do it so that it doesn't destroy any of what is good about her?"

Marcia shook her head and said, "Good officer or not, I'd say you ought to be going by policy. That's all I can suggest."

Questions

1. How would you advise Lieutenant Patterson to proceed in the matter of Officer Nelson?

2. Do you feel that Patterson's failure to take action thus far affects his ability to take action now? Why, or why not?

The Problem Employee and Employee Problems

16

In so complex a thing as human nature, we must consider it hard to find rules without exception.

George Eliot

The broad effects which can be obtained by punishment in man and beast are the increase of fear, the sharpening of the sense of cunning, the mastery of the desires; so it is that punishment tames man, but does not make him "better."

Friedrich Nietzsche

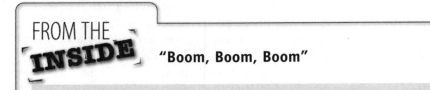

"Boom, Boom, Boom"

It's a maximum-security penitentiary with eight towers. Tower Four is in the middle of the north fence line, and as it happens the officer there has a direct view down the north corridor of the institution as well. The towers are linked by an open intercom to each other and to the control center, so that everyone on the perimeter can know what is happening on other key posts. The officer in Tower Four, Frank Haskins, is in his late forties, average in performance, but has been increasingly outspoken against the administration. He also has personal grudges against a number of other line personnel, including Manuel Contreras.

It is a pleasant summer day; there was rain earlier, but the yard has just opened and a few inmates are straggling out. Tower duty can be boring, and so the officers on those posts tend to pay attention to any interesting diversions.

A sound starts coming over the open intercom. "Boom, Boom, Boom." As if a person is simulating the sound of a gun going off. Tower personnel know it has to be coming from one of the towers or the control center, but no one can identify the voice. Binoculars are picked up, and Towers Three and Five (the corners on either side of Four's fence line) see a strange thing.

Haskins has his M-14 rifle mounted to his shoulder and he is leaning partially out the tower window. He clearly is aiming his rifle at inmates on the yard. As he does so, more "Boom, Boom" sounds come over the intercom, synchronized with re-aiming of his rifle. No bullets. No actual firing. Just the sound of Frank simulating shooting inmates.

Then, before anyone in a tower can get on the phone and report what is going on, Frank turns, so that he is aiming directly down the north corridor window. Manuel Contreras is standing at the door to H Unit, waiting to be buzzed in. And as the other tower officers watch, Frank clearly says, "Boom, Boom, Boom" as he aims directly at Contreras.

■ Is There Such a Person?

Most experienced managers could quickly offer several real-life examples of problem employees over the years. The "From the Inside" vignette certainly would be one. (In the real incident, Frank was relieved of duty at once and eventually given a disability retirement on the basis of mental illness.) Indeed, it would be a rare institution that did not have many problem employees at once, although hopefully not as acute as that one.

In actuality, however, no one knows how many, or how few, problem employees there really are in a given institution. A person may be a true problem employee, but the manager may see only the apparent problem. Also, one ought to consider the manager, his or her leadership style, and any established expectations communicated (or not communicated) to the individual. These, too, may be part of the problem that an individual employee displays.

Usually the problem employee will be neither a very good nor a very poor performer. Rarely can outstanding workers (that is, those exceeding management's expectations in terms of quality of work, amount of work, and interpersonal relations) be considered problems. In addition, the worker whose performance has been chronically substandard and generally unacceptable should no longer be there to present a problem. This presumes, of course, that staff who turn out to be simply unable to do their jobs are properly "weeded out" after being given every reasonable opportunity to learn.

Rather, the problem employee is usually a worker whose performance, both functionally and interpersonally, falls in the mid-range and perhaps a bit short of the average. The problem employee is tough to instruct and correct, tough to motivate, and generally troublesome to handle. There are days when employees in this category make their supervisor wish for nothing but inmate problems to deal with!

Problem employee types sometimes are labeled with terms such as "know-it-all," "wise guy," "blabbermouth," and "complainer." Such terminology tends to zero in on one or two narrow behavioral characteristics and magnify them to the extent that they overshadow the whole person. Also, categorizing a person tends to lead to the assumption that the person is in fact that way. The label thus tends to channel a relationship with the individual along certain narrow lines. It is easy to be led into automatically reacting to certain people in certain given ways.

For instance, you need something from the chief of security but you know that "that stubborn so-and-so" is difficult to deal with. You may readily adopt a hard-line approach in making your request without realizing you are doing so. Conversely, you may know that one of your employees is undergoing personal stress because of a serious illness in the family. Your approach to that individual may be gentler and you may be inclined to give the person more latitude in his or her job performance.

Actually, though, people are all different, and every employee is different. The majority of staff behave in somewhat the same fashion, owing largely to the rules, methods, and procedures they are required to follow in the correctional environment. But this common behavior in no way means that they are alike. Each has a particular personality, and each is a collection of attitudes and feelings that add up to a unique individual.

Some supervisors are unaware of, or perhaps ignore, the differences between staff that mark some as so-called problem employees. They may unconsciously decide that it is easier to stand on the authority of the supervisory position and use orders and ultimatums for handling the problem employee.

The problem employee, however, can be a particularly interesting challenge for the supervisor. Yes, a particular employee may be difficult to get along with. But a manager who learns to get along with one so-called problem employee and skills at relating with people will improve all the way around.

People (whether employees, inmates, or others) cannot be pigeonholed or categorized in any fair manner. Even the use of the term "problem employee" is unfair. Although it is a broad category, it results in grouping people together under that single label. This process often leads to a gradual narrowing of the categories and eventually to stereotyping. In addition to being hazardously unreliable, stereotyping is unfair: It almost always constitutes unwarranted generalization, and as G. K. Chesterton said, "All generalizations are dangerous, including this one."

Be on guard against the tendency to apply labels to employees, especially labels with negative connotations such as *stubborn, grouchy, lazy, undependable,* or *dull.* When they are used, it erects obstacles that may obscure more favorable, but perhaps less obvious, characteristics that are also present. Also, as discussed in the chapter on performance appraisal, when applying such labels managers are rendering personality judgments that most of them are unqualified to make.

It is easy to be unwittingly led into accepting a label generated by someone else. A new supervisor coming in from the outside may get a rundown on all employees from the departing supervisor: "this one is cooperative, that one is a grouch, this one is headstrong and insists on doing things the wrong way, that one is a crybaby," and so on. It is bad enough that managers are inclined naturally to form their own shaky judgments, let alone compounding the problem by accepting another supervisor's similarly shaky judgments. At the very least managers need to form their own impressions of employees through personal observation and interaction. When categorizing or labeling employees, a manager is positioning him- or herself to think of the person as actually being that way, and consequently to expect the person to behave according to that label.

■ Dealing with the Problem Employee

Ups and Downs

Recognize that everyone (including managers) has good days and bad days. The mood of the day, how things have been going, how a person feels physically, and many other factors have a bearing on how a person comes across to others at any particular time. Most of the people who work in any given department have their good days and their bad days. They are just as likely to be subject to the ongoing pressures of the job as they go through the day. And if you have worked in an institution for any length of time, you know that one nose-to-nose confrontation with an inmate can affect the entire day, and not for the good. Make sure to form impressions of employee abilities from longer-term observations and interactions.

As in relations with all employees at all times, try to be empathic. Try putting yourself in the employee's position, and try to appreciate fully why that person feels and acts in such a way. When there are visible signs of problems in an employee's conduct or job performance, seriously consider reasons why others (including yourself) may behave under similar circumstances. What causes coolness, distance, irritability, or other less-than-desirable behavior?

In dealing with the problem employee, recognize initially that direct, forceful attempts to alter behavior will probably fail. For instance, meeting deepseated stubbornness head-on with hard-nosed determination is more likely to be destructive than constructive. Recognize that very often managers cannot change so-called difficult people but that often it is possible to accept and perhaps even utilize their peculiarities.

Time to Be Troublesome

Ironically, just as it is with inmates and work programs that keep them gainfully occupied, one of the keys to dealing with the problem employee is to keep the person constructively occupied. The act of doing meaningful work, with energies directed into obviously useful channels, is by far the best cure for many problem employees. Activity is certainly not the cure for everything, but it can remove some problems, such as irritability, boredom, or frustration owing to inactivity, and it can "keep the lid on" other problems. When people have time on their hands, they tend to dwell on themselves and their troubles. Things that bother employees are more likely to come to the surface and be magnified when they are idle or underutilized.

Whenever possible, use employees in the capacities for which they are best suited. Everyone is a collection of numerous qualities; not everyone does everything with the same degree of success or is equally suited to all assignments. Managers should recognize that the needs of the institution come first and it is often necessary to place people in less-than-ideal assignments. But there is nevertheless a degree of management responsibility to use employees in the manner to which they are best suited. When employees are utilized such that their talents and preferences are applied to best advantage, problems are less likely to develop.

Many employees find time in which to be troublesome because they are not sufficiently challenged by their work. Some simply do not care for what they are doing and indeed there are many tasks in a correctional institution that few people would enjoy doing day in and day out. (Think about the work of a receiving officer, which in part involves searching the private parts and orifices of each inmate who enters and leaves the facility.) Part of the solution to the problem of keeping employees constructively occupied could include offering increased responsibility to those who are able to assume it and rotation of job duties, when possible. This has the effect of spreading around the least desirable tasks, so they have minimum negative impact on any few select persons. Many security departments base their entire staff assignment system on this premise, with highly structured rosters that ensure that all staff work each shift and each post over a period of time.

The Whole Person

In the long run, it is practically impossible to separate the person off the job from the person on the job. To a greater or lesser extent, people bring their outside problems onto the job and carry their job-related problems off the job. With some employees this crossover is minimal. With others, however, a small crisis in either facet of their existence can affect attitudes and behavior in the other. Thus the employee who becomes sullen and withdrawn may have done so either because of some job-related experience or because of something that happened off the job. If someone has become "moody" or "stubborn" or invited the application of some similar label, that alone should cause to wonder what is behind the behavior.

The behavior projected by an employee can be an emotional defense against treatment received on the job. Because people are different (some distinctly so) the same approach will not work with everyone. As a supervisor, make it a practice to remain on friendly terms with everyone—friendly, but impersonal and businesslike—conscientiously trying not to play favorites while avoiding getting "too close" to employees. To some staff this will be appropriate behavior; it will be seen as the mark of a good supervisor. However, some employees will see this same behavior as artificial and perhaps label this as "cold" or "phony" (it is not just the supervisors who do the labeling.)

Make an effort to get to know every employee, openly expressing interest in him or her as an individual and inquiring into his or her personal interests. Again, to some employees this behavior will exemplify a good supervisor. To others, however, it will seem nosy—inquiring into things that are "none of your business." A good manager makes it a practice to circulate about the group during the workday or tour his or her part of the institution, in order to simply show people he or she is there, available, and interested. When done consistently and constructively, this behavior will be accepted by most employees as appropriate supervisory behavior. Others, though, may see it as representing a distrustful attitude because management is constantly "checking up" on them.

Whatever a manager does in making a sincere effort to be a good supervisor, a few employees are likely to react negatively. These negative reactors are likely to be "problem employees."

■ Seven Guidelines

The following general guidelines can be useful as a starting point in dealing with problem employees.

Be a Listener

Make it clear that you are always available to hear what is bothering your employees. Display an open attitude, conscientiously avoiding the tendency to shut out possible unpleasantries because you "don't want to hear them." Many employees' doubts, fears, and complaints are created or magnified by a closed attitude on the part of the supervisor, so your obvious willingness to listen will go a long way toward putting some troubles to rest.

Be Reasonable

Always be patient, fair, and consistent, but retain sufficient latitude in your behavior to allow for individual differences among people. Use agency rules as they were intended, stressing corrective aspects rather than punishment. Apply disciplinary action when truly deserved, but do not use disciplinary threats to attempt to force change by employees.

Be Respectful

Recognize and respect individual feelings. Further, recognize that a feeling, as such, is neither right nor wrong—it is simply there. What a person does with a feeling may be right or wrong, but the feeling itself cannot be helped. Do not ever say, "You shouldn't feel that way." Respect people's feelings, and restrict your supervisory interest to what each employee does with those feelings.

Be Level-Headed

Problem employees are frequently ready and willing to argue in defense of their feelings or beliefs. However, by arguing with an employee you simply solidify that person in a defensive position and reduce the chances of effective communication of any kind.

Be Flexible

If possible, let your supposedly stubborn or resistant employees try something their own way. As a supervisor you are interested first in results and only secondarily in how those results are achieved (as long as they are achieved by reasonable, permissible methods). There is no better way to clear the air with the employee who "knows better" than to provide the flexibility for that person to try it that way and either succeed or fail. In other words, the employee who appears stubborn or resistant may not be so by nature but may rather be reacting to authoritarian leadership. More participative leadership might be the answer.

Be Attentive to Complaints

Pay special attention to the chronic complainers, those employees who seem to grouch and grumble all through the day, and spread their gloom and doom to anyone who will listen. Chronic complaining is, of course, a sign of several potential problems and also breeds new problems of its own. The chronic complainer can affect departmental morale and drag down the entire work group. You should make every effort to find out what is behind the complaining. It may be wise to even consider altering assignments such that a complainer is semi-isolated or at least limited in the opportunity to spread complaints.

Be a Relationship Builder

Give each employee some special attention. The supervisor-employee relationship remains at the heart of the supervisor's job, and each employee deserves to be recognized as an individual as well as a producer of output. Honest recognition as individuals is all that some so-called problem employees really need to enable them to stop being problems.

■ A Special Case: The Dead-End Employee

The "dead-end employee" is discussed in this chapter because this person is caught up in a set of circumstances that can lead to problems. The dead-end employee can go no further in the organization. A move to a supervisory position may not be possible because he or she lacks the basic qualifications. Further promotion and significant pay raises may not be possible because the employee is already at the top of the grade. The supervisor may be limited organizationally from making any significant changes that would enhance job satisfaction for the employee. In short, the dead-end employee is blocked from growth and advancement in all channels.

This type of employee presents a special motivation problem. There are no more material rewards left with which to prevent creeping dissatisfaction, and other rewards, the true motivators that should be inherent in the job, are limited. This is a common situation in large correctional institutions, where promotional opportunities are limited, the ranks are crowded with well-qualified people, and the opportunities to restructure the work are limited.

It is unfortunate that many dead-end employees become problems because these employees very often have the most to offer to the organization. Their experience often surpasses that of others in the department. Yet because their professional advancement is stymied, the organization may not be getting the full benefit of their abilities and experience. It falls to the supervisor to deal with the problem by appealing to the individual through true motivating forces that stress job factors rather than environmental factors.

In dealing with the dead-end employee:

- Consult the employee on various problems and aspects of the department's work. Ask for advice. It is possible that an employee with years of experience in the same capacity has a great deal to offer and will react favorably to the opportunity to offer it.
- Give the employee a bit of additional responsibility when possible, and let the person earn the opportunity to be more responsible. Some freedom and flexibility may be seen as recognition of a sort for the employee's past experience and contributions.
- Delegate special one-time assignments. Again, years of experience may have prepared the employee to handle special jobs above and beyond ordinary assignments.
- Use the dead-end employee as a teacher. The experienced employee may be quite valuable in one-on-one situations, helping to orient new employees or teaching present employees new and different tasks.
- Point the dead-end employee toward certain prestige assignments such as committee assignments, special projects, attendance at an occasional seminar or educational program, or the coordination of a social activity such as a retirement party or other gathering.

Note that all of the foregoing suggestions deal with ways of putting interest, challenge, variety, and responsibility into the work itself. Many of them do not involve formal "job description" duties, but collateral or informal tasks.

In dealing with the dead-end employee, special attention must be given to true motivating forces. That is because the potential dissatisfiers (environmental factors such as wages, fringe benefits, and working conditions) are present in force. If the employee has come to regard an occasional pay increase as deserved reward for putting up with the same old nonsense, when the top of the scale has been reached and pay raises stop, then dissatisfaction will begin. All the suggestions made relative to the dead-end employee are intended to help the person find sufficient motivation in the work itself and avoid the weight of the dissatisfying factors that most often are beyond the control of the supervisor.

There are other potential solutions to the problem of the dead-end employee. Maybe it is possible to transfer the person to a completely different assignment. Perhaps a rotational scheme can be arranged, in which several employees trade assignments on a regular basis. Also, the dead-end employee may be cross-trained on several other jobs within the department, and thus be given a chance to do a variety of work and become more valuable to the department.

Perhaps it is unfair to discuss the dead-end employee in a chapter on "problem employees," since many such employees may present no problem at all. However, it is to the supervisor's advantage to recognize the dead-end employee as at least a slightly special case that a bit of conscientious supervisory attention can keep from becoming a real problem.

■ Absenteeism

As well as being a significant problem in its own right (particularly in departments that operate on rigidly scheduled rosters) absenteeism is usually a symptom of other problems. To the supervisor, absenteeism is an immediate problem of sometimes significant dimensions. Even a modest percentage of absenteeism in the department is likely to upset schedules and cause a great deal of supervisory time and effort to be devoted to juggling personnel and assignments to cover necessary tasks.

The employee who develops a pattern of chronic absenteeism becomes a "problem employee." Chronic absenteeism, in turn, should suggest that this employee may have problems (personal or otherwise) that are being at least partly revealed by their effects on work attendance.

To some extent absenteeism plagues supervisors and managers in every industry and in every conceivable organizational setting, not just corrections. Of course a great deal of absenteeism is legitimate. People get sick and have other difficulties that sometimes keep them away from work. A considerable amount of absenteeism, however, is not legitimate—at least not to the extent to which managers define legitimate in terms of the organization. In addition, there is no sure way of determining how much absenteeism is legitimate and how much is not.

While it is impossible to estimate with any degree of accuracy, it is no stretch to say that absenteeism costs the country billions of dollars each year in lost output and other costs. When someone fails to show up for work, usually one of two things happens: (1) either the employee's work goes undone that day or (2) someone else must be assigned to do the absent employee's work.

In both cases significant cost can be involved. Perhaps sick leave benefits are paid to the absentee. Wages certainly are paid to a replacement employee (often at an overtime rate). Continuity and efficiency are likely to be lost because the employee was not there to perform his or her duties. Numerous rippling inefficiencies occur that ultimately have a dollar impact on the organization. In the correctional setting, as in the case of departments like security, absenteeism within a fixed-post, fixed-shift structure can be a major disruption.

To a degree, absenteeism is governed by employee attitude. Employee attitude is in turn influenced by numerous factors. An employee experiencing a negative turn in attitude is likely to discover that unwarranted absence, once indulged in, is just that much easier the next time around. This can be especially true if absenteeism is in effect aided by silent or tolerant supervision.

Controlling Absenteeism

Since there are legitimate reasons for employees to be absent, absenteeism is a problem that can never be completely cured. However, absenteeism can be reduced and controlled through conscientious supervisory attention. The following eight guidelines can be useful in controlling absenteeism:

1. In dealing with employees, stress that employment is a two-way street. Sick leave is an employee benefit with an associated cost, one that is provided for use only when needed. It is a privilege, not a right, and as such should not be abused. Let your employees know that absence should not be taken for granted, and let them know what it does to your department in terms of added cost and lost output.

2. Let your employees know that their attendance is important to the operation of the department. Openly publicize your concern for absenteeism and its effects on department performance.

3. Start new employees the right way, including the expectation of regular attendance in their orientation. Make sure they clearly understand all the rules governing absenteeism and the use of sick leave benefits.

4. Keep accurate attendance records, and do so with the employee's knowledge. If possible, do not put yourself in the position of having to have another department (for instance, personnel or payroll) research an employee's attendance record when a question arises.

5. Have absentees report to you when returning to work. This generally will not bother legitimate absentees, but it puts a certain amount of pressure on the healthy "stay-aways." In any case, even assuming that a particular absence may indeed be legitimate, you should be sufficiently interested in your employees' well-being to briefly check with someone returning from a day or two of absence. You may want to insist that the employee speak to you personally when they are calling in sick, in order to put additional pressure on the individual whose reasons may be questionable.

6. Do not allow your system to reward for absenteeism. Take, for instance, departments in which employees rotate to provide weekend coverage, or the security department where rotating shift and weekend coverage often is built into the work roster. There may be people who seem to experience "illness" only when scheduled to work on Saturday or Sunday. If possible, arrange your scheduling such that a person who calls in sick on a scheduled Saturday or Sunday will be rescheduled to work that day on the following weekend so the employee cannot avoid the fair share of weekend duty through the use of sick leave.

7. Discuss unusual patterns of absence with the employees involved. If someone's supposed illness or personal problem always creates a long weekend or stretches a holiday into two days, at least make it known that you are aware of the pattern and feel perhaps it would be more than coincidence should this pattern continue. If your personnel regulations allow, insist that the employee bring in a written explanation from a doctor for persistent absences.

8. Use incentives available to you as a supervisor to discourage absenteeism. It would make sense to delegate a special assignment or a particularly interesting or appealing task to someone you can reasonably count on to show up regularly for work. Make it known that this measure of dependability is one of your rea-

sons for selecting the person you chose. Also, make appropriate use of employee attendance records at performance appraisal time. While attendance is not likely to weigh heaviest in a performance appraisal (unless, of course, it is exceptionally poor) you can certainly apply attendance in its proper relationship with other appraisal factors in either extending or withholding praise and reward.

Morale and Motivation

Employee morale and individual motivation to perform are key factors in a department's rate of absenteeism. Generally, low morale and lack of individual motivation will encourage increased absenteeism. Some people stay away from work because they are ill and would do so in any case. However, many absences happen when employees are feeling "on the fence"—neither especially sick nor particularly well. If this "blah" feeling happens to coincide with a "blah" attitude toward the job, the employee will stay home as long as sick-time benefits remain available.

All the factors having a bearing on the employee's attitude toward the organization, the job, and the work—the motivators and dissatisfiers discussed in Chapter 13—can influence attendance. Some of the most frequent causes of unwarranted absences are boredom, repetition, lack of interest, lack of challenge, and the inability to see positive results from one's efforts. Anything the supervisor can do to improve the chances of employee self-motivation will also be positive steps toward reducing and controlling absenteeism.

■ The Troubled Employee

Employees with personal problems—problems people cannot help but bring to work with them—are rarely able to do their best work. In rare cases—Officer Haskins, for instance—they can present safety and security risks equal to that of any inmate. Supervisors may be in a difficult position relative to the troubled employee. Getting the work of the department done, and done well, is the manager's responsibility. How employees perform their duties in completing the department's tasks is also the manager's business. An appropriately people-oriented supervisor should be interested in the employee as a whole person, but the employee's private life and personal problems are not within that scope. Aside from certain dependency issues that merit a referral to other resources, they represent an area management cannot enter without specific invitation.

In dealing with the apparently troubled employee, a good manager does not prod and does not push. Availability to the employee and a willingness to listen are the main options. A manager may have to go as far as to provide the time, the place, and the opportunity for the employee to talk, without specifically asking the employee to "open up." Quite often, if the manager's openness is evident the troubled employee will respond in kind.

In relating to the troubled employee:

- Listen, but be aware at all times of the temptation to give advice. Some of the most useless statements that possibly could be made begin with, "If I were you. . . ." Also, although many troubled employees could use advice, it is usually advice that the manager is unqualified to deliver. The best that can be done under most circumstances is gently to suggest that the employee seek help from qualified professionals.

- Be patient, and show concern for the employee as an individual. Although managers should naturally be concerned with an individual's impairment as a productive employee, do not parade this before the troubled person. Rather, be patient and understanding. When possible, be patient to the extent of easing off on tight deadlines and extra work requirements until the person is able to work through a problem.
- Do not argue, and do not criticize an employee for holding certain feelings or reflecting certain attitudes. Avoid passing judgment on the employee based on what you are seeing and hearing.
- Be discreet, and let nothing a troubled employee tells you go beyond you. Be extra cautious if an employee tends toward opening up to the extent of revealing information that is extremely personal and private. While it often does a person good to be able to simply talk to someone else about a problem, there is a risk of saying too much and afterward feeling extremely uncomfortable about having done so. If you can, try to demonstrate that you sympathize and understand without allowing the employee to go too far. Always provide assurance that what you have heard in such an exchange is safe with you.
- Reassure, and when you are honestly able to do so, provide the employee with reasonable confidence about important issues. These might include security of the employee's job, possibility of reducing undue pressure while problems get worked out, and the presence of a friendly and sympathetic ear when needed. You need not even know the nature of the employee's outside problem to supply very real assistance by reducing the job-related pressures on the individual.

In dealing generally with the troubled employee, a good supervisor will listen honestly and sympathetically and do what can be done to reduce pressure on the employee. It is equally important, though, to leave giving specific advice related to the problem to persons qualified to deal with such matters.

Employee Assistance Programs

A typical agency employee assistance program (EAP) motivates employees in need of assistance to accept early counseling. The intent of the program is to help them regain their productive capability, reduce absenteeism, minimize grievances, reduce the need for disciplinary action, and improve morale. As typically organized, the EAP is a confidential program available to all agency employees with alcohol, drug, or emotional problems.

Originating in the 1970s, EAPs are almost universally considered a standard program throughout government as well as in private industry. With an EAP in place, organizations can help troubled employees find sources of appropriate help at hand.

A well-functioning EAP can help increase work quality, reduce productivity losses, and control tardiness, absenteeism, and other undesirable conditions that affect job performance when employees' personal problems carry over into the work environment. A great many kinds of personal problems—drug abuse, alcoholism, and compulsive gambling, as well as marital, legal, financial, and emotional difficulties—can harm an employee's performance. The EAP is intended to provide employees who are troubled by such problems with personal, confidential assistance.

Ordinarily, the EAP consists of a network of service providers, or a referral system to such providers, available to employees in need of specific kinds of assistance. For the sake of complete confidentiality, actual treatment services are almost never provided by a component of the institution or agency whose employees the program serves. All employees will know about the EAP, but the institution's only direct involvement will be through the services provided by the organization's EAP coordinator.

Supervisors in organizations that operate EAPs should be trained to recognize job-performance problems that might result from personal difficulties. When a manager encounters such circumstances with a particular employee, it is then possible to suggest (perhaps as an alternative to normal disciplinary processes) that the employee consider visiting the EAP coordinator.

An employee might enter the program independently, directly approaching the EAP coordinator without referral. Most EAPs encourage self-referrals from employees who believe they are heading for trouble, if not already in trouble. Beyond management referral and self-referral, third-party referral is also possible—perhaps from a family member, clergyman, personal physician, or other concerned party.

Although employees' lives outside of the workplace are truly none of the supervisor's legitimate concern, employees' work performance indeed is. Having learned the more common signs of serious personal problems (at least to the extent of affecting job performance) managers then are able to tell employees how to gain access to the EAP should they desire help. The supervisor neither commands nor directs an employee to engage the services of the EAP. The supervisor simply recognizes the signs and reminds employees of the availability of the EAP.

However, in some organizations an employee's refusal to comply with a supervisory referral to an EAP can be taken as an inference that the person is not trying to intervene in and correct performance problems that triggered the referral. While agency policy differs on the consequences of such a refusal, it ultimately could be used in the disciplinary or termination process.

Over a period of time, as employee acceptance of EAPs has grown, self-referrals have tended to increase relative to other kinds of referrals, and the focus of many programs gradually has moved from crisis intervention to preventive care. A mature, well-functioning EAP can be the most comprehensive resource available for dealing with the troubled employee.

■ The Real "Problem"

The true people-problem of the supervisor is not the problem employee as such but rather the basic challenge presented by the vast differences to be encountered among people. Everybody is different, so there is no single "right way" of dealing with all employees. Should you have twelve employees reporting to you it is conceivable that for you there may be twelve different "right ways" of dealing with these employees.

A good general rule for supervision is to give as much (or more) time to a new employee as would be required by a new major assignment. Get to know your employees even better than you know your work. Regardless of the function you supervise, employees are your greatest resource. It is ultimately the humane and understanding use of the human resource that will determine success as a manager.

EXERCISES

Exercise 16-1: The Great Stone Face

Six months ago you were working as an accountant in a small state correctional facility. Today, you are supervisor of a business office with 20 employees in the agency's headquarters. Most of your staff have been there for years.

One of your employees, Paul Steiner, is assigned to maintain a complex accounting system for your correctional industry operation that requires considerable extra training. None of your other employees is able to do Paul's job because your predecessor never trained anyone else in the job.

Paul strikes you as good at what he does. However, he has days (at least one a week) on which he refuses to speak with his coworkers. His silences are well known; the other employees refer to him as "the great stone face." Since his regular duties require contact with other employees, when one of his silent moods strikes the department's work flow is impaired and people begin to complain.

On several occasions you have given Paul the opportunity to talk with you, but so far he has given you no clue as to any difficulty that might be behind his moods. All of your indirect offers of help have been ignored and when directly asked if anything is bothering him, he ducks the question.

Instructions

Develop a tentative approach to the overall problem presented by Paul's behavior. Consider not only the "problem employee" specifically but also the effects of his mood changes on the department. (An "if/then" approach is suggested. For instance, "If I try this particular direction and such a response is forthcoming, then I'll go on to try. . . ."

As circumstances permit, share your thoughts on the problem in a discussion group.

Exercise 16-2: The First-Class Grouch

"As your assistant, I'm certainly not trying to tell you what to do," said Carol Ames. "You're the boss, and I'm only pointing out—again—a problem that's leading us into lots of grief."

"I know," Administrative Services Manager Sam Henson said with more than a trace of annoyance. "I'm trying to take it the way you mean it. I've heard it from several people and I know we've got a problem with Nancy. I just don't know how to deal with it, that's all."

"It has to be dealt with," Carol said. "The receptionist/switchboard operator is in a position to leave a first and lasting impression on a lot of people. For months now, she's been generating an endless trail of complaints. I've heard from visitors, staff, and the headquarters office and a lot of other people—just about anyone you care to name—about her curt, rude treatment of them. It's been going on for way too long, and it's getting worse. And now she's starting to mix up phone numbers—routing calls to the wrong offices."

Sam said, "I know. I had hoped that whatever was bugging her would pass. But it hasn't. She's gone from bad to worse. And it's too bad—she's been here a long time, and this is only relatively recent."

"One of us needs to talk with her. Or at least make some attempt to find out what's wrong."

Sam spread his hands, palms up, and said, "I've tried to talk with her. Just a week ago I gave her a chance to talk in private. I even asked if I could help out in any way, but . . ." He shrugged helplessly.

"But what?"

"She told me nothing was wrong, or something like that. I got the impression that she was telling me—kind of roundabout—to mind my own business."

"Well, something is wrong," Carol said, "and we need to do something about it. She's coming across as a first-class grouch and the whole institution is suffering."

Instructions

Develop a tentative approach for dealing with the apparent attitude problem presented by the receptionist/switchboard operator. Make certain you provide for reasonable opportunity for correction of behavior and that you account for:

- Possible ways of assisting the employee with "the problem"
- The necessarily progressive nature of any disciplinary action considered
- The needs of the department

The Supervisor and the Human Resource Department

17

Chapter Objectives

- Introduce the human resource department as a vital staff function that exists to support operating management and the employees of an organization.
- Describe the major functions of a human resource department.
- Briefly review the evolution of the human resource function.
- Outline the functions of human resources and indicate how these functions relate to the role of the supervisor.
- Describe a number of action steps the supervisor can take to ensure that he or she will obtain appropriate service from human resources when needed.
- Suggest what the supervisor can do to establish a working relationship with human resources that will lead to improved human resource service to the organization.
- Emphasize the essential *service* nature of the human resource function.

I use not only all the brains I have, but all I can borrow.

Anonymous

With the help of a surgeon he may yet recover, and prove an ass.

William Shakespeare, *A Midsummer Night's Dream*, V, i, 318

FROM THE INSIDE

The Importance of the Background Check

The personnel officer was interviewing an applicant for the job of unit secretary. She seemed bright, motivated, and qualified. Her application had a few job history gaps in it, but she had what seemed to be an adequate explanation. Because of a shortage in this job category, she was hired provisionally, conditional on clearing a background check.

About 3 months later (backlogs being what they are in agencies doing law enforcement background checks) the results of the investigation came back. The personnel officer called her supervisor and asked him to come to the office and bring the woman with him. He said they were going to have to fire the woman, but didn't say any more over the phone.

When the two arrived, the personnel officer asked the woman if she had neglected to tell them everything they needed to know about her work record. She said that she didn't have anything else she could think of to add. She was then asked if she had accurately related any arrests or convictions in her background. She hesitated, and then said, "Well, maybe not."

The personnel officer then told her that she was being terminated, and would have to leave the institution at once (under escort). Her background investigation had disclosed that the employment gaps she had not fully explained were times when she was working as a prostitute, and that she had numerous arrests for prostitution.

■ "Personnel" Equals People

Supervisors are charged with the task of facilitating the work performance of a number of people. In this role they are expected to ensure that the efforts of a department or work group are applied toward attainment of the organization's objectives. This must be done in such a way that the group functions more effectively with the manager than it would without him or her.

As regulatory and statutory requirements relating to employment continue to increase and become more complex, managers cannot be expected to know every applicable rule. They need help, and the place where the supervisor can find help in abundance for issues of this type is the personnel (human resources) department.

Throughout this book the terms "personnel" and "human resources" have largely been used in an interchangeable manner. Other designations that may be encountered are *employee relations* and *labor-management relations*. These latter two labels have also found use as descriptors of subfunctions of modern human resources, with *employee relations* relatively referring to dealing with employee problems and problem employees, and *labor-management relations* referring to dealing with unions. In any case, *human resources management* is today's more all-encompassing title for what goes on in the department that most organizations once called *personnel*.

Management is frequently described as getting things done through people. People do the hands-on work and other people supervise them, and still other people oversee the supervisors and managers. The human resources department is the staff function that assures managers get the best people and get the best possible performance from them—and in some cases (as in the "From the Inside" anecdote in this chapter), remove them from the organization as well.

Even the most sophisticated piece of equipment requires periodic maintenance to ensure its continued functioning. So, too, do the human beings who run a correctional institution require regular maintenance. And given that the "human machine" is generally unpredictable and varies considerably from person to person in numerous dimensions, the supervisor's "maintenance" function will encompass many activities.

There are many places in a correctional organization where the supervisor can go for help with various tasks and problems. For people problems, however, and for some straightforward people-related tasks that cannot yet be described as problems, the supervisor's greatest source of assistance is the human resource department. It remains only for the supervisor to take steps to access that assistance. It is to the individual supervisor's distinct advantage to know exactly what should be expected from human resources and how to get it when needed.

■ A Vital Staff Function

The mission of this staff department is to assist line managers in acquiring, maintaining, and retaining employees so that the objectives of the organization may be fulfilled. As a critical staff function, human resources does none of the actual work of the correctional facility. Rather, human resources facilitates the work of the organization by concerning itself with the organization's most importance resource.

As a support department, human resources should be prepared to offer a variety of employee-related services in a number of ways. Human resources staff should anticipate numerous kinds of difficulties and needs, and should communicate the availability of assistance throughout the organization. For example, a personnel policy manual that dispenses advice and guidance in employee matters, and top management's typical instructions to supervisors to seek individualized guidance from human resources staff, are both essentially "advertising" for human resource services.

Human resources staff should advise those in the ranks of management about the services they provide. However, the human resource department cannot anticipate every specific need of each individual supervisor or manager. To truly put the human resource department to work, the supervisor must be prepared to take his or her needs to them and expect answers or assistance.

The human resource department has long been of increasing value to the organization at large and the individual supervisor in particular. Its value has increased because rational, necessary responses to a number of forces (both external and internal to the organization) have created additional organizational tasks. Two major forces have been the expansion in the number and kinds of tasks that have fallen to human resources, and the proliferation of laws affecting various aspects of employment.

The privatization of certain aspects of corrections has added yet another dimension to this issue. Personnel management laws that formerly applied only to private business apply to private correctional organizations, whereas a totally different statutory and regulatory structure may apply to public correctional facilities. Sorting out these issues is not something in which the chief correctional supervisor, food service administrator, or power plant manager can or should be expert.

The reasons for a separate personnel management function are discussed next.

An Increase in Employee-Related Tasks

Like the majority of departments in a modern organization, there was a time when human resources did not exist. And also like other departments, this one arose to fill a need. The development of the human resources function in government, and in corrections in general, roughly paralleled trends in the private sector.

In the private sector, the earliest human resource departments were commonly known as employment offices. They were created as businesses grew large enough to see the advantages of centralizing a great deal of the process of acquiring employees. Employment and employment-related record keeping made up all the work of the employment office. When wage and hour laws came into being, the "employment office" absorbed a great deal of the concern for establishing standard rates of pay and monitoring their application relative to hours worked. This began the compensation function.

In response to new laws and other pressures (both internal and external), organizations began to provide compensation in forms other than wages. At this point the "employment office" took over the administration of what became known as "fringe benefits." And as organizations responded to labor legislation and to labor unions themselves, labor relations functions were added to the growing list of activities that shared a common theme, all had something to do with acquiring, maintaining, or retaining employees.

Other people activities were added as needed over time. What had once been the employment office became "personnel," literally, "the body of people employed by the agency." During the last two decades the term *personnel* began to be replaced by the term *human resources*, but the essential meaning remains the same. All the while, the human resources function grew in value as it took on an increasing number of employee-related functions.

A Proliferation of Employment Laws

A number of laws were the primary cause of a great deal of the increase in employee-related tasks described in the foregoing section. For example, the establishment of Social Security, worker's compensation, and unemployment benefits all created important tasks for human resources staff. A great deal of labor relations activity was brought about by these laws. In addition, various anti-discrimination laws, including the Civil Rights Act of 1964, the Age Discrimination in Employment Act, and the Equal Pay Act, brought with them an increased volume of work for human resources.

Antidiscrimination laws have forever changed the way most organizations work. This category of legislation has created a strongly law-oriented environment in which lawsuits and other formal discrimination complaints have become routine human resources business. This legislation has also turned employee recruitment in general, and specific processes such as performance evaluation and disciplinary action, into legal minefields filled with traps and pitfalls for the unwary. And in the process they have created more work for human resources and have created myriad reasons for the individual supervisor to turn to human resources on more occasions.

A Tendency Toward Decentralization

There is a tendency toward decentralization in correctional administration, in the form of what most systems call "unit management." Unit management calls for increased first-line manager responsibility and moving decision-making as close as possible to the bottom of the hierarchy. Unit management means that certain decisions that might once have been made by an associate warden (like, for instance, inmate classification decisions or deciding how far to proceed in a particular disciplinary action against an employee) have been moved down to the level of the supervisor. As more employee-related decisions migrate to the first line of management in this fashion, the more the first-line manager—the individual supervisor—has to depend on the guidance and support of the human resource department. Thus the present tendency toward decentralization in the form of unit management has increased the value of human resources to the supervisor.

■ Learning About the Human Resource Department

To be able to get the most out of the organization's human resource department it is first necessary to understand the nature of the human resources function. The manager also should know how human resources relates organizationally, and be familiar with the specific functions performed by the organization's human resources department.

Staff Versus Line

Human resources has already been described in these pages as a staff function. As opposed to a line activity (a function in which people actually perform the work of the organization, such as correctional officers or food service staff), a staff function enhances and supports the performance of the organization's work. The organization's work is more effectively accomplished with the staff function than without it.

The distinction between line and staff is critical to appreciate because a staff function cannot legitimately make decisions that are the province of line management. Operating decisions belong to operating managers; they must be made within the chains of command of the line departments. The business manager of a correctional facility is not the person who actually makes the decision to expend money for new floor buffers in a housing unit; the unit manager or perhaps the safety and sanitation officer make that decision. The primary purpose of human resources in enhancing and supporting work performance is to recommend courses of action that are (1) consistent with legislation, regulation, and principles of fairness, and (2) in the best interests of the organization as a whole.

It is not unusual for some managers to blame human resources staff for decisions that did not reflect a choice they would have made themselves. Complaints such as "Personnel made me do it," or "This is the human resources department's decision," are not uncommon from managers whose preferred decisions are altered because of human resources' recommendations. However, human resources generally does not (and should never) have the authority to overrule line management in any matter, personnel or otherwise. If, as occasionally is the case, a personnel-related decision made by line management must be reversed for the good of the organization, that reversal is by higher line management.

Human resources may reach out and bring higher management into the process when a supervisor insists on pursuing a decision that human resources has recommended against (for cause). But line management at some level actually must make the decision, not human resources staff. Whether line management readily listens to its advisors in human resources depends on many factors, including their estimation of the importance of the human resources function and the human resources department's track record in making solid recommendations. But in the end, the decision is management's to make.

This situation explains the choice of the second quote at the beginning of this chapter. Human resources staff can provide the expertise needed (the surgeon's function), but the individual manager still is in a position of making choices with that information (and in the process could wind up looking like Shakespeare's proverbial ass.)

The Human Resource Reporting Relationship

In virtually all agencies, the human resources department reports to top management. Generally, human resources should report to the level that has authority over all of the organization's line or operating functions. At the institutional level, depending on the particular organizational scheme employed, the human resource manager might report to either the warden or the associate warden.

Human resources must be in a position to serve all of the organization's departments equally. Independence and impartiality are essential for human resources services to be provided to the whole organization. If the human resources function is organized directly under the warden or a single associate warden, there is a relatively low risk that the organizational structure itself will bias the provision of human resources services toward one department or portion of the organization. If the human resources manager reports to one of several associate wardens, as is the case in many prisons, then extra care should be exercised to ensure that the other departments supervised by that associate warden do not receive any form of preferential treatment by human resources staff. In such situations, the warden must be alert that this dynamic does not develop.

Also be wary of the occasional practice of duplicating human resources functions within the same facility. For example, there is an argument to be made for the health services department to be delegated some human resources functions (medical staff recruitment, for instance) while the institution's regular human resources office serves all other departments. Although there arguably are advantages to be gained from this approach, splitting or subdividing other human resources activities tends to create duplication of effort while increasing the organization's exposure to legal risks.

The Human Resource Functions

There are almost as many possible combinations of human resource functions as there are human resources departments. A great many activities that may be generally described as "administrative" can find their way into the human resource department. However, this section will address the significant activities or groups of functions that are often identified as the tasks of human resources.

The basic human resources functions are:

- Employment, often referred to as recruiting. This is the overall process of acquiring employees are advertising and otherwise soliciting applicants, screening applicants, referring candidates to managers, checking references, extending offers of employment, and bringing employees into the organization.
- Compensation, or wage and salary administration. This is the process of creating and/or maintaining a wage structure and ensuring that it is administered fairly and consistently. Related to compensation, as well as to other human resources task groupings, are job evaluation, the creation of job descriptions, and maintenance of a system of employee performance evaluation. When the correctional agency develops these standards and procedures, it ordinarily is the local human resources department's responsibility to administer them.
- Benefits administration. This activity is a natural offshoot of wage and salary administration, since benefits are actually a part of an employee's total compensation. Benefits administration consists of maintaining the organization's benefit structure and assisting employees in understanding and accessing their benefits.
- Employee relations. This activity may be generally described as dealing with problem employees and employees who have problems. It may range from handling employee complaints or appeals through processing disciplinary actions to arranging employee recognition and recreation activities.

One can see the foregoing four general activities at work in essentially every human resource function regardless of size and overall scope. In a very large institution or correctional agency these may be separate sections within the department. Each may have its own chief and its own staff, perhaps including subdivisions of those functions. In a very small institution, these are likely to be the tasks of a single person with multiple duties.

One additional basic function that may be encountered in the human relations arena is labor-management concerns. This may be a functional title that identifies a whole department in a very large institution, or simply a function of the human resource department. It is also a relatively generic label that applies to the maintenance of a continuing relationship with a bargaining unit, that is, a labor union. Again, depending on size, labor-management relations ordinarily is a subdivision of human resources or simply one of several responsibilities assigned to one person.

Other activities that might be found within human resources include:

- Employee Assistance Program. Often this is part of human resources, although it also can be managed by the institution's chief of mental health.
- Training, for both managers and rank-and-file employees. With the exception of continuing education for some specialties such as medical staff, formal training functions most often are part of human resources.
- Payroll. In the past this often has been part of "personnel," but payroll functions also may be organized under the business office. However, a working relationship between human resources and business office personnel has always been essential. In recent years, integrated personnel/payroll systems have started a shift of the payroll function back toward human resources.

■ Putting the Human Resource Department to Work

The first, simplest, and most valuable advice to be offered for getting the most out of the human resource department involves the age-old, two-step process of initiation and follow-up. It is but a slight variation on a practice followed by most successful supervisors.

The successful supervisor knows that any task worth assigning is worth establishing a specific deadline. The "do-it-when-you-have-a-chance" or "when-you-think-about-it" approach breeds procrastination, delay, and inaction. Any well-thought-out, specific assignment must be accompanied by a target for completion and a deadline which (though it may be generous or even loose) leaves no doubt as to expected completion. And when that deadline arrives and no results have been forthcoming, the supervisor then exercises the most important part of the total process—faithful follow-up. Faithful follow-up is the key; the supervisor who always waits a week beyond the deadline is behaviorally telling the employees that they always have at least an extra week.

Anything needed from the human resource department should be addressed in a similar manner. Relative to the human resources department, the process might be summarized as follows:

- Make certain the issue is part of human resources' responsibilities, and determine, if possible, who in human resources would be the best person to approach on the topic.

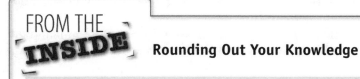

Rounding Out Your Knowledge

Using the foregoing paragraphs as a guide, determine exactly which functions are performed by your agency's human resource department. Further, take steps to attach a person's name to each function. Managers should seek to be in a position to understand how the human resources department is organized, to know generally who does what, who reports to whom, and who bears overall responsibility. Moreover, be certain to know the organizational relationships of sometimes-human resources functions (like, for example, employee assistance functions) that may belong elsewhere in the organization, to ensure that specific issues are referred to the correct department.

Next, take the time to make a list of the supervisory functions or activities that can lead to seeking information or assistance from human resources. The lists of most supervisors may have a great deal in common in that they might include:

- Employment—finding a sufficient number of qualified candidates from whom to fill an open position.
- Benefits—providing information in response to employees' questions about benefits.
- Compensation—providing information responding to employees' pay questions.
- Employee problems—determining where to send a particular employee who is having difficulty with a specific problem.
- Job descriptions/job evaluations—determining how to proceed in assessing the grade or pay step of any particular position.
- Policy interpretation—determining the appropriate interpretation of personnel policy for any particular situation.
- Disciplinary actions—determining how to proceed in dealing with what appear to be violations of work rules.
- Performance problems—determining how to proceed in dealing with employees whose work performance is consistently below the agency's standards.
- Performance appraisals—securing guidance in completing appropriate performance appraisals and finding out how much one can depend on human resources to coordinate the overall appraisal process.

The preceding list can doubtless be expanded by each supervisor who may refer to it. One helpful method of expanding your list includes leafing through your agency's personnel policy manual and employee handbook. That process will bring to mind additional areas of concern encountered in the workplace.

- Refine the question or need such that the inquiry is sufficiently specific to permit a specific response.
- When making contact and conveying needs to the appropriate person, if the answer is not immediately available ask when an answer will be supplied.
- If the promised reply date occurs later than the legitimate need date, negotiate a deadline agreeable to both you and human resources.
- If the agreed-upon deadline arrives and there has been no answer, follow up politely and diplomatically, but follow up faithfully. Never let an unanswered deadline pass without following up.

This process should be applied not only to problems, issues, and concerns that a supervisor would consider taking to human resources. It should also (and especially) be applied to questions and concerns that employees bring to you as a manager. If an employee's question in any way involves human resource concerns and if you are unable to respond appropriately, then you need to take the question to human resources as though it were your own.

Some Specific Action Steps

Any number of supervisory needs present opportunities to put the human resource department to work. The more frequently encountered of these include the following.

Finding New Employees

There are any number of points in the employment process at which the supervisor and human resources must work together. Fulfill the line management end of the working relationship, and expect human resources staff to fulfill theirs. Tracking vacancies in each department ordinarily is a human resources function. But supervisors also need to monitor vacancies and give the human resources department prompt notice when they become aware that a vacancy is about to occur.

Screening prospective job candidates is another area in which supervisors and human resources staff must work together. For example, if none of the five candidates human resources has supplied for a particular position is truly appropriate, ask for more. Do not settle for only what is given if that is not enough. As to the manager's part of the arrangement, do not continue to call for more applicants in the search for the "perfect" candidate if two or three who meet the posted requirements of the job have been identified.

Also, stay in touch with human resources concerning extending offers, checking references, and scheduling preemployment physical examinations and starting dates. Do not be unreasonable: recognize that these activities take time. Making this level of interest and attention known will encourage completion of the process.

Bringing Job Descriptions Up to Date

The supervisor ordinarily has a significant responsibility in maintaining current job descriptions for the department. The human resources department usually also has the responsibility for associating a pay level with each job and for maintaining official files of up-to-date job descriptions.

The manager's part is best done within the department. To the extent possible, job descriptions should be written by those who do the work and those who supervise the work. However, if a manager deliberately involves human resources, it will contribute consistency to job descriptions and ensure each job is placed on the proper pay scale. Once again, the manager's visible interest in the process will encourage timely completion of human resources activities.

Disciplining Employees

Regardless of the extent of human resource involvement in processing a disciplinary action, it is not the human resources department that decides upon the action. The human resources department disciplines nobody (except employees of the human resources department, as necessary). Any employee deserving of disciplinary action must be disciplined through his or her immediate chain of command. That having been said, most organizations require the supervisor to submit proposed disciplinary actions through the human resources department before implementation. Whether or not this is true for a specific organization, the manager needs to make the best possible assessment, and it would be wise to go to human resources for advice in advance. Managers should expect to receive sound advice—whether in the form of a single recommendation (complete with rationale for doing so) or as multiple alternatives, each with its own explanation of possible consequences.

The decision is technically still the manager's, and if it is a poor decision he or she probably will bear the brunt of the consequences. Do not let human resources avoid responsibility by failing to provide specific direction. Insist on complete human resources input in deciding upon disciplinary action.

Evaluating Employees

One of the most important tasks of the supervisor is appraising employee performance. No formal system of performance appraisal can function consistently throughout the agency without the central guidance usually provided by human resources staff. Although managers can certainly evaluate employees' performance without the assistance of human resources, they can do a much more consistently acceptable job of evaluation with human resources involvement.

Most of human resource's involvement in evaluation should occur automatically as far as the supervisor is concerned. The human resources department should provide forms, instructions, schedules, and reminders throughout the process. However, if the human resources department is not always on top of the supervisor's employee appraisal needs, then the supervisor should be ready to:

- Keep track of scheduled review dates and ask human resources for forms and timetables if not supplied automatically.
- Ask for periodic instruction in how to apply rating criteria, especially if criteria have changed and refresher instruction is not supplied.
- Keep human resources advised of how changing job requirements may be affecting the application of criteria based on previous requirements.
- Periodically ask human resources for rating profile information (if the agency or institution maintains it) that reveals patterns in the supervisor's rating practices. This data can show whether those patterns are changing with time, and perhaps also shows how this specific supervisor's rating patterns compare with those of other supervisors.

Dealing with Training Needs

If the human resource department has responsibility for any kinds of employee training (and in most correctional agencies there is a better-than-even chance that this is so), do not wait for needed training to come to the department's employees. If there are training needs in the work group, take them to human resources. If, for example, several employees require training in basic inmate management techniques, take a well-developed request for meeting that need to human resources. Be ready to negotiate a timetable for providing the training, and offer to become personally involved in the training. (With appropriate human resources involvement, every supervisor is a potentially valuable instructor in some topic.)

The foregoing suggestions are offered primarily to convey a general idea to the supervisor. The human resource department exists as a service function for all employees. It remains for the supervisor to take each legitimate personnel-related need to the human resources department and to ask for (and expect) an honest response.

■ Wanted: Well-Considered Input

The most effective human resource departments are not one-way dispensers of information and assistance. They interact with managers and respond to the needs of the organization's workforce. However, the human resources department can go only so far in anticipating needs and meeting them. To be fully effective, the human resources department must learn of employee needs from employees and supervisors and must in turn go to top management with solid proposals for meeting the most pressing needs.

Some employees (although usually a minority) take their own questions, concerns, and suggestions to the human resource department. However, many employees will never do so for themselves. Their supervisors are the conduit for meeting those information needs (needs that also should be made known to top management if they constituted emerging patterns).

Among the kinds of information that the supervisor should pass to human resources are:

- Reactions to various personnel policies, especially when policies seem to have become less appropriate under changing conditions.
- Employee attitudes concerning pay and benefits, especially perceptions of inequities and alleged instances of unfair treatment.
- Complaints (and compliments as well) about employee services such as locker room conditions, quality of uniforms provided, or other related issues.
- Comments on the appropriateness of various employee benefits, and perceptions of benefits needed or desired as opposed to those presently given.
- Potential changes in methods of acquiring, maintaining, or retaining employees that might afford the agency a competitive edge in those areas.

■ Understanding Why as Well as What

It is relatively easy to determine what the human resources department does within the organization. However, it is necessary to go beyond *what* and develop an appreciation of *why* this department does what it does. It also is useful to know why the human resources operation sometimes must take a position in opposition to a line department's position.

Consider the case of the supervisor who appeals to the human resources department to help resolve a seemingly unending series of difficulties by agreeing to the termination of a particular employee. As the supervisor says, "I simply can't do any more with this person. He's chronically late in spite of all my warnings. His absenteeism disrupts staffing. He uses up his sick time as fast as he earns it. His attitude is absolutely terrible. He's been overly aggressive with inmates and rude to inmate families during visits. And the way he talks to me borders on insubordination most of the time. His people skills are marginal and he's a disruptive influence when he's working housing units. I've been patient longer than anyone has the right to expect of me, but nothing has changed. I want to terminate him." Upon questioning, the human resources staff learn that there is very little in the way of a written record to back up the supervisor's assertions.

As sometimes occurs in such circumstances, the human resource practitioner hearing the supervisor's request briefly reviews the employee's background and immediately recommends against termination. This may understandably disappoint and upset the supervisor and leave him or her displeased with the human resources department. He or she may complain, with some justification, that HR staff ought to be more supportive in the effort to get rid of an unsatisfactory employee. He or she may well view human resources as obstructive and adopt an adversarial position. The manager perhaps will even attempt to solicit the assistance of higher management in opposing the human resource position.

Why would human resources be automatically protective of an employee like that described above? The differences lie first in the supervisor's perspective versus the human resource perspective, and second, in the frequently cited employee personnel file—"the record."

The supervisor's perspective is unit or departmental. The supervisor is legitimately focused on the good of the operation at that level, and the employee in question threatens it. However, human resources must view the issues in two ways that conflict with the supervisor's perspective. In micro terms, human resources must be concerned with the rights of the individual employee. In macro terms, human resources must be concerned with the good of the total organization. The organization is of course no more than the sum total of a number of individuals. But in focusing on the total organization, the human resources perspective is different than that of the supervisor—who is more concerned with more than one employee but much less than the total workforce.

Then there is "the record" to consider. In the foregoing case all of the supervisor's warnings concerning tardiness were unrecorded oral warnings.

Likewise, there are no warnings on absenteeism, except for a single written warning that is too old to give weight to in a current disciplinary action. No other warnings appear in the personnel file. True, in the mind of the supervisor this employee has always been less than satisfactory. But the personnel file includes several performance evaluations which (although not glowing with praise) suggest at least minimally acceptable performance. In short, there is no basis for termination except in the supervisor's mind.

In regard to problems such as those just described, the human resources department is:

- Defending the rights of the individual, not only because doing so arises from a sense of fairness but also because there are many laws requiring the agency to do so.
- Protecting the agency from numerous legal risks.

Whenever there is a chance that an employee problem or complaint will be taken outside of the agency, it is even more important to think in a single vein—if it is not in "the record," it never happened. There are instances of sustainable termination for major infractions that were accompanied by immediate discharge (although these days even many of these actions are successfully challenged). However in most instances a discharge must be backed up with a written trail describing all that occurred. It is legally necessary to be able to demonstrate that the employee was given every reasonable opportunity to correct the offending behavior or improve the unsatisfactory performance.

■ Emphasis on Service

As a staff function, human resources is organized as a service activity. Service departments provide no inmate supervision or productivity in the organization. They do not directly advance the work of the organization. However, they support the performance of the organization's work and in a practical sense become necessary. For example, if a pure service such as the facility maintenance shop did not exist, the physical plant would gradually self-destruct. Similarly, without human resources to see to the maintenance of the workforce, the overall suitability and capability of that workforce would steadily erode.

Recognize human resources for what it is: an essential service function required to help the organization run as efficiently as possible. Learn what the human resources department does, and especially learn why the department does what it does. Provide input to the human resource department. Forge a continuing working relationship with the human resources department, setting a clear expectation of service from this essential department. Challenge the human resources department to do more, to do better, and to continually improve service. Doing so will put the human resources department to work for employees at every level in the organization.

EXERCISES

Exercise 17-1: Where Can Human Resources Help?

Review the human resource functions as outlined in this chapter. Within each major function, identify the subfunctions that involve you as a supervisor. For example, under the employment or recruiting function you will probably have no role in locating or screening applicants, but you will be actively involved in interviewing job candidates. As another example, you will have no active responsibility for administering benefits, but you will often be expected to come up with answers to your employees' questions about benefits.

When you have generated a complete list of human resource subfunctions that require your active involvement, determine who in your human resource department is most involved with each and associate one or two names with each subfunction. Most supervisors' lists will be likely to include selection, interviewing, answering benefit questions, doing performance evaluations, and taking disciplinary action. This can become the basis of a reference list to help familiarize you with the individual specialties and strengths of the organization's human resource practitioners.

As an additional activity, review your list of pertinent human resource subfunctions against Chapter 29, "The Supervisor and the Law." For each human resource subfunction with which you are involved, indicate the major areas of legislation (for example, "antidiscrimination laws") and the specific laws (for example, "Immigration Reform and Control Act") that are of concern to your performance of the subfunction.

Exercise 17-2: A Favor or a Trap?

You are the receiving and discharge department supervisor in a medium-security facility. One morning well before the start of your department's normal working hours you were enjoying a cup of coffee in the employee's lounge/locker room, sorting out some notes on your plans for the day, when you were approached by one of your employees. The employee, Cerese Carter, one of your two or three most senior employees in terms of service, seated herself across from you and said, "There's something going on in the department that you need to know about, and I've waited way too long to tell you." Cerese proceeded to tell you ("In strictest confidence, please, I know you'll understand why.") that another long-term employee, Marge Jenson, had been making a great many derogatory comments about you throughout the department and generally questioning your competence.

For nearly 10 minutes Cerese showered you with criticism of you, your management style, and your approach to individual employees, all attributed to Marge Jenson. On exhausting her litany, Cerese proclaimed that she did not ordinarily "carry tales" but that she felt you "had a right to know, for the good of the department—but please don't tell her I said anything."

Cerese's comments were filled with "she saids" and "she dids" (she being Marge) and generally twice-told tales without connection to specific incidents.

But something extremely disturbing clicked in your mind while you were listening. Recently your posted departmental schedule had been altered, without your knowledge, in a way indicating that someone had tried to copy your handwriting and forge your initials. Two separate, seemingly unconnected comments by Cerese together revealed that one of only two people could have altered your schedule. Those two people were Marge Jenson—and Cerese Carter herself.

As Cerese finally fell silent you were left with an intense feeling of disappointment. You wondered if you could ever again fully trust two of your key employees.

Questions:

1. What should be your immediate response to Cerese Carter? Why?

2. Do you believe you have the basis on which to proceed with disciplinary action?

3. How can the human resource department help you in these circumstances?

Exercise 17-3:

Review the "From the Inside" vignettes at the beginning of Chapters 15 and 16, and discuss the issues raised from a human relations department point of view.

The Supervisor and the Task

IV

Ethics and Ethical Standards

18

Chapter Objectives

- Understand the nature and scope of potential ethical problems in corrections.
- Define ethics within the context of the modern correctional organization.
- Review the formal code of conduct of the American Correctional Association.
- Introduce the major areas of concern in the area of professional ethics and the generally accepted ethical standards of conduct within corrections.
- Describe the areas of ethical concern having the greatest impact on the role of the correctional manager.
- Describe the "appearance test" as applicable to potentially unethical behavior and examine the handling of conflicts of interest.
- Review the manager's responsibility for communicating and modeling ethical standards and conduct to his or her employees.

I look upon the simple and childish virtues of veracity and honesty as the root of all that is sublime in character.

Ralph Waldo Emerson

When a man's knowledge is sufficient to attain, and his virtue is not sufficient to enable him to hold, whatever he may have gained, he will lose again.

Confucius

Ethical Dilemmas in Corrections

The warden's wife is a realtor. As employees are transferred to the facility, she would like to try to sell them homes in the area. She asks the personnel officer to provide her with the names of such employees in advance, so she can get to them before they contact other realtors. The personnel officer is concerned that to do so would compromise the privacy rights of the incoming staff, and give her an unfair competitive advantage.

A caseworker happens to notice a busload of new inmates coming in and recognizes a former school classmate, who is there on a 90-day court-ordered study. While never close to this individual in school, he notifies his supervisor and writes a memo to the file, with the intent of ensuring that he not be involved in any of the case decisions regarding that inmate.

A relatively new correctional officer on outside patrol notices some pheasant hunters in the state-owned fields adjacent to the institution. When he tells them hunting is not permitted, they offer him $50 to let them continue hunting. He refuses and orders them to leave. Later in the day, he discovers a second group of hunters in another restricted area. When he tries to remove them, he learns they are friends of the assistant superintendent of the facility, who has told them they can hunt there.

A case management supervisor engages in a highly personal relationship with a secretary under his supervision. As this affair proceeds, she receives what others perceive to be unusually favorable treatment in the workplace.

Agency policy permits inmates to receive abortions. An associate warden of a facility housing female inmates has religious views that prohibit him from being involved in any way in providing abortions to inmates. When an occasion arises where the paperwork relating to an inmate abortion comes to him for approval, the administrator refuses to be involved.

A headquarters administrator in a large agency is involved in private consulting work—a situation that agency policy permits under most circumstances. His outside work takes a turn and begins to involve other organizations with which his agency has financial ties. Concerns are raised over whether this constitutes a conflict of interest, even though his duties do not involve any control over agency finances.

A correctional officer is attracted to a female inmate. He offers to bring in a quarter pound of marijuana in exchange for sexual favors. She turns out to be a DEA street informant in custody on an unrelated charge. She advises prison authorities, wears a "wire," and the officer is caught in the act as he gives her the drugs and tries to engage in the sex act. Prosecution follows termination.

A newly promoted administrator claims $500 worth of miscellaneous moving expenses on his government-reimbursed transfer. This is permitted by policy if receipts are presented. In doing so, he prepares several fraudulent "receipts" for nonexistent expenses. Unwisely, he uses the same typewriter and same kind of paper to type them. Noticed by an alert clerk

FROM THE
INSIDE

processing the reimbursement transaction, a promising 15-year career ends in disgrace.

An inmate "befriends" a correctional officer who has some personal vulnerabilities. She becomes infatuated and arranges to be assigned to escort him on a community visit to a dying relative. They abscond to a nearby city for a tryst and he remains at large for several weeks. The episode ends when he is apprehended, and she is terminated and prosecuted for aiding in an escape.

A camp administrator who is an up-and-coming prospect for a warden's job is in financial difficulty because of his child's high medical expenses. In a position to influence approval of various inmate furloughs, he begins to accept money and other favors from inmates in exchange for authorizing furloughs that otherwise would not be approved. When he refuses to comply with one such request, the inmate involved "snitches him off." The result is termination in lieu of prosecution.

The food service supervisor in charge of purchasing for a large penitentiary is steered to a particular supplier by an inmate. He makes large purchases of institutional food items from what turns out to be a mob-connected vendor. After being paid in various forms for his complicity in steering government business to the inmate's associate, the arrangement is discovered. Termination of an otherwise promising manager is the outcome.

A correctional supervisor speaks a bit too freely and explicitly in front of inmates about his marital problems. He comments that his wife won't give him certain kinds of sexual favors. Word gets around in the inmate population. On a late night tour of a housing unit, an inmate offers to provide some "comfort" in the form of oral sex. One such event leads to others, and eventual discovery. Termination (and today in many jurisdictions, prosecution) ensues.

Each of the situations mentioned in "Ethical Dilemmas in Corrections" actually occurred, and will be referred to later in this chapter. They, and the additional examples used elsewhere in this chapter, represent syntheses of real-world situations encountered by the authors over the years. They illustrate the wide range of dilemmas in which employees can find themselves and which require conducting themselves in an ethical and honest manner. The first six vignettes in the list represent situations where some ethical decision was required. The second six show what can happen in the real world when an employee makes the wrong choice. Some of them involved choices that eventually led to outright illegalities. But all of them started with a decision about ethical behavior in the workplace.

It is worthy to note that not only supervisors encounter ethical challenges in the workplace. Many of the examples used involve line personnel. The reason they are included is not only that they may also be encountered by managers, but because managers need to be ready to deal with this kind of issue in their subordinates.

■ Honesty

This chapter does not address an abstract issue. In fact, corrections can be an ethical morass, with numerous questionable or tempting situations presenting themselves to employees in virtually every job category.

But any discussion of ethics has to start with a simple concept—honesty. The dictionary tells us that honesty encompasses such concepts as "integrity, uprightness, truthfulness, sincerity."[1]

This is not a mysterious concept. It is, however, one that has been seriously eroded in today's culture. We read of large corporations (the now-infamous Enron comes to mind, but it certainly is not the only one) in which top executives seemingly feathered their own nests while apparently defrauding line staff and investors. We read about the mutual fund industry (long thought to be a safe haven for the small investor) being manipulated for insider's gain. Closer to home, we hear about correctional officers bringing drugs and other contraband into institutions.

Not that dishonesty and greed are new phenomena, but the pervasiveness of reports such as these is disheartening. The seeming prevalence of such dishonest acts also may set a tone in the workforce that honesty is not the best policy—that only the suckers play by the rules; that the folks that bend (and even break) the rules are the ones who come out ahead.

The management challenge that this presents is simply stated—set and maintain high standards for honesty in your part of the organization, no matter where that is. This starts with personal honesty, of course. Employees who see their boss lying or taking shortcuts (and who may even know of his or her dishonest acts) will decide that it is all right for them to do likewise. And they also learn a little bit about what the organization is really like, and from the noncorrectional world. As mentioned in an earlier chapter, one of the authors' has a daughter (recently graduated from college) who has taken her first real-world job. Several other new employees came on board at the same time she did, having been offered positions on a new project. However, they immediately were shunted to another type of work (with different, less interesting duties and lower pay), with the promise that the new operation would start soon. Three months later, they had not been trained and the person who was designated as their supervisor was no longer involved in the project. It seemed apparent that when they were hired, the organization was not even close to activating this new operation. It seemed clear to these three employees (and everyone else who was aware of the situation) that they had been lied to repeatedly about the program's implementation timetable. Two things resulted from this episode. The people directly affected were very unhappy and one of them is looking for other jobs. But in addition, every other employee in the organization has learned a lesson about being wary in trusting what the company says.

Setting and maintaining high standards of honesty also extends to the way you as a supervisor interact with your staff. Naturally, if you lie to them yourself, the message is a very direct one. If you laugh or shrug off lightly accounts of dishonest behavior on the part of others, employees will learn that you do not take such things seriously. If, on the other hand, you deal honestly with them and cast a disapproving view of questionable situations you are sending a much different message.

And your conduct matters a great deal as well. If you are seen as someone who is trying to "push the envelope" on travel reimbursements, they will notice. If they don't see you putting in an honest day's work, they'll take cues from that. Your personal honesty, displayed day in and day out, will do a lot to set the tone for your department.

Now the harsh reality is that there are times in corrections when a lie has to be told, or the truth must be concealed. A hostage negotiator can't reveal some things to a hostage-taker. (Actually, hostage negotiators may be told very little about tactical preparations just to remove the potential problem of even an inadvertent disclosure.) No one would argue that the name of a confidential informant should be revealed. It may be necessary to conceal the timing and substance of certain personnel moves. It even may be a matter of legitimate security to conceal or even present false facts on some matters.

But these are special cases, and employees will understand them. In the important day-to-day operation of the institution or agency, such dishonesty has no place. Deal with employees in a scrupulously honest fashion, and they (for the most part) will reflect that trait in their dealings with both staff and inmates. And that honest foundation is really the heart of ethical professional conduct throughout an organization.

■ Correctional Ethics Framework

Turning to the *New Shorter Oxford Dictionary* for a definition, we find that ethics is: "The moral principles or system of a particular leader or school of thought; the moral principles by which any particular person is guided; the rules of conduct recognized in a particular profession or area of human life."[2]

For the purpose of this chapter, *ethics* is the latter component of the definition—the rules of conduct recognized in a particular profession or area of human life. This places ethics in the context of the work organization. But whether a person ascribes to this definition or another, in all definitions of ethics there is mention of morals or morality. Indeed, one of the briefest but most useful descriptions of ethics is *moral code*.

Whether written or unwritten, every organization has a "system or code of morals" based on how its leaders want to relate to others. Directly or indirectly, this code sets standards for how the organization's employees should behave as well. An increasing number of organizations, within corrections and otherwise, are issuing formal ethical standards of conduct for their employees.

Ethical codes can arise from formal and informal sources. Agencies often have published codes of conduct. In most instances, people admitted to membership in a profession (mental health and medical personnel, for instance) agree to ascribe to specific ethical standards of conduct as a condition of entry to their profession.

Moreover, individuals have personal standards of conduct, although for most people these standards must be inferred from their behavior. (And as far as individual standards are concerned, one's observable behavior is usually a far more accurate indication of true ethical standards than one's own words might be in attempting to articulate these standards.) Whether individual or organization in origin, behavior is generally governed by these ethical standards of conduct plus legal requirements in the form of applicable laws and regulations.

The American Correctional Association has published the following formal code of conduct that serves as a model for our profession. Because of its comprehensive and highly realistic content, portions of that document form the basis for the rest of this chapter.

1. Members shall respect and protect the civil and legal rights of all individuals.

2. Members shall treat every professional situation with concern for the welfare of the individuals involved and with no intent to personal gain.

3. Members shall maintain relationships with colleagues to promote mutual respect within the profession and improve the quality of service.

4. Members shall make public criticism of their colleagues or their agencies only when warranted, verifiable, and constructive.

5. Members shall respect the importance of all disciplines within the criminal justice system and work to improve cooperation with each segment.

6. Members shall honor the public's right to information and share information with the public to the extent permitted by law subject to individuals' right to privacy.

7. Members shall respect and protect the right of the public to be safeguarded from criminal activity.

8. Members shall refrain from using their positions to secure personal privileges or advantages.

9. Members shall refrain from allowing personal interest to impair objectivity in the performance of duty while acting in an official capacity.

10. Members shall refrain from entering into any formal or informal activity or agreement which presents a conflict of interest or is inconsistent with the conscientious performance of duties.

11. Members shall refrain from accepting any gifts, services, or favors that are or appear to be improper or imply an obligation inconsistent with the free and objective exercise of professional duties.

12. Members shall clearly differentiate between personal views/statements and views/statements/positions made on behalf of the agency or association.

13. Members shall report to appropriate authorities any corrupt or unethical behaviors in which there is sufficient evidence to justify review.

14. Members shall refrain from discriminating against any individual because of race, gender, creed, national origin, religious affiliation, age, disability, or any other type of prohibited discrimination.

15. Members shall preserve the integrity of private information; they shall refrain from seeking information on individuals beyond that which is necessary to implement responsibilities and perform their duties; members shall refrain from revealing nonpublic information unless expressly authorized to do so.

16. Members shall make all appointments, promotions, and dismissals in accordance with established civil service rules, applicable contract agreements, and individual merit, rather than furtherance of personal interests.

17. Members shall respect, promote, and contribute to a workplace that is safe, healthy, and free of harassment in any form.[3]

The remaining portions of this chapter will address these areas and show their importance to a professional manager. Some are addressed in more detail than others, but all are important. (The fact that these examples do not all involve managers does not mean that managers are not vulnerable in those areas—they certainly can be and the information in this section should serve as a body of cautionary material.) But using these examples also should prepare supervisory staff to deal with these issues as they arise with their line personnel.

Use of Official Position for Personal Advantage

The title says it all; no employee at any level should use his or her official position for any personal gain. Admittedly, this is a major category that represents a lot of "gray" territory, and it overlaps with other areas of ethical concern. In particular, areas of vulnerability involve outside business activities, contacts with vendors, confused personal relationships, and certain assumed "privileges of rank" —all of which can entangle managers in this problem to varying degrees.

This issue starts most simply with maintaining integrity with regard to the durable assets of the institution or agency. Every employee is responsible in a general way to protect these assets against loss, theft, and misuse. But agency property also may not be stolen, used for personal benefit, loaned, sold, given away, or disposed of in any manner without appropriate authorization. Material that is declared surplus, obsolete, or scrap must ordinarily be disposed of according to specific agency or governmental policies.

Typical abuses include unauthorized appropriation or use of tangible assets such as computers and copiers and other office equipment. This can be seen also in misuse of other types of equipment and tools, vehicles, supplies, reports and records, computer software and data, and facilities. Intangible assets such as intellectual property, trademarks and copyrights, proprietary information including computer programs, confidential data (including business plans for private correctional firms), and such must be protected as well.

But this area of ethical conduct is much broader and represents a wide array of real-world problems. Recall one of the lead-in examples—the hospital administrator who falsified moving expenses. This situation actually boiled down to theft by deception, and the case was handled as such. Other examples abound, however. Some agencies reimburse moving expenses by paying employees for the net weight of their U-Haul truck loads, in lieu of using a commercial mover. Stories are rampant of barrels of water being moved cross country to increase the eventual payout to the person doing his or her own moving. Employees send serviceable equipment out to the dump and then retrieve it after hours. A machinist in the power house uses agency equipment to make parts for his personal mechanical projects. The print shop manager prints up brochures for his church. Each of these is another form of fraud, deception, and theft of government resources by staff.

A few detailed examples may be useful, so think again about the warden's wife who is a realtor. On first blush, the warden's wife is not an employee, so she is not bound by any particular corrections-related ethical code. However, the warden himself is likely to benefit if his wife sells more homes. The peril comes if he instructs the personnel officer to provide this information to his wife so she has a competitive advantage by using privileged information. An argument can be made that at some point the agency newsletter provides information about promotions and transfers, so she would get the information anyway. But by getting the names in advance of other realtors, she has a competitive advantage. The warden may not be able to keep his wife from selling houses to incoming employees, but he should not allow the personnel officer to provide this information. (We will discuss next the problems created for the personnel officer in this scenario, which are significant as well. And as a further complication not delved into here, think about the incoming employee's dilemma. He or she is forced to choose between using the warden's wife and some other realtor; will this have any career impact?)

Next, take the case of the administrator who has an agency vehicle assigned to her for ease in performing her duties. She is permitted to take the car home, so she can return to the institution on short notice. Is it also all right for her to take the car shopping downtown? Is it all right to take her children to school in it? Is it all right to take it on a weekend getaway with her husband? Is it all right to go visit her brother in a distant city for a few days? The rationale for allowing personal use gets thinner as these hypothetical situations progress. And at some point (beyond just looking at agency policy on this topic) a conscientious manager needs to consider when a clear appearance of personal advantage begins to emerge.

The "appearance test" also would apply to an agency information systems manager who works closely with a local computer company to supply the institution's automated data system needs. The company offers to give the manager an extremely deep discount on a computer for his personal needs. In deciding whether to take the deal, the manager must ask if this is an offer that would be made to any other customer, or whether it may appear that it is being given to him in exchange for the institution's business. Even if the offer would be available to anyone, the appearance problem persists. And if the offer is only because he does business with the company, it clearly is impermissible.

Relationships with inmates and their families inherently involve abuse of official position. Most agencies explicitly prohibit employees from having any relationship with inmates or their families, other than to conduct their professional duties while in contact with them. Because of the manipulation potential when such contacts go beyond professional bounds, there is no debate in the corrections community on this subject, yet it comes up repeatedly.

Relationships among employees also are an ongoing concern. The issue of workplace romances and other emotional entanglements is inextricably bound to issues of sexual harassment. This is an area of increasingly vigorous litigation, both in terms of individual and organizational defendants being charged with (either conducting or knowing of and tolerating) inappropriate conduct. No agency can afford to ignore the need for strict regulation of such relationships, but there is constantly visible evidence of just that kind of inaction.

The person contemplating actions that fall in this ethical area needs to examine agency policy (and his or her motives) very carefully. The appearance test is critical. Think about how you would explain it to the average taxpayer. Imagine explaining it to the legislative committee that controls your agency budget. Visualize how this situation would be portrayed in a newspaper article. The rule of thumb here, as in so many of these cases, is, "when in doubt, don't do it."

Conflicts of Interest

Conflicts of interest constitute something of an overarching category of ethical concern; many other ethics problems either start with or inextricably involve this issue. Conflicts exist when employee loyalty is divided between an individual's organizational responsibilities and any outside interest. A potential conflict of interest is present whenever an objective observer of one's actions would have cause to wonder whether the observed actions are motivated solely by organizational concerns or external (usually personal) concerns.

An organization's employees most certainly have the right to engage in outside financial, business, or other activities, as long as these activities do not interfere with the conscientious performance of their duties. It is necessary to avoid both actual conflict of interest and any behavior that creates the appearance of conflict of interest. A merely perceived conflict of interest is genuine to the outside party.

Think back to the computer specialist mentioned in an earlier example. The computer deal just looks too cozy to explain away unless it is absolutely clear that anyone walking in the door of the computer firm could get the same deal. Unless that clearly was the case, this situation passes neither the appearance nor the actual conflict test. Let's look at a few of the other instances already cited.

The case management administrator who was carrying on an affair with someone he supervised created an immense conflict of interest when it came to assignments and promotions within the department. Beyond the morale implications for other secretaries, this also creates vulnerability for a potential future sexual harassment claim. Indeed, any of the previous examples with sexual overtones creates a highly visible conflict of interest for the employee when it comes to decisions about the involved inmate or employee.

The case of the administrator with outside consulting interests is more complicated. He started out with agency approval for this activity, which came into question when financial ties began to arise between his agency and his clients. Examining the facts, it was clear that this employee had no duties relating to finance. The client relationships preexisted his headquarters assignment, so there was no way he could have received the consulting job because of his position. In headquarters, he had no way of influencing the decisions about whether agency money flowed to a client of his. He had filed a memo withdrawing himself from any discussions or decisions that might come up in the future that would appear to present a conflict. In this actual instance, a full ethics investigation resulted in a decision that there was no conflict.

Finally, the example of the caseworker who recognizes an old school acquaintance provides a simple model for how to respond to many such cases. He notified his supervisor at once. He documented the facts, letting everyone in the chain of command know that he should not be involved in the decisions regarding the inmate's case. He made a conscious effort to avoid the inmate while the forensic study was completed.

Conflicts of interest are tricky. What may seem like one to you is clearly not one to me (and usually those views depend on whose interests are involved.) The safe rule in such instances uses the "appearance" test—if there is the slightest possibility that anyone would view this as a conflict, then avoid the activity.

Here are a few guidelines to avoiding problems in this area:

- Employees who are in positions involving financial matters should avoid placing business with any firm in which their family or close business or personal associates have a direct or indirect interest (usually financial).
- No employee should derive personal financial gain from transactions involving the organization unless the organization is advised of—and approves of—the arrangement in writing.
- Employees should conduct every aspect of a personal business venture totally outside of the workplace, and on nonwork time. For example, the employee who, while at work, solicits orders for cosmetics, food containers, jewelry, and so on, is in violation of ethical standards. The person who (on work time and using the organization's equipment) makes photocopies for a part-time activity is similarly in violation technically.
- Employees should not allow any situation to arise in which they have hiring authority or supervisory responsibility over a relative.
- Employees should not solicit, offer, accept, or provide any consideration that could be construed as conflicting with the agency's interests. This would include exchanging things such as meals, gifts, loans, entertainment, or transportation with anyone dealing with the agency or institution.
- Employees should not accept gifts exceeding the maximum value established by the agency; never accept gifts of cash in any amount.
- Employees must safeguard inmate and staff information against improper access or use for financial gain by unauthorized interests.
- Those in authority to do so should require vendors and contractors to be aware of the organization's ethical standards, and maintain impartial relationships within those bounds.
- Employees should not endorse any product or outside service on behalf of the organization (such as security devices, food service equipment, self-defense items, etc.).
- If serving as a member of any community organization or in a public office (if permitted), employees should abstain from any discussion or decision affecting their employing agency. If in doubt, they should always disclose the situation and seek resolution of an actual or potential conflict of interest before taking what might later be deemed an improper action.

Questions concerning a potential conflict of interest can usually be addressed with the agency's human resource department.

■ Finally, in some organizations managers are asked to sign a conflict-of-interest statement either indicating the presence of potential conflicts or the absence of such. This statement may be renewed periodically.

Partisan Political Activity

It is not uncommon to find that an organization's code places a limitation on employees who participate in elective government activities. The typical restriction is that they may only hold nonpartisan offices, such as a school board seat. Even in such cases, it is necessary to ensure that employees are not seen as representing the institution or agency.

However, partisan political issues are a difficult subject in many jurisdictions, particularly as one moves up the career ladder. In many states, counties, and local government units, selections for top management positions are associated in some way with partisan political connections. Clearly, most state corrections director positions are political appointees, and as a result their jobs are quite vulnerable. But political factors can intrude, even lower in the chain of command and in lower levels of government.

Examples of this kind of situation are numerous. Managers and line employees are "asked" to sell tickets to political fundraising functions. They are "asked" to contribute their own funds to party coffers. They are "asked" to contribute personal time stuffing envelopes for a candidate. They may even be assigned to do partisan political work while on duty.

Some states have legislated on this topic, and the federal system is essentially free of such influences. But in many systems, if you want to get a certain job, a promotion, or other considerations in the system, having the right political connections is the fastest route. This is not to say that people of proven ability cannot get ahead even in these agencies. But in many instances being compliant with these political pressures enhances the opportunities that are available.

There is no way around it—these are tough situations. If a manager has moved up in the organization to the level where he or she receives this kind of pressure, that person has to make a decision about how he or she wants to make a living and how he or she wants to live his or her life. If you believe that your moral or ethical principles would be compromised by participating in a specific activity, don't do it. But be ready to suffer the career consequences.

And while it is not often discussed, if you do participate in such partisan activities, be ready to suffer a different set of possible consequences. That is because in many jurisdictions it is illegal for government employees to do the things listed three paragraphs above. While the climate and culture of an organization may make it seem par for the course to engage in these activities, it is possible that doing so may violate state or federal law.

To help employees in these areas, public agencies may include specifics in their codes of conduct, such as the following:

■ An employee speaking out on public issues must avoid the impression or appearance of speaking for the organization.

- Employees who hold public office must do so as individuals, not as representatives of the organization. They must pursue the duties of such office in a manner that does not conflict with institution or agency responsibilities.
- No agency funds or other resources (including facilities) may be used to support any political activity.
- No employee may be reimbursed in any manner for political activity or expenditures.

Acceptance of Gifts or Favors

Speaking bluntly, the correctional environment is ripe with opportunities for inmates, their families, and their associates to try to corrupt staff in material ways. This can take the form of money, gifts, trips, discounts, sexual favors, and many other enticements. It is the rare employee who does not encounter an opportunity, some time in his or her career, to stray down this dangerous path.

The example of our computer manager comes to mind again, but the list at the beginning of the chapter contains others as well. The food service administrator who received gifts and money to steer institution food purchases to an inmate's mob associates ultimately learned a very hard lesson. The camp administrator who took bribes to approve furloughs threw away a promising career.

Inmates will try to entice you, particularly if they perceive any personal vulnerabilities. It is extremely unwise to discuss personal problems (marital, medical, financial, etc.) on the job, and particularly when inmates can overhear such discussions. Even mentioning personal woes to other staff can result in inmates learning about them through subsequent unguarded conversations. And when a motivated inmate has information about some possible weakness on the part of an employee, that person becomes a potential target for compromise.

Having an inmate approach you with an offer of a bribe can take subtle forms; don't expect an outright offer. It may be an intimation that they know someone on the outside who can help with some "advice." It may be a friend of theirs who can help out with a part-time job. It may be a business deal on the outside that the inmate can't take advantage of, but is too good to pass up. It may not even be from the inmate. It may come in the form of a phone call from an inmate's family member or associate.

It even may come out of the clear blue sky, like it did to the correctional officer who refused a bribe to allow hunters on restricted state property. And while he did the right thing in refusing, he did not, however, do everything right. He was young, inexperienced, and working independently on patrol. He did not think it was necessary to inform his supervisors and did not memorialize the encounter with these hunters. Had they later made allegations against him for something connected to this incident, he would have been poorly prepared to rebut them.

Prudence in Public Statements

Fortunately, most managers will not be in a position to make formal statements on behalf of the agency. The term "fortunately" is used deliberately, because anyone who has been a public spokesperson for an agency knows how weighty a task that is.

But it is possible, and indeed likely, that most correctional managers will occasionally be asked about their employment and their views on corrections by members of the public. And in those instances, it is vital to separate personal views from those of the agency.

This topic is related to the proper use of public and protected information categories, which will be discussed next. Suffice it to say that information gained on duty that is not in the public domain has no place in the conversational arena outside the workplace.

Diligent Adherence to Public Records Requirements

This aspect of the code requires correctional employees to be diligent in keeping records that have public safety implications. The correctional officer who does not file an incident report against a deserving inmate violates this precept. A case manager who slants a progress report to make an inmate look more favorable to a parole board also does so, even if it is because he likes the inmate and not because anything of value has passed between them. The work supervisor who deliberately misrepresents an inmate's work record because of fear of inmate retaliation had breached this responsibility.

These are areas that can be easily overlooked in the daily ebb and flow of an institution or community supervision setting. But cumulatively, they have an immense bearing on the picture the criminal justice system develops about offenders. Any deliberate misrepresentation, omission, or insertion of improper or inaccurate information violates this standard.

Reporting Unethical Conduct

Here is the other area where the young officer on patrol chasing hunters was remiss. Instead of raising the issue of his superior allowing hunters in an impermissible area, he kept quiet. Now one might ask, who would he have gone to, and to what effect? These are reasonable questions, particularly when coupled with his status as a relatively new, low-ranked employee. The fact is, when he revealed this incident years later, he expressed the view that he thought at the time that the assistant superintendent had the latitude to waive the rule.

If that was an honest answer, then so be it. But in many instances, an employee becomes aware of an unethical (sometimes illegal) act on the part of co-workers, and knows very well that it is wrong. In such a case, the person has a moral, ethical, and often legal, responsibility to report the misconduct. So easy to say, but so hard to do in practice.

That is because as a matter of practical reality, the expectation that this employee will disclose the improper conduct collides with another part of the law enforcement world. That is the view that officers never "snitch" on other officers. How individual employees deal with this conflict says a great deal about their character.

Return for a moment to the personnel officer who has been asked by the warden's wife to provide employee information. This clearly is internal agency information until it is published in an openly distributed newsletter. Suppose the warden tells him to go ahead and provide the information. What should he do? Tell the warden that he won't do it and hope for the best? Call his headquarters superior (a staff position working for the warden's boss), relate the situation, and hope for the best? Do it, risking that an incoming employee will complain about the situation and thus disclose the way the names are being released, and hope for the best?

The point is that in some cases there are no good answers and certainly no easy ones either. A person of good character will refuse to do the improper thing and trust that the system will vindicate him or her.

Take the example of the associate warden who refused to participate in inmate abortions for religious reasons. This is a real case, recounted not so much as an example of how to deal with informing on coworkers as for the ("damned if you don't, damned [in his mind, literally] if you do") dilemma. Following his refusal he was punitively transferred, filed litigation claiming religious persecution, and raised this issue with legislators, who passed restrictions on the agency's ability to force staff with such religious beliefs to participate in abortions. (Not unexpectedly, those actions further aggravated his relationship with his agency superiors.) After a new agency head learned of the case, he ultimately settled the lawsuit in a very agreeable way for the manager. But for years his career was in limbo, he was ostracized by many coworkers, and underwent immense personal stress with no assurance of a positive outcome. Sometimes hard-core ethical issues can lead to long, complicated, life-changing situations, and not always with a happy outcome.

Employees who find themselves confronted with illegal or unethical behavior on the part of their coworkers have no easy choice. It is easy to say, "do the right thing," despite the personal and career consequences. But one may need further motivation to do the right thing. For those, it is worth noting that in many situations a person who knows about malfeasance and does nothing to stop or report it becomes complicit, and thus subject to sanctions as well. The conclusion is that, in the absence of more noble motivations, it is better to report the act and suffer some limited consequences than to not report it and suffer even more serious effects if your knowledge is later revealed.

Nondiscrimination

This is an area that is covered extensively in employment law, and little more need be said here. Regrettably, discrimination on the basis of race, sex, creed, or national origin is hard to extinguish. Old prejudices, patterns of thinking, and organizational biases are far harder to overcome than they should be. It is not only unethical, but illegal for corrections professionals to engage in conduct that in any way reflects those biases and prejudices.

Information Integrity

Certain categories of information must be protected within a framework established by statute and agency regulation. To that end, an agency's ethical standards may set forth the following principles:

- Employees and former employees may not disclose personnel, inmate, or proprietary information except as specifically authorized.

- All agency property and information in one's possession must be surrendered upon termination of employment.
- Employees may not use, either directly or indirectly, information acquired through employment with the organization for personal gain or the gain of others.
- Official information is for official purposes only. Any variation from that theme places the person involved in jeopardy, both ethically and in some cases legally. Here again recall that poor personnel officer, dealing with a persistent warden's wife.

But there are other sides to this issue as well. Many actual instances of this kind are inadvertent, but even such a lapse can have serious consequences in the correctional setting. A receiving department officer who mentions to an inmate the time of an impending transfer can be setting up the escorting officers for an ambush-based escape attempt. An employee who casually mentions the timing of a shipment from or to the industrial area may be contributing to an escape plan or the introduction of contraband.

There are, however, instances where misuse of information is calculated. A correctional officer who uses visiting room records to contact the attractive girl-friend of an inmate is misusing official information, as well as departing from the professional role he should be playing. A personnel clerk who tips off a friend that he or she did or did not get the promotion is violating this concept. A caseworker who tells an inmate the substance of a highly confidential discussion of a parole hearing panel likewise is erring seriously.

A very specific area of concern is inmate HIV status. This is a highly sensitive matter. Lapses in this area can result in not only violation on an inmate's privacy rights, but in come cases can result in assault, injury, and even murder of the immune-compromised offender. This information (HIV status) is disclosed on a need-to-know basis for a reason, and every employee, not just managers, must be acutely aware of this fact.

It also is easy for someone in contact with a notorious or celebrity inmate to want to talk about that person with friends or acquaintances. It's sort of fun, after all, to tell people how you are dealing personally with someone who others only know from the media. No matter what the rationale, such loose talk is inappropriate. It constitutes a violation of inmate personal privacy at a minimum. But it may violate statutory restrictions that exist in almost every jurisdiction as well.

Impartial Personnel Decisions

This is an area that is ripe for manipulation in some systems, particularly those in which there may not be well-formulated personnel policies. The mechanics of selection, training, and promotion of agency personnel are addressed elsewhere in this book. In those discussions it is assumed that politics (personal or partisan) play no part in those decisions. In most systems, an additional factor against the use of partisan or personal factors in promotions is the existence of collective bargaining agreements that prescribe these practices in great detail.

But the reality of it is that interpersonal politics can tempt decision-makers and partisan factors can influence decisions as well. Informal and poorly documented promotional practices can allow personal factors to enter decision-making. Corrections history is replete with examples of referral systems that involve local politicians providing workers to facilities of all types.

When a manager is confronted with these kinds of practices, the same advice applies as noted earlier in connection with reporting unethical conduct. These practices are largely illegal and surely unethical. Adhere as closely as possible to prescribed agency promotion policy. Document copiously. And if confronted with pressure to promote someone for extra-policy reasons, do the right thing.

■ Additional Issues

Fortunately for correctional managers, most anticipated ethics issues will be addressed by policies established by the agency. Most agencies have very clear prohibitions against staff having any unprofessional relationships with inmates or their families. In most jurisdictions, sexual relations between law enforcement and correctional personnel and those in their custody is a criminal matter. A facility housing female offenders in which abortions are performed will ordinarily have a policy exempting staff who are opposed to abortion for religious reasons from participation in such procedures. Published codes of conduct are common at many levels. Moreover, when ethics issues arise that may not be clearly covered by policy, or involve exceptions or extraordinary circumstances, some agencies have ethics officers, whose job is to assist in this kind of situation.

All employees are responsible for understanding and complying with the laws and regulations applicable to their jobs. Each manager is responsible for ensuring that employees have all the information they need to enable them to do so. To that end, organizations should publish their ethical standards of conduct and disseminate them to all employees. These standards also should be distributed externally as appropriate to vendors, contractors, third-party agents, and others as necessary to advise these entities what to expect in business dealings with agency personnel.

It also is advisable to put in place a system for reporting alleged or potential violations of the organization's ethical standards. Employees are urged to report what they believe may be a violation of these ethical standards. Reports should be immediate, thorough, and directed to either the individual's immediate supervisor or the chief human resource officer. For potential violations that might appear especially sensitive and for those of such a nature that direct reporting might compromise the reporting employee, it would be advisable for the organization to establish an "employee ethics hotline" number. An employee may use this number to make an anonymous report or to request guidance in describing or addressing a potential violation.

Ultimately, it falls to the individual to conduct him- or herself ethically. But it is up to managers to ensure that all employees receive, review, and understand the organization's ethical standards of conduct. Consequently the ethical standards of conduct should be a regular subject of both new employee orientation and continuing education. In terms of managing a department or group, the manager should strive to model ethical behavior in all aspects of job performance. Thorough orientation and education notwithstanding, there is probably no more effective influence in shaping employee behavior than the manager's visible behavior. If the manager visibly observes and conscientiously adheres to the ethical standards of conduct, his or her employees are more likely to do the same. The manager's continued demonstration of ethical behavior is one of the most important dimensions of successful ethical management.

Employees at all levels have the continuing responsibility to display complete integrity in all aspects of their work activity. Integrity influences the reputations of people as individuals, and individual reputations together ultimately determine the reputation of the organization. Indeed, no set of ethical standards can ever replace a balanced combination of sound judgment, common sense, and personal integrity.

As a final comment on this subject, private corrections presents a slightly different mix of issues, because the profit motive can potentially intervene between management and sound correctional practice. Managers in private facilities are acutely aware that their operations must be run both as a business and a prison. Many of the statutory and regulatory safeguards built into public agency operations are not applicable to private prisons. Not all contracts impose stringent controls over vulnerable areas.

For this reason, ethical conduct in the private sector involves avoiding some additional issues, as follows:

- Skewing conduct and progress reports to lengthen inmate stays, thus supporting higher billable population levels.
- Payment of gratuities or in-kind favors to government officials in exchange for contracts.
- Duplicate billing, or billing for services not actually rendered.
- Questionable timing of releases and commitments to maximize user-days.
- Filing false or erroneous cost reports for agency reimbursement, when contracts so provide.
- Utilizing incentives that violate antikickback regulations or other similar statutes, or engaging in questionable financial arrangements between the institution and vendors supplying the facility.
- Knowingly failing to provide covered services or necessary care.

EXERCISES

Exercise 18-1: Is the Boss Always the Boss?

Carl Mason is business manager of a very large, medium-security private prison. He reports to Robert Green, associate warden for operations. Green, formerly business manager at the prison, has been with the company since its inception more than 12 years ago, and has been in his present position for 5 years. Mason is the third business manager reporting to Green in less than 5 years. He has never heard a predecessor's opinion of Green, but his own is that Green tends to micromanage. He seems to hang on to as much control as possible within the department.

As a result of some recent dramatic systems changes, Mason's information services section expanded and as the upgraded automation took effect, general accounting was eventually overstaffed. Mason accepted the eventual necessity of reducing staff in the accounting group. However, a budget deadline was coming up and he had not been able to achieve the needed reduction through attrition. It looked as though he might have to lay off two people unless he was lucky enough to have someone resign or retire. The only person anywhere near retirement was Ned Kline, who often expressed his intention to work until age 65. Kline was 9 months shy of turning 62.

Green requested a plan for bringing staff down to the required level, along with Mason's personnel budget projections. Mason submitted two accountant names for layoff: Ralph Brown and Jane Manriguez. Green responded by suggesting instead that Mason lay off Kline and another member of the accounting staff, Jerry Victor. Mason did not agree, and he asked for Green's reasons. Green responded that he believed Kline and Victor to be the two least productive people in accounting and that Kline had a "chronic attitude problem."

Mason disputed both of these suggestions, pointing out that he had gone by straight seniority in his recommendations although he had been given no criteria for designating employees for layoff. Mason considered Victor capable, and Victor was third from the bottom in seniority. Brown and Manriguez were the logical ones to go on a last-in, first-out basis. He generally agreed with Green about Kline's productivity and attitude, but he felt he had to point out that Kline was always known as a complainer but the issue was never addressed. In fact, Kline's personnel file held years of "satisfactory" performance evaluations and not a single criticism or complaint. If keeping Kline this long had been a management mistake, Mason reasoned, it was unfair as well as risky to get rid of him at his present age.

Green told Mason to do what he thought was right; he was only offering suggestions. Mason knew there was no love lost between Green and Victor. They had had occasional differences on business matters, and when they communicated at all it was curtly. Mason felt that Green was using the layoff to get rid of people he did not particularly like. One week later Green asked Mason if he had revised his recommendations. The layoffs were to take place in stages, with a pay period between individual layoffs. One had to go at the end of the current week.

Mason indicated that his first choice to go was Brown. Green repeated his earlier recommendation to layoff Kline and Victor. He felt that Kline would not be hurt because he was vested in the retirement plan and was known to own some rental property on the side. On Wednesday, two days before the target for the first layoff, Mason prepared a termination notice for Brown. He went to Green for the required signature, but Green refused to sign. He said to Mason, "What would you do if I gave you a direct order to lay off Kline and Victor?" Mason answered, "I don't know."

Green said he was going to keep the notice until the following day and do some thinking. When Mason went to see Green the next day he found that Green had not signed the layoff notice for Brown, but had rather prepared—and already signed as higher management—termination notices for Kline and Victor. He showed them to Mason and said, "I feel that it's in the best interests of the institution if these two are the people to leave. They're the least capable employees in your department, and when push comes to shove, the finance operation is me. I'm sorry you chose to ignore my suggestions, so I'm giving you a direct order: Sign these two layoff notices and get Victor out of here this week and Kline at the end of the next pay period."

Instructions

Imagine yourself in the position of Carl Mason and consider what ethical dilemmas are presented, what courses of action might be open to you, and what you believe you would have to do if indeed you were placed in this position.

Exercise 18-2: What Is Appropriate and What Is Not?

Address the following situations in essay form. In each case describe what you believe is ethically appropriate and what is not, and why. Also, be sure to describe in as much detail as necessary what you believe should be done about each situation.

1. Inmate-patient Smith, visibly upset, charged into the prison hospital administrator's office to register a complaint about comments he heard on the hospital's elevator. It seems that the institution doctor was discussing a particular case with a physician's assistant. The doctor described the patient as "imagining things" and "not nearly as sick as he thinks he is." The other laughed softly and agreed, saying, "We might want a psych consult for that one." They left the elevator on the same floor as Smith, just a few steps ahead of him. Mr. Smith returned to the elevator and went straight to the administrators' office.

2. Christmas was just 2 days away. Bob Williams, the business manager for the private corrections facility, was in his office when a heavy package arrived. It was an ornate wooden wine crate containing a dozen bottles of relatively expensive wine. A note attached to the crate said, "A little something for you and your capable assistant Ms. Brown." It was signed by a vendor who did a small amount of business with the institution, and was regularly proposing to take on more. The same day, one of the institution's department heads, Betty Young, dropped in on Williams to wish him Merry Christmas and leave him a beribboned bottle of champagne, as she was doing for every department head.

Exercise 18-3: The Introductory Scenarios

Take any of the twelve scenarios in the "From the Inside" feature at the beginning of the chapter and discuss them in detail.

- What was the first sign of a potential problem in this scenario?
- Were there things that the employee ignored that would have prevented the full development of an ethical problem?
- What alternate course of action might have been taken?
- Were there things in this scenario that pointed to poor supervision, or places where timely supervisory intervention could have presented a problem?

ENDNOTES

1. *The New Shorter Oxford English Dictionary,* 4th edition, *s.v.* "honesty."
2. *The New Shorter Oxford English Dictionary,* 4th edition, *s.v.* "ethics."
3. ACA Web site (www.aca.org). Adopted by the ACA Board of Governors and Delegate Assembly, August 1994.

Decisions, Decisions

19

Chapter Objectives

- Establish a direct relationship between the amount of effort going into a decision and the potential consequences of that decision.
- Identify the elements of the basic decision-making process and describe the steps followed in rational decision-making.
- Emphasize follow-up on implementation, which has traditionally been the weakest part of the overall decision process.
- Define constraints, both absolute and practical, and identify the various forms in which they appear.
- Establish perspectives on risk, uncertainty, and judgment in decision-making.
- Define the "no-decision option" and review its implications.
- Review decision-making authority and responsibility.

As a rule . . . the person who has the most information will have the greatest success in life.

Benjamin Disraeli

All decisions should be made as low as possible in the organization. The Charge of the Light Brigade was ordered by an officer who wasn't there looking at the territory.

Robert Townsend

"It's My Decision"

The warden and his new executive assistant were transferred to the maximum-security prison about the same time. The institution was seemingly poised on the verge of having to make major operational and program changes to accommodate an increasingly difficult inmate population—in effect move to a "super-max" operation.

As they discussed the problems involved in managing a place like this—with many dangerous prisoners and immense potential for escapes and assaults—they came to the topic of critical decisions. The agency had a strong tradition of oversight through headquarters and regional administrators. With that tradition came an understanding that wardens were sometimes told by their superiors to take actions with which they did not necessarily agree.

The executive put the question to the warden, "Do you think you'll be told whether to make this change, and how to do it?"

The warden replied, "Nobody else in the agency will ever tell me what to do down here. They may suggest. They may intimate. They may even cajole more than a little. But when it comes down to it, nobody there wants to take the responsibility if things go wrong. So the decision always is going to be mine, and mine alone."

■ A Fact of Life

Decision-making is a fact of life for correctional managers. Indeed, every employee makes decisions every day, sometimes a great many of them. Some of them are small and inconsequential and are made automatically, or very nearly so. Others are larger, potentially significant, and require considerably more time and effort. Like the warden above said, some of them will be weighty and lonely.

Generally, the potential consequences govern the amount of effort that goes into making any particular decision. This stands to reason. Decisions involving appreciable risk usually are not (and should not) be made lightly, nor can they usually be made with speed and ease. A decision to authorize an evening educational class for inmates who work during the day is far different than that made when determining the best way to resolve a hostage situation. Lee Iacocca said, "If I had to run up in one word what makes a good manager, I'd say decisiveness. You can use the fanciest computers to gather the numbers, but in the end you have to set a timetable and act."

Most people are self-programmed to make some decisions, especially small decisions of little consequence and those made regularly. Most managers probably do not consciously think of making a decision about what to have for lunch or which shoes to wear to work. Nevertheless, they are following the basic decision-making process. They are receiving information, forming and comparing alternatives, selecting one of these, and translating those choices into action.

The more often managers make certain kinds of decisions, the better they get at it. Decisions in new areas require more effort to do the job right. Take the manager of the security department of a new institution that is about to open. The job likely could include ordering chemical agents and munitions for the armory and towers. With no other guidance (in the form of headquarters procurement specifications, for instance) the supervisor might gather information about several brands and carefully evaluate them for cost and other features. This could take a significant amount of time, and might require switching to a second or perhaps even a third brand later on as information was accumulated based on experience during training exercises or even in actual usage. However, chances are that each time the ordering process was repeated the task would become quicker and easier. Eventually the manager might reach a point where the decision to order these chemicals and munitions became no more than the simple act of reordering a product that had already been accepted as meeting the department's needs.

In a similar way, people become self-programmed to make many periodic decisions. Regardless of preprogramming, however, and regardless of the existence of comprehensive policies and procedures, supervisors face numerous situations that require original decision-making. Policies and procedures never cover everything. What, then, to do when a necessary decision cannot be made "by the book," as often is the case in the correctional setting? What do you do when no one is immediately available to tell you what you should do?

When a situation requiring action is left up to you (and there are plenty of these in corrections) you have to do something. A supervisory position requires the incumbent to exercise judgment as to whether the situation is in fact his or her problem, and to exercise judgment in arriving at a solution.

■ The Basic Decision-Making Process

The six elements of the basic decision-making process are (1) identifying the problem, (2) gathering information, (3) analyzing information and arranging it into alternatives, (4) choosing an alternative, (5) implementing the chosen alternative, and (6) following up on implementation.

Regardless of the scope or potential impact of any particular decision, most of the foregoing elements are present to some extent in every decision-making situation. In minor situations the problem or need may be self-evident and require no attention beyond simple recognition. This is still a time to gather information, analyze it, form alternatives, and make a choice. Of course in simple, repetitive situations (again, for instance, consider deciding what shoes to wear to work) preprogramming may compress these steps considerably. Implementation is always present, since a decision is nothing without it—the person must eventually make a choice and put the shoes on. Follow-up is also always present, fleeting though it may be—do the shoes really match the outfit, and does that little scuff mark show?

As decision-making situations grow in scope and complexity, however, the importance of each element of the process looms greater.

Identifying the Problem

Before tackling any problem requiring a decision, it is necessary to consider whether the situation is a real problem. One has to be sure that something is based on fact and not the result of opinion, misinterpretation, or bias. Do not simply "jump into" every apparent problem that comes along. Investigate to determine whether it is indeed a problem deserving attention.

Look also at the nature of the apparent problem. Is it a unique situation, not previously encountered and certainly not covered by existing policies and procedures? Is it a common, recurring problem, one for which specific procedures exist? Is it perhaps self-solving, one of those rare but nevertheless genuine situations that correct themselves when left alone?

Also be mindful of whether the issue being confronted really is the problem or if it simply is a symptom of the real problem, which is hidden. Finding a cut bar is a problem all right, but it also is an indicator of a larger problem—an attempt to escape, to break into the commissary, or to engage in some other unauthorized activity. It also suggests that problems in tool control and inmate supervision may exist. Being led astray by symptoms, which are the obvious, surface indications of trouble, is a hazard inherent in supervisory decision-making. Effective treatment of a symptom may lead to immediate improvement. However, the symptom may soon reappear, suggesting that what seemed to be the problem may not be the problem at all.

If, for instance, a backlog occurs in the sentence computation process, you may feel the need to do something about it. Perhaps you authorize overtime or give the employee involved some temporary help and in a few days the backlog is gone and the computation specialist is current with incoming work. A few days later, however, you notice that a backlog is again growing; there is a 3-day, then a 5-day accumulation of uncomputed sentences. In reducing the backlog you treated a symptom. The return of the backlog suggests the problem is elsewhere, perhaps in staffing itself, distribution of work, inmate commitment patterns, or somewhere else.

It is not always clear if a symptom is only a symptom until it has been dealt with once and it returns. However, by scratching beneath the superficial indications of trouble it often is possible to determine where the real difficulty lies before deciding on corrective action.

Gathering Information

Volumes have been written on this and the next stage of the decision-making process. Elaborate quantitative approaches to decision making concentrate on the collection of information and the arrangement of this information into alternative choices.

The task is essentially to gather enough information to give some degree of assurance that the decision is being made consistent with reasonable risks. The broad objective in gathering information is to gather everything that has a bearing on the decision-to-be. Gather everything that can reasonably be collected, considering the time and effort that can or should be devoted to the decision. This is research and observing with a specific purpose in mind—to get as much information as needed (or can be obtained) to help make an intelligent decision.

Nobody can say how much information is enough, and a manager must recognize that "enough" information may never be available. It is only possible to suggest that to the extent possible, information-gathering efforts should be consistent with the potential impact of the decision to be made.

If you happen to be selecting a new booking camera for the receiving and discharge area, you may spend a great deal of time talking with salespeople. You will study the prices, operating costs, and features of many different machines. You might talk to staff in other institutions who use various cameras. Your activity could consume a number of hours spread out over days or weeks, and this effort may be justified because of the amount of money at risk. However, if you happen to be trying to decide what brand of pencils to buy for the office, you had best not spend three days gathering data about brands and prices.

Setting aside easy examples, in a riot or hostage situation, the issues are the same, and yet very different. In all likelihood you will be required to make life-and-death decisions. But in doing so you will be confronted with the fact that you have no way of gathering all of the critical information you think you need to make your decision. You may not know exactly how many hostage-takers there are, where they are located, how they are armed, their mental state, and many other relevant factors. (We ignore in this discussion the various technical and negotiation strategies that a crisis manager might use to gain some of that information. That is because it would be a rare case when all of these information elements are known with such certainty as to make a decision-maker comfortable when the time came to make the final decision.) In such a case, it is almost certain that you will wish you had more information before you make the decision.

In gathering information for decisions, work with facts and not with opinions. A great deal of so-called information comes to a manager directly or indirectly from other people. In gathering these decision-making building blocks, take care to separate the factual from the subjective.

Buying pencils probably is a no-brainer. But while researching the booking camera issue, you can ask staff at other institutions for specific figures on their patterns of film usage, frequency of equipment breakdown, and other factors that are important to your operation. Facts and opinions both may be useful, but the latter will more readily lead astray.

Analyzing Information and Arranging It into Alternatives

This, along with gathering information, constitutes a cyclic process. A manager finds this is a process of arranging and evaluating bits and pieces of information while still collecting more information. The tendency is to "fill in the gaps" as the process continues, going back for more information whenever questions arise or weaknesses arise in the data. This cyclic process can go on for quite some time in decisions of considerable impact. Try to obtain as much information as possible when the potential consequences are significant.

However, this is the stage in which the process begins to break down in the hands of some so-called decision makers. Projecting the impression of being conscientious, cautious, and thorough, some people continue gathering more and more information and keep refining the alternatives further and further, somehow seeming to take forever getting around to actually making the decision. As already suggested, however, in a decision of importance, the decision-maker will never truly have "enough" information. It may not be possible to reach the stage where there is a 100 percent comfort level about taking a risk based on the information in hand. However, there is no way to "if" a problem to a solution. Somewhere along the way the manager must take a stand, accept some risk, and decide. Of course, the higher the degree of risk, the more the information desired, but even in a true life-and-death crisis, there will come a time when (often still lacking all of the desired information) a decision will have to be made.

Any number of different kinds of information will greet a manager when researching a problem. In buying a booking camera you will learn about cost, size, operations, service and maintenance, warranties, color, and a dozen other factors. You may even learn about how computers and digitalizing processes can allow you to store inmate pictures on computer disks and then use them for nonbooking purposes. The cost and organizational implications of a decision like this are high. But even in low-cost activities, like buying pencils, price, brand, color, hardness, and other elements are relevant. It is the same in almost every decision-making situation. A number of factors are present, and all may or may not have a bearing on the final choice.

The organization's needs, the department's needs, the inmates' needs, and the manager's needs and preferences all have a bearing on the decisions that eventually are made. However, it is not often possible to achieve complete satisfaction of all needs and preferences.

For instance, you might like a booking camera that gives you the lowest operating cost and the longest warranty, but it turns out that machine A has the lowest operating cost and machine C has the best warranty. You then find yourself in the position of having to make smaller decisions along the way relative to your decision factors themselves. You are led to discover that trade-offs are necessary between and among decision factors and that you will settle for less in regard to the factors that mean the least to you for the sake of achieving satisfaction on certain other factors.

You may be confronted with the need to send a response team in to handle an uprising without knowing the full extent of the problem. You know that doing so will likely increase the risk to the responding personnel, because if the inmate forces are greater than you estimate, the staff may be captured, injured, or killed. Against that possibility is the knowledge that, left unchecked at an early stage, the uprising likely will spread, involving more inmates, more territory inside the facility, and more staff in other areas that now may be secure—tough decisions to make, and tough consequences either way.

Ultimately, decision-related factors will be arranged into a number of alternatives, from among which one will be selected. The number of alternatives may be clearly limited and well defined. Sometimes, however, the number of alternatives is up to you. If there are five brands of 12-gauge shotgun ammunition on the market and eight makes of booking cameras available, the decision-maker should not be limited to looking at only one or two possibilities. If the decision is to send the response team in now or wait for state police backup, at least this kind of binary situation is clear (if unappealing).

Of course, most managers cannot afford to perform an exhaustive analysis of every possible choice. Rather, they will scan the field of available choices using some broad criteria and concentrate on developing perhaps three or four of the more appealing choices as specific alternatives. Several possibilities may emerge along the way, only to be discarded because certain constraints rule them as unsuitable (more about constraints later).

Choosing an Alternative

Left with several reasonably well-defined alternatives, make the decision by picking the one that apparently best fills relevant functional needs and meets applicable preferences. It is a matter of comparing the possible choices and taking the one that "comes to the top" (both objectively and subjectively) for the decision maker. This solution should be able to withstand the following tests:

- Does this answer deal with cause (the problem) rather than simply with effect (a symptom)?
- Does this answer have the effect of creating policy or establishing precedent, and as such should it be formally expressed as a guide for future decisions?
- Will this solution, if implemented, have adverse effects on any other aspects of the prison's operation?

Implementing the Chosen Alternative

Implementation is action, putting the decision to work. Without implementation a decision is no decision at all, it is simply an academic exercise in "What if we did this?" A true decision is both choice and action.

From this point forward most of what was said in Chapter 5 about the basic management functions could be repeated. Planning, organizing, directing, coordinating, and controlling all come into active play in the implementation of a decision and in the follow-up on that implementation.

Following Up on Implementation

Follow-up is generally the weakest part of the decision-making process. Employees resist change for many reasons. Old habits persist, new habits are difficult to form, and, most important here, conditions surrounding the decision may change. The moment specific information is obtained and committed to paper, it starts to become obsolete. Needs and preferences change, products change, people change, and the environment changes. Time goes on, and the more of it that passes (and with significant decision making the lag between choice and complete implementation can be considerable) the more change is likely to accrue. The manager will need to clarify instructions, assess timing, and make adjustments as necessary, and in general supervise the entire implementation effort.

During implementation, be open to employee suggestions. Learn from the way implementation appears to be going. Do not be afraid to admit a mistake. Be willing to reverse or withdraw the decision if experience shows it is turning out to have been a bad one. Sometimes, through no one's fault, an alternative that appeared best on paper will go sour in practice, and it makes little sense to continue pushing a poor choice. On the other hand, however, if partial implementation only assures that the decision was the best that could be made under the circumstances, then stick to it and see it all the way through.

■ Constraints

In decision-making, constraints are those limiting conditions that rule out certain alternatives as possible or desirable choices. Once identified, constraints signal how far one may go in considering certain alternatives; they show what is practical and what is not.

The constraints commonly encountered in supervisory decision-making involve time, money, quality, personalities, and politics. Other factors such as limitations of physical space and shortages of various material resources may also appear as constraints. However, since it is possible to build space and buy material, these seeming constraints reduce themselves to limitations of money as well.

Time

Everything done in the organization takes time. It takes time to process a request for an FBI fingerprint check, escort an inmate to an outside hospital, order and obtain ammunition from a supplier, design and build a new segregation unit, organize to react to a riot, perform a disciplinary hearing, and accomplish thousands of other activities.

Time may be a critical factor in a decision, or it may not matter at all. If you are still wondering, for instance, about what brand of pencils to buy, it may not matter at all that you need to wait an extra week to get the best deal on the preferred brand. However, if you are deciding on alternative procedures for reacting to a hostage situation you may be limited to considering only those alternatives that can be successfully brought to bear within a tightly restricted amount of time. As it relates to health, safety, and the quality of inmate care (as it relates, in fact, to life and death), time is a key constraint in many supervisory decisions made in correctional operations.

It is necessary to assess the decision situation for the importance of time in the outcome. Once realistic timing is determined, the alternatives shaping up as well outside the limit begin to rule themselves out. In situations in which it may not be critical, time can be considered flexible. Timeliness can be traded off against other factors. When time does not matter, the best choice may be to complete a given task more slowly for the sake of greater convenience or lower cost. When time is of the essence, get the best information you can (including, if possible, advice from others who have been in such situations) and then make the best choice you can from the available solid options.

Money

Money—the full cost of implementing a decision—is a common constraint. Often the limits are clear, defined by the amount of money available.

For instance, when you go to lunch with just five dollars in your pocket you know that all meals costing more than five dollars are ruled out by financial constraints. The money constraint can also be put there by someone who is making decisions of another kind or from another perspective. For example, you would like to buy a new laptop computer for your department but your immediate supervisor says, "No, it's not in the budget, we have no money available. You'll have to do without it for another year." Although there might be money available in the organization, someone else has made a decision on the relative worth of your proposal versus other uses for the money. In such a case you are fully as constrained as though the money did not exist at all.

In short, any time an otherwise workable alternative is more costly than available resources allow, that alternative is effectively ruled out.

Quality

Quality—as represented by effective safety, security, and inmate programs—is always a primary consideration in the operation of a correctional facility. It can be just as great a constraint as time or money—sometimes overriding all others. Generally, when managers make decisions affecting inmate programs and services they are faced with the necessity of considering only those alternatives that do not compromise the quality of the facility's safety and security systems. Often the "quick and dirty" solution to a problem is clearly the best in terms of time, money, or both. But if quality of the organization's core functions (staff and inmate safety and security of the institution) is likely to suffer then the solution is unacceptable.

Personalities

Although personalities may not seem to present legitimate constraints on decision alternatives, they nevertheless must be reckoned with. It is important to try to avoid getting tied too tightly to the belief that "people shouldn't be that way." The fact of the matter is that people are "that way," and the constraints presented by personalities can be fully as limiting as absolute financial constraints.

There are managers who will not buy a particular product that analysis shows to be the best available simply because they do not like the salesperson. Some employees will not work well with other employees because of personality clashes, so an alternative that might otherwise be the best solution may not work at all. Also, some people are more resistant to change than others—some are more solidly attached to old habits or have strong tendencies toward doing things their own way as opposed to anyone else's way.

Although personality constraints are usually not the most powerful forces shaping a decision, they are nevertheless important. People are themselves a key factor to consider in analyzing alternatives. Approval and implementation are accomplished only through people, and people's acceptance of the decision and their ability to work with it are fully as important as the elements of the decision itself. This "people" consideration flows into the labor-management arena as well. If there are clear historical or current indications that the union will oppose a course of action, that fact alone will have implications for choosing, modifying, or abandoning that option.

Politics

Politics may not appear to constitute a legitimate constraint in a decision situation. However, the political implications of a decision are as real as the personality factors, and they often have far greater impact.

This book cannot effectively address how partisan politics might affect high-level administrators in state agencies (in terms of not only programmatic decisions but also job tenure). But that factor certainly is a reality in many correctional systems, and it would be naïve to ignore it.

The other kind of politics—interpersonal/organizational—is a reality as well. In the ideal, correctional agencies are organizations of people (employees) with a common purpose. The agency serves society, which itself is a body of people having certain wants, needs, and preferences. Within both corrections and society are many subgroups organized along various social, economic, professional, vocational, and other lines, and each of these groups has its own desires. Hardly a supervisor has not said at one time or another, "Politics shouldn't matter in running the organization," but in reality political considerations do matter.

Politics (the word carries a negative connotation for many in the profession) is frequently described as "the art of the possible." Specifically in correctional organizations it is the art of the possible given the actions and characteristics of all the various groups involved. For instance, a particular decision that serves the financial interests of the agency may alienate the employees. A decision that may appear good for the state's budget, a personnel freeze, for example, may be seen as bad for the facility and would thus alienate its staff. Within the organization, a decision that serves the needs of the security department may unavoidably make the job of the food service department more difficult and thus alienate those employees. A decision that serves the needs of the business office may legitimately upset relations with the maintenance staff.

This is not to say that political considerations should rule or that "pressure groups" should get their way simply because they are larger, stronger, or more vocal than other groups. However, it is necessary to consider political implications because a decision that is made in the face of the strongest opposition, whether that opposition is justified or not, generally requires far more in the way of resources and effort to implement.

The foregoing examples show that playing "politics" often is a realistic acknowledgement of the fact that managers have to find ways to accommodate the legitimate needs and concerns of others in the organization, or even outside the organization. In short, it is the price to be paid for functioning within a network of people and organizational entities.

Absolute and Practical Constraints

Constraints may be absolute or practical in nature. An absolute constraint clearly defines a limit beyond which the manager cannot go.

For instance, if you are choosing a booking camera and $2,000 is the absolute extent of what you can spend, then $2,000 is the limit for any camera you seriously consider. More often than not, however, constraints present themselves by limiting the practicality of a decision. A certain condition becomes a constraint because it renders an alternative impractical under the circumstances.

As another example, if you decide it is important to have two particular functions located adjacent to each other in the booking area you might be faced with considerable expense for altering physical layout. You may then look at other possibilities: there may be space available across the hall, in an adjacent room, or on the second floor of the receiving and discharge building. You can function under any of these alternatives, but you will obviously do better with some than with others. It may be that the closer together you bring the two functions (saving staff escort time in the process), the more financial resources you will consume implementing the decision. Given that finances are limited but you are still trying to achieve reasonable operating efficiency, you develop trade-offs and settle perhaps for the second or third most desirable choice.

This kind of decision may be driven by structure of the budget; there may be no construction/renovation funds but ample salary or overtime funds. Many constraints, then, are in effect not saying "You absolutely can't do this" but are rather saying "You're going to pass a point where you shouldn't do this because it's no longer practical"—like locating one function next door even though it costs a fortune, or locating it elsewhere because it costs nothing in terms of construction or renovation.

A manager needs to assess realistically the constraints inherent in the situation, recognize the inherent limitations, and focus on realistic, practical alternatives. Alternatives that lie beyond the bounds of identified constraints are really not alternatives at all.

■ Risk, Uncertainty, and Judgment

Some decision-making theorists speak of something called "perfect information." This is a state that exists only when the manager knows all there is to know about all aspects of every available alternative. Some would argue that if there truly were such a thing as perfect information there would be no decision at all: the "decision" would have made itself because the only true alternative would be self-evident.

Because there is no such thing as perfect information, there are always elements of risk and uncertainty in a decision-making situation. Risk is there because something may be lost (be it time, money, effectiveness, or perhaps life itself) if the wrong decision is made. Uncertainty also exists. Since the manager does not know everything about all aspects of the situation and has no guarantees that things will come out right, he or she does not know that the final choice is the right one.

One major objective in the decision-making process is to minimize risk and uncertainty by learning as much as is practical about each decision-making situation. Since risk and uncertainty are always present, there is always the need for judgment in decision-making. Decisions do not make themselves: People make decisions. All efforts at gathering and analyzing information, as well as all sophisticated quantitative decision-making techniques, are no more than efforts to reduce the extent of pure judgment required in decision-making.

For example, someone lines up three booking cameras for you and asks, "Which one do you want?" You might make a purely judgmental decision by pointing and saying, "I'll take the middle one." The probability that you have made the "right" decision is literally one-third, based on a random choice of three available alternatives. Consequently, you experience a two-thirds chance of being "wrong." However, suppose you analyze all the data you can obtain about the three cameras and allow your judgment to be influenced by quantitative information. Although you might never be absolutely sure you are making the right decision you could well reduce the chance of being wrong to considerably less than two-thirds.

In the final analysis many decisions will be right or wrong because of human judgment—regardless of the amount of quantitative information involved. It is not the manager's objective to try to eliminate judgment from decision making. This cannot be done. Rather, refine that judgment by learning as much as it is practical to learn about the alternatives.

The No-Decision Option

In even the simplest of decision-making situations there are always at least two choices: to decide or not to decide. The no-decision option entails selecting the latter choice. In effect this is a decision to "decide not to decide." This does not have to be a consciously made decision, but it can, and very often does, occur by default through procrastination. Taking no action on a problem amounts to the exercise of the no-decision option.

Appreciate that whether the no-decision option is exercised by choice or through procrastination it is still a decision. Frequently it is the decision with the most potentially far-reaching consequences; making no decision in a riot situation would be disastrous. But in less dramatic circumstances, all too often managers adopt an attitude (either consciously or unconsciously) that suggests, "If I'm really quiet maybe it'll go away." Sometimes it does indeed go away and things get better. However, things usually do not get better. To cite one of the corollaries of the well-known Murphy's Law: "Left unto themselves, things invariably go from bad to worse."

The Range of Decisions

Decision-making situations may range from the highly (but rarely totally) objective, with plenty of factual information on which to decide, to the purely subjective or totally judgmental decision. The decision-making process described in this chapter essentially applies to all decisions. The differences lie in the kinds of information with which a manager must deal in developing alternatives.

It usually is possible to be more comfortable with so-called objective decisions. Facts, figures, and other data are available to work with. It is possible to compare prices, statistics, hours, positions, or some other specific indicators and to make choices. Orderly approaches to decision-making are possible, and improved facility at making such decisions comes with practice in conscientiously following the steps of the process through decision after decision.

The highly subjective decision is another matter. Little or no data are available. One must make choices based on rules and regulations, policies, procedures, and precedents—all of which form a backdrop but do not give clear guidance. Very often the core decision-making process requires instead a basic sense of what is right or wrong, fair or unfair, or logical or illogical. Many personnel-related decisions fall in the area of the subjective, and while the basic decision-making process does not apply nearly as specifically as with most objective decisions, improved decision-making ability again comes largely through experience.

Although decisions may range from the mostly objective to the wholly subjective, the character of any particular decision may be greatly influenced by the conditions under which it must be made. The single circumstance that probably has the greatest effect on the character of any particular decision is the imposition of pressure.

Decision-making pressure is often felt as a limitation of the time available in which to investigate properly and render a decision. It is one thing to face a situation in which there is more than adequate time to develop and assess all workable alternatives. Working on the annual budget—deciding in a relatively calm environment how to prioritize funding—could be an example of this kind of decision. It is another matter entirely to realize that undesirable consequences will result if a decision is not made by a deadline, and that this deadline does not leave time enough for reasonable investigation. Responding to a hostage-taker's ultimatum is in this category.

Unfortunately many supervisory decisions are pressure decisions (albeit less critical than the last example). Managers have to accept the fact that limited time will squeeze them into a less-than-desirable pattern of analysis and action. In this regard, time may constrain not only the alternatives but also the entire decision-making process.

■ Responsibility and Leadership

An anonymous quotation goes, "It's all right to pull decisions out of a hat as long as you're wearing the hat." While this statement is only partially true (it is not usually all right to pull decisions out of a hat) the point about responsibility is well made. If a manager has been given the authority to make a decision, an equivalent level of responsibility should accompany that authority. In making decisions for the department a manager must be consistent with their duties as a supervisor. The manager is responsible for the output and actions of all assigned employees. However, it is not appropriate to make a decision with which employees from other departments (or the supervisors of other departments) must comply. In accepting a given amount of responsibility the manager ordinarily acquires decision-making authority consistent with that responsibility and within his or her sphere of authority.

Also on the subject of authority, when delegating certain decision-making powers to some employees, be sure to extend authority and responsibility in equivalent amounts. Authority and responsibility are each weakened, if not negated entirely, by the absence of the other. How much decision-making authority is delegated to employees will be a direct reflection of the supervisor's leadership style. Generally the autocratic or authoritarian leader will prefer to retain all such authority while the participative leader will involve the employees in shaping and choosing decision alternatives.

◼ No Magic Formula

It is possible to create valid guidelines for making many decisions. Managers do it all the time: they generate rules, regulations, policies, and procedures to guide decisions. Institution emergency plans for responding to riots, escapes, and other crises are examples of how some structured guidance can be provided for the most complex of situations.

However, in spite of what a few determined bureaucratic leaders may believe, it is not possible to anticipate all contingencies and pre-make all decisions. It is not possible to cover everything with rules that say, in effect, "When this situation arises, apply this remedy." A riot response plan does not tell what to do in every type of riot; it only provides broad response parameters within which the on-site crisis manager makes specific decisions. Good supervisory decisions always will remain a matter of arriving at a proper emphasis on all decision elements through judgment based on facts and figures, knowledge, experience, advice, intuition, and insight.

EXERCISES

Exercise 19-1: The New Booking Camera

You are the receiving and discharge supervisor, and you must select a new booking camera for the department. The need for the camera has already been established, but you do not yet know which particular machine to select.

You have investigated several cameras and discarded a number of them because they do not take the kind of pictures you need or their costs were far above what your equipment budget would allow. You are left with three cameras from which to choose. All three will give you the kinds of photos you want, and all three will fit in the available space and are compatible with available power sources.

Camera A is available for $1,500. It must be purchased outright; it is not available on a lease. It will last at least 5 years in normal operation.

Camera B costs $3,000. It is also available on an 8-year lease at $500 per year. The estimated useful life of camera B is 8 years.

Camera C costs $2,500. This camera is also available on a lease at a cost of $450 per year for a term of 6 years. Camera C has an estimated useful life of 6 years.

Questions

Based on the very limited information you are given above, which of the three cameras might you select? Why?

Would you alter your decision if you were constrained by the necessity to limit expenditures because of a severe fiscal problem in the agency?

What additional information would you like to have before making a decision? (In other words, what important information has been omitted?)

Exercise 19-2: Deciding Under Pressure

You are facilities manager of a large correctional complex that includes four separate facilities. Several times over the past few months the power plant supervisor (whose plant supplies steam and electricity to all four prisons in the complex) has mentioned to you that the maintenance workload and asbestos abatement problems on steam lines throughout the complex have been increasing and that he needs more help. His requests have never been more specific than "more help," nor have they been very strongly stated, so you have not looked into the situation.

However, on Monday of this week the supervisor came to you and said: "The plumbing crew needs an additional pipe fitter and I need more funds now to deal with the abatement problem. I'm tired of waiting and tired of being overworked and worrying about a lawsuit on this asbestos thing. If something isn't done this week, you can find yourself somebody else to run the power plant."

Questions

1. List at least three courses of action that may be open to you. What are the apparent disadvantages of each course of action you listed?

2. What hazards are presented by the way the "pressure" was introduced, and as facilities manager how are you "at risk" in this situation?

3. How do you want to present this situation to your immediate supervisor? Are there any risks to you or your career?

Managing Change: Resistance Is Where You Find It

20

Chapter Objectives

- Establish a perspective on "change" as an unavoidable feature of the correctional environment.
- To the extent possible, learn to differentiate between inflexibility and resistance.
- Review the more prominent reasons why people resist change.
- Consider the effects of change on the pursuit of careers in corrections.
- Identify likely sources of employee resistance to change.
- Establish guidelines for the supervisor to consider in managing change and minimizing employee resistance.

Have no fear of change as such and, on the other hand, no liking for it merely for its own sake.

Robert Moses

Progress occurs when courageous, skillful leaders seize the opportunity to change things for the better.

Harry S. Truman

Necessary Changes

The new deputy warden transferred in with a mission. He had been told that the warden he would be working for was essentially retired in place, and had little motivation to manage beyond the acceptable minimums. The message to the deputy was, "Get the place shipshape no matter what it takes."

Upon touring the institution, it was clear that a lot of work had to be done. It wasn't even clear that acceptable minimums were being achieved. Sanitation was a major concern. Inmate personal property was far in excess of acceptable levels. Staff wore their uniforms dirty and/or in poor condition. There was a sense that the institution was on the "pay-me-no-mind" list and that no one in the agency cared what went on there. Consequently staff members were lackadaisical and poorly motivated. His conversation with the warden after this tour made it clear that he had a free hand, as long as he didn't require the warden to get more involved than he now was.

The new deputy came in like a whirlwind. He toured housing units every day. He railed on sanitation, scraping accumulated dirt and wax out of floor corners with his fingernail clipper, to show staff and inmates that the tile could come clean. He spent time every day in food service, checking sanitation, looking at temperatures on cooler and freezer thermometers, prowling the steam line with a vengeance. He issued a standing order that any time there was a forced cell move in segregation that he was to be there to see that it was done properly. He talked with line staff everywhere in the institution about their duties. He attended every labor-management meeting and his level of participation showed he was listening carefully. He also made sure he was available to inmates, standing where they could talk to him at mealtime, taking questions and fielding complaints. He insisted the unit managers and caseworkers do the same. He carried a note pad, made notes, called department heads, followed up on calls, and generally plagued his line managers with attention to detail. He held staff regular meetings that had meaningful content. He established new, productive ties with union leadership.

■ The Nature of Change

It has been said over and over again that the only constant in this world is change, and often managers do feel (with good reason) that there are few permanent features in the everyday world. In the example in "From the Inside," the new deputy visited a multitude of changes on the staff of the institution in a short time. The stresses were significant, but the communications aspects of this administrator's style were such that over time, the facility and its staff benefited greatly.

In many ways, change has become the dominant force in modern lives. Stability is a thing of the past. No longer is it possible to depend from one generation to another on the same reliable characteristics of living, the same technology, and the same social structures and values. Consider, for example, the technology of daily living. Compare the era of the Roman Empire with the period of Colonial America and consider how little true change occurred in some 2,000 years.

Methods of transportation remained about the same—if one wanted to go somewhere one did so in a conveyance with two or four wheels pulled by an animal. Means of communication differed little, at least in effectiveness if not in availability. (It is true that the impact of the printing press was felt in Colonial America, but its effects were not widespread because the general literacy rate remained low.) Whether one lived in the days of the Caesars or in the days of America's founders, cooking, heating, lighting, and plumbing would have been accomplished in about the same way. In both societies slavery was an accepted practice. It is perhaps a sad commentary on human development to point out that one area in which noticeable advancements were made was warfare (since gunpowder was introduced along the way and refined in several "practical" ways). It seems mankind had learned how to end life more efficiently without learning how to improve it appreciably.

Over the years, technology—the total storehouse of knowledge—remained fairly constant until the late 1820s and the advent of the railroad locomotive. The invention of the locomotive improved transportation and began to draw people closer together in practical terms. In a very real sense this development launched a period of steady technological growth that lasted more than a century.

That increase, however, pales by comparison with the rate of knowledge growth experienced since about 1945. With the splitting of the atom and giant strides in electronics, technology has literally exploded. Recent technological growth has been such that it cannot be satisfactorily measured. New knowledge is accruing so rapidly on so many fronts that nothing will "stand still" long enough to be measured. Now, we blandly accept that in the relatively brief time that has elapsed since the mid 1940s, total knowledge has grown by a factor of thousands!

This is true of corrections as well. The availability of many types of sophisticated electronic perimeter detection systems (while foreshadowed by what old-timers will remember as a "snitch wire" on the wall) could hardly have been envisioned 25 years ago. We have electronic monitoring for nondangerous offenders. We benefit from elaborate computerized records systems and motion detection circuitry in closed circuit television monitors for isolated areas. Remote medical diagnostic procedures are available, such that inmates do not have to leave the facility to be examined by a physician. There is teleconferencing technology for use in remote court hearings. These and many other new applications of modern technology have given corrections new opportunities to manage facilities more effectively and cost efficiently.

The point is that most of what managers work with—and most of what they have, use, and even enjoy in a technical sense—was developed within the lifetimes of the generations now alive. And just as technology is not stable, neither are careers. Managers find themselves faced with constant change as a way of life. They must continually upgrade their knowledge and skills simply to stay even with the advances made in their chosen field.

People tend to equate stability with security. However, this is undeniably among the first generations on earth to experience such massive change within a single lifetime. In the past there was security in being unchanging and inflexible—in adopting a set of values, a pattern of living, and an approach to work, and pursuing these for life. However, this behavior is no longer completely appropriate in the workplace. The times in which we live suggest that job security now lies in the ability to learn and be flexible and adaptable.

■ Inflexibility or Resistance?

Failure to keep up with legitimate change in a chosen profession can severely diminish a manager's effectiveness. In the early stages of a career, it is entirely likely that a great deal of what both managers and employees do will change dramatically in the next 20 years. (The aforementioned "snitch wire" really was still close to state-of-the-art in the early 1970s when the authors were starting their careers.) Today's methods, techniques, and requirements will give way to new and generally more sophisticated counterparts. It will become more and more likely that the work a person is called on to perform in the latter stages of a career will bear little resemblance to the duties performed when they first entered corrections. This suggests that flexibility could be essential to a manager's continued usefulness to the agency.

Most people also are resistant to change to some extent. Sometimes resistance is rightly founded. Restraint is a stabilizer, a needed counterbalance to frivolous disruption and change for the sake of change. However, a great deal of resistance has its basis in human nature alone and forces the basic needs of the individual to take precedence over the common good.

Think for a moment of what your own reactions may have been 25 years ago to a few seemingly wild, "crackpot" ideas. Actually telling whether an offender is in his house or not from 20 miles away? ("Impossible.") Automatically activate a closed circuit television monitor in a control room when there is movement in the field of view of the camera a half mile away? ("Never work.") Depend on some microwave gadget to tell me if an inmate is about to hit the fence? (Not on your life! We need live officers in towers to do that!) A machine that can tell me if drugs have been stored in that inmate locker? (Come on, what have you been smoking?) Yet all of these and many more are realities today.

Most people are not instant believers in drastically new techniques or in the workability of marvelous new gadgets. Often it is because the average person does not have sufficient grasp of the principles on which these advancements are based. Just as often it is because they are reluctant to accept principles or theories that have yet to be proved. Perhaps they fail to appreciate that a great deal of today's technology that now is taken for granted was, in its day, subject to resistance of almost destructive proportions. When the Wright Brothers were still trying to get their plane off the ground, many supposedly knowledgeable people of the day, engineers and scientists among them, were publicly labeling powered flight as impossible. When the Wrights tried to sell aircraft to the U.S.

military, nobody was interested because they saw no possible combat application. More than once the telephone was branded an out-and-out fraud, with one critic going so far as to say that "even if it were possible to transmit the human voice over metallic wires, the thing would be of no practical value." Likewise, the automobile and the railroad locomotive had their detractors.

Many people view such past attacks on today's accepted technology as on the order of someone saying to Columbus: "Don't try it, Chris, you'll sail off the edge." However, when the initial response to a seemingly far-out idea is "it can't be done," we are reacting in the same way.

Consider how many ideas suggested on the job are disposed of on the spot with supervisory reactions like, "There's no budget for that"; "The warden won't like it"; "The security staff would kill me"; "They won't let us do it"; and "The old way's good enough for me." There are so many ready-made reasons "why not to" that it often requires an extremely compelling *why* to break through with a new idea.

In addition to constant technological change and continuing social and legislative change, in the last several decades two significant trends have been shaping and accelerating change in corrections. They are the courts and privatization.

The scope of litigation affecting correctional operations has greatly broadened over the last three decades. From a former "hands-off" attitude toward corrections, the courts (and particularly federal courts) have become much more involved in the oversight and actual operation of many correctional organizations. A great many positive gains have been made in the conditions of confinement at many prisons as a result of this escalation of judicial activity. However, this trend clearly has had an impact on the day-to-day management of many correctional facilities, and has reached deep into many agencies to affect the way line managers do their jobs.

Privatization is the second significant trend having an impact on corrections. The debate on this issue is complex and significant. To look at this, one must set aside the philosophical debate over the role of government in the correctional function in today's society. Private corrections is inarguably proving to be valuable in that it offers an important population relief function for today's criminal justice system. But it also provides a contrasting view of how a traditionally public function can be performed. This challenges traditional correctional administrators to review how they do their jobs and manage their resources. If a private correctional organization can provide adequate services and programs to inmates at a reduced cost, one must ask whether some of the economies that result in those cost savings cannot be implemented in the traditional institution. Questions stemming from private corrections can be threatening to those in public corrections, as particularly evidenced by the resistance of labor organizations to privatization of correctional functions.

■ Why Resist Change?

People resist change primarily because it disturbs their equilibrium and threatens their sense of security. For the most part people seek equilibrium with their surroundings—that balance of values, activities, and environment with which they can be most comfortable. To some people that equilibrium is a comfortable, dependable, rut. To others, equilibrium is a pattern of change, but it is their change, ordered in their way. With equilibrium comes a measure of security. People shift and move against daily forces in small ways to reestablish continually and maintain equilibrium just as surely as water continually seeks its own level. However, change from outside the person intrudes; equilibrium is disturbed, security is threatened, and resistance results.

In the workplace, most instances of resistance can be traced to one or more of the following general causes:

- Organizational changes in which departments are altered or interdepartmental relationships or management reporting relationships are changed.
- Management changes in which new management assumes control of an organization or department.
- New methods and procedures indicating that people are expected to do new and different work or accomplish old tasks in new ways.
- Job restructuring requiring that tasks be added to or be deleted from people's jobs.
- New equipment representing new technology or technological departures from equipment previously used.

For a moment, accept the assertion that change disturbs equilibrium and threatens security. Then go a step further and accept one additional factor that seems borne out in practice—people really most fear the unknown. It is the unknown that actually disturbs equilibrium and threatens security. When employees seem to be resisting change it is usually because all of the implications of the change are not known, understood, or appreciated.

However, most available ways of improving supervisory effectiveness also involve changing the way things are accomplished in the department. To have improvement there usually needs to be change—change in ways of doing things and especially change in attitude. In the last analysis the success of any particular change depends almost completely on employee attitude.

Consideration of change, then, returns again to consideration of the supervisor's approach in dealing with employees.

■ The Supervisor's Approach

As a supervisor interested in implementing a particular change, there are three avenues along which to approach employees. The supervisor can (1) tell them what to do, (2) convince them to do it, or (3) involve them in planning for the change.

Tell Them

Specific orders—commands—have been described as one of the marks of the autocratic or authoritarian leader. The supervisor is the boss, a giver of orders who either makes a decision and orders its implementation or relays without expansion or clarification the orders that come down from above.

The authoritarian approach is sometimes necessary. Sometimes it is the only option available under the circumstances—there is no time to convince or build a consensus when dealing with a riot. However, the "tell them" approach is the approach most likely to generate resistance and should be used in only those rare instances when it is the only means available.

Convince Them

In most instances, though (including those in which the change in question is a nonnegotiable edict from the upper reaches of the agency), there is room for explanation and persuasion. At the very least the manager can try to make each employee aware of the reasons for the change and the necessity for its implementation. The manager may have to champion the cause of something clearly distasteful (to them personally as well as to some employees). They do so because it may be good for the institution overall, or good for the inmates, or even perhaps because it is mandated by new regulations of some sort. Employees may not like what they are called on to do, but they are more likely to respond as needed if they know and understand the "why" of the change.

Employees deserve information, and information serves the manager well because it often removes the shadow of the unknown. Few if any changes cannot be approached by this "selling" means, and the authoritarian "tell them" approach should be reserved for those, hopefully infrequent, occasions when someone clearly cannot be "sold."

Involve Them

Whenever possible (and especially as it affects the way they perform their assigned tasks) involve employees in shaping the details of the change. It is beyond question that employees are far more likely to understand and comply when they have a role in determining the form and substance of the change.

A manager considering new equipment and with sufficient lead time will get the input of the people who will have to work with the equipment once it is in place. If the facility is to be expanded or be remodeled and the department will change its physical space and arrangement, it would be important to obtain employees' opinions on where things should be located and how work should flow. Through involvement, change can become a positive force. The involved employees will be more likely to comply because the change is partly "theirs."

In *A Passion for Excellence*, Tom Peters comments, "People who are part of the team, who 'own' the company and 'own' their job, regularly perform a thousand percent better than the rest."[1] This dynamic is not confined to the business world, it works just as effectively in the correctional setting.

And there is another potential benefit to involvement as well. Employees know the work in ways that a supervisor may perhaps never know it. A manager supervises a number of tasks, some of which he or she may once have performed. However, employees perform those tasks every day in hands-on fashion. Thus they know the details of the work far better than management and are in a much better position to provide the basis for positive change in task performance.

It is suggested here, as elsewhere, that participative or consultative approaches to management are the best ways of getting things done through employees. The most effective ways of reducing or removing the fear of the unknown make full use of communication and involvement.

Guidelines for Effective Management of Change

Plan thoroughly. Fully evaluate the potential change, examining all implications in regard to its potential impact on your department and the total organization.

Communicate fully. Fully explain the change, starting well in advance, and assure that your employees are not taken by surprise. To the extent possible, make it two-way communication; pave the way for employee involvement by soliciting their comments or suggestions.

Convince employees. As necessary, take steps to convince your employees of the value and benefits of the proposed change. When possible, appeal to employee self-interest. Let them know how they stand to benefit from the change and how it may perhaps make their work easier.

Involve employees when possible. Recognizing that it is not possible to involve employees in all matters—a line manager cannot do much about a mandate from above. But involvement is nevertheless possible on many occasions. Be especially aware of the value of line employees as a source of job knowledge. Tap this source not only for the acceptance of change but for the development of genuine improvements as well.

Monitor implementation. As with the implementation of any decision, monitor the implementation of any change. This is especially true for those involving employee task performance—increase supervision until the new way is established as part of the accepted work pattern. A new work method often is dependent for its success on willing adoption by individual employees. Just as often it can be introduced in a burst of enthusiasm only to die of its own weight as the novelty wears off and old habits return. New habits are not easily formed, and employees need the help that a manager can furnish through conscientious follow-up.

Resistance to change will never be completely eliminated. People possess differing degrees of flexibility and exhibit varying degrees of acceptance of ideas that are not purely their own. However, involvement helps, and most employees are willing to cooperate and genuinely want to contribute. Beyond involvement, however, communication is the key. Full knowledge and understanding of what is happening and why is the strongest force the supervisor can bring to bear on the problems of resistance to change.

EXERCISES

Exercise 20-1: New Perimeter Systems

The director of a large prison system was attending a conference at which vendors were showing new perimeter detection systems. Most perimeters in this administrator's system were secured by towers, although some had the venerable "snitch wire" systems on top of their walls. Liking what he saw, the director made a decision that all new institutions (including penitentiaries) built by the agency would have perimeter detection systems, augmented by roving vehicular patrols. Attendant to that decision was the design change that no more institutions would be built with walls—only fences. After all, a perimeter patrol officer in a truck could see through a fence, but not a wall.

Wardens in the system, who at that time still mostly had risen from the ranks of the security staff, resisted this proposition. How could you trust an electronic gizmo to replace a tower officer? You could see if a taut-wire system was physically intact, but how did you know if buried sensors were operable? Everybody knows that walls are what you need for high-security operations—fences are for lower security institutions. There was more than a minor ripple of discontent.

Questions

1. Does it seem that the director's new-found confidence in new technology was completely communicated to lower management?

2. What concrete steps might have been taken to get the word out that this change had corresponding benefits?

Exercise 20-2: Surprise!

On Monday morning when prison industries employees arrived at the facility they immediately noticed the absence of the superintendent of industries. This was not unusual; she was frequently absent on Monday. However, while she rarely failed to call the factory when she would not be there, on this day she still had not called by noon.

Shortly after lunch the supervisors of the various sections of the factory were summoned to the warden's office. There they were told that the superintendent was no longer employed by the institution. They (the supervisors) were told to look after things for the current week, and that a new superintendent already had been hired and would be starting the following Monday. All the supervisors were told about their new supervisor was that it was somebody from outside of the institution.

Questions

1. What was right or wrong about the manner in which the change in superintendent was made?

2. What would you suppose to be the attitudes of the industries staff upon hearing of the change?

3. With what attitudes do you suppose the staff will receive the new manager?

4. In what other ways might this change have been approached?

ENDNOTES

1. Peters, Thomas J. and Nancy Austin. *A Passion for Excellence*. New York: Warner Books, 1989.

Communication: Not by Spoken Word Alone

21

I have made this letter rather long because I have not had time to make it shorter.

Blaise Pascal

I had heard you were a very great man, but I don't think so. I heard your speech and understood every word you said.

Davy Crockett to Daniel Webster (attributed)

Excerpts from Real Government Correspondence

"In order to accomplish a rational, coordinated program of management and tenure adjustment, in accord with Bureau goals, the various frameworks in which functional programs are accomplished must, to the greatest extent possible, and on a periodic basis, be objectively defined, analyzed, and put into proper perspective."

"It may be concluded that multivalued decision problems are so common in economics that objectives and criteria of decisions are best formulated in a way that takes uncertainty explicitly into account; this can be done, for example, by subjecting the economic optimum to the restriction of avoiding immoderate possible losses, or by formulating it as minimizing maximum possible losses."

"This office's activities during the year were primarily continuing their primary functions of education of the people to acquaint them of their needs, problems, and alternate problem solutions, in order that they can make wise decisions in planning and implementing a total program that will best meet the needs of the people, now and in the future."

Huh?!

■ The Written Word

Letters, memoranda, and other written communications are essential in the operation of any organization. Indeed, in many instances documentation is critical to a correctional manager's survival. But at the same time it seems that managers have to put up with far more paperwork than they care to handle. This is certainly true in corrections, where the paper tiger has become a beast of considerable proportions. But a great deal of this paper is nevertheless necessary. Many organizations function quite well in spite of hefty amounts of paperwork, but just try to run an organization without paper.

However, the written word possesses a serious drawback. A piece of writing is essentially a one-way communication, providing no opportunity for immediate feedback in the development and transmission of the message. Once committed to writing, one is unable to amend, correct, clarify, or defend the contents in response to the reader's reaction.

Because of the one-way nature of this means of communication, the need for clarity in writing becomes critical. However, clarity is the attribute most often lacking in written communications in the organizational setting. The examples of actual public agency writing cited in "From the Inside" illustrate this point quite well.

■ Sources of Help

Far too many letters and memos resemble low-grade Thanksgiving turkeys: they are hefty and meaty looking, but when cut into they are mostly just stuffing. This chapter will present some guidelines aimed at helping take some of the "stuffing" out of letters and memos. However, although these guidelines can help improve writing clarity, no single chapter in a book will make someone a "good" writer, especially if they have basic difficulties with grammar, punctuation, and usage. To become an effective writer two things are needed: (1) the desire to improve and (2) the help provided by practically oriented resources such as writing teachers and good reference works on writing.

Numerous books on writing techniques are available. Several are described in the recommended readings at the end of this book, but if only one of them were to be available, first consideration should be given to *The Elements of Style* by William Strunk, Jr., and E. B. White. This book has fewer than 100 pages, yet contains more solid, usable advice per page than any other book available. It is a great place to start when a manager decides to get help in improving professional writing. Another particularly good reference on how *not* to write in "bureaucratese" is *Gobbledy Gook Has Gotta Go,* a U.S. Department of the Interior publication.[1]

■ Guidelines for Better Letters and Memos

Written communications perform several important functions. They are used to advise (or inform), explain, request, convince, and provide permanent records. Many agencies have specific requirements for the format or style of written materials. In addition, many managers have their own specific (but often unwritten) expectations regarding style and content for material originating in their department or institution. As a writer, be aware of those formal and informal guidelines for written work. Use the advice in this chapter with those factors in mind. If a supervisor will not accept written material with contractions in it, then by all means adjust to that condition. If first-person usage is not acceptable, don't use it. But within the constraints of the particular situation, use of the following guidelines will help a manager improve his or her writing in a minimum amount of time.

Write for a Specific Audience

A particular letter or memo may be going to one person, or it may be intended to be read by several people. The author will need to decide who, specifically, it is being written to. The person who will receive the communication (ordinarily the person for whom the message is principally intended) is your main audience. However, there may also be a sizeable secondary audience—others who will receive, read, and perhaps make use of the communication.

Many managers seem to believe they should write in such a way that anyone picking up a particular document will "get the message." However, this is a difficult task at best. It becomes nearly impossible when there is a sizeable secondary audience including people of widely varying backgrounds and different degrees of familiarity with the subject.

Since the author of a document really cannot write for everybody who might read the letter or memo, write specifically for the primary audience. If there is difficulty identifying the primary audience, sift through the likely recipients of the message with this question in mind: Who of all these people needs this information for decision-making purposes? Often the primary audience will consist of a single person, but it could just as well be two, three, or more people.

If you are a shift commander writing about the need for specific change in policy on recreation yard coverage, perhaps all of the other shift commanders (and maybe even the heads of some other departments) should be aware of the issue. However, it would be your immediate superior, the chief of security, who would be the primary audience because it is in that position in which the decision-making authority concerning policy is located. On the other hand, if the chief of security is releasing a new policy with which all shift commanders are expected to comply, then the memo announcing the policy will have all shift commanders as its primary audience.

Use what is known about the primary audience in deciding how to structure the message. Can it be done on a friendly, first-name basis? Must it be a formal letter, or will a brief, casual note suffice? Does this person seem to prefer detail, or would a concise overview be enough? Knowledge of the primary audience will suggest how to communicate.

Avoid Unneeded Words

Volumes could be written about this subject. However, understanding and exercising one simple concept—that of the "zero word"—will go a long way toward removing excess words from the written message.

Every word in a given piece of writing can be placed in one of three categories: Necessary, optional, or zero.

A necessary word is one that is essential to getting the basic message across. An optional word, as the name suggests, can be used optionally to qualify or modify a necessary word or phrase. A zero word contributes nothing to the message and should be removed.

Consider the sentence, "Harvey is certainly an exceptionally intelligent man." It contains only three necessary words: Harvey is intelligent. Note, however, that even with all zero words and optional words removed what remains is still a sentence.

The word "exceptionally" is the only optional word in the sentence. In getting the message across, it may be important to say that "Harvey is exceptionally intelligent" rather than simply, "Harvey is intelligent." While this is perfectly acceptable, watch out for the excess use of such modifiers and qualifiers. After a while they not only become tiresome but they also begin to lose their impact.

The sentence includes three zero words: "certainly," "an," and "man." At least they are zero under normal circumstances—assuming that Harvey is a man. Still, if Harvey were anything else (a dog, for instance), the sentence would say so. The word "an" is there for structural reasons, and "certainly" is certainly unnecessary. That is because in terms of what the writer is trying to convey, Harvey either is or is not intelligent and "certainly" does not make it any more binding. Zero words abound in most business communication. However, they are relatively easy to get rid of with conscientious editing.

This is not to say, however, that the zero word infests all writing to the same extent. In many uses of written language, writers are attempting to tell stories or create moods or impressions. However, writing an interoffice memo is not writing a poem, a novel, or even a textbook. The primary objective is to get a message across with clarity. A memo can be correct in every sense of grammar and usage although stripped of every zero word.

Select a sample of your writing and go hunting for zero words. If in doubt about a word, sound out the sentence without it. If the sentence remains a sentence and continues to carry the message you wish it to carry, then the word is probably a zero word. Chances are you will find a surprising number of zero words, including many uses of *the*, *that*, *of*, *and* other simple words.

Often a writer will use unnecessary words in groups, applying multiple-word phrases to do the work that could be done by one or two words. This is especially common in business correspondence in which some phrases have reached cliché proportions. Consider these examples:

- The use of "due to the fact that" when the writer could simply say "because"
- Saying "be in a position to" when all that is needed is "can"
- Saying "in the state of California" when "in California" says the same
- Using the stuffy "with reference to" when the job can be done by "about."

Avoid roundabout phrases in your writing. They simply add bulk to your communication without adding clarity. In fact, such words not only fail to add clarity but they also can actually harm your message by surrounding and obscuring your real meaning.

Use Simple Words

Almost every technical and professional field has its own jargon—the unique technical terminology of a special activity or group. However, to some people, jargon is nothing more than confused, unintelligible language.

It is one thing if the author happens to be a correctional doctor writing to an audience of one who happens also to be a correctional doctor. In this case it is possible to get away with the free use of the language of the specific field being discussed. However, as a group, correctional employees usually include other highly educated, specialized professionals, as well as secretaries, tradespersons, and other workers at various levels in the organization. Many of these staff in different fields have their own "languages."

Medical, mental health, and other technical professionals are among the worst offenders when it comes to sprinkling their correspondence with jargon. However, the "in" language of a field should not be allowed to cross departmental lines to any considerable extent. As already suggested, doctor-to-doctor may be a safe channel for the use of jargon. However, doctor-to-business manager or psychiatrist-to-correctional officer are channels calling for completely different approaches. Again, consider the primary audience (in this case, the recipient's background and familiarity with the subject) in preparing to write.

If technical professionals are the worst offenders when it comes to jargon, then administrative personnel, consultants, and other managerial professionals are the worst offenders when it comes to made-up words. Management is cluttered with once-sensible words to which have been appended the suffixes -ize and -wise, hence the tendency to stylize, prioritize, and regularize and speak of things as timewise or wordwise, for example.

Consider, for instance, the increasing number of people who show an apparent fondness for the likes of "operationalization." Interestingly, this clumsy word has in recent years, through use, become a "legitimate" word by its inclusion in several dictionaries. However, this legitimacy makes it no less awkward and overblown.

Coined words and barbarisms of legitimate words are a great deal of the clutter that clouds communication. These can also lead to trouble when such a language twist creates an unintended meaning. In a classic example from the field of medicine, a young administrative staffer, assigned to study systems and staffing in a hospital pathology department, referred on paper to the staff of the department of pathology as the "pathological staff." The department's director, who was sensitive to the analysis in the first place, flew into a rage after reading the word pathological (literally referring to the department's employees as the sick staff) and it was days before the remainder of the report was read.

Edit and Rewrite

During editing and rewriting, the zero words, the roundabout phrases, and other verbal stuffing should come out of any correspondence. There are few pieces of writing that cannot be improved by careful editing or perhaps rewriting. Very few people (and this statement includes professional writers) can go from thought to a completely effective finished message in a single try. In fact, professional writers probably do far more editing and rewriting than do most writers of day-to-day business correspondence.

Therein lies the problem. A great deal of what is wrong with writing is wrong simply because the writer does not put enough time into it. As this chapter's opening quotation suggests, given a particular message to get across it generally takes more time to write a shorter letter than it does to write a longer one.

If you are thinking that better writing is too time consuming, that you would have to double your correspondence time to provide that extra reading through every letter and memo you write, think also of the cost of misunderstanding. Have you ever had to spend valuable time and effort smoothing out some problem that developed because a written message was misunderstood? You can edit many memos in the time it takes to solve a couple of knotty problems arising from missed communication.

■ Changing Old Habits

In day-to-day writing many people are often unconsciously still trying to please dear Mrs. Smith who taught them English "way back when." Even in the 1940s, 1950s, and 1960s, students were taught how to write letters that sounded as though they were lifted from a Victorian secretarial handbook.

Most of what has been said so far in this chapter is "legal" in terms of what Mrs. Smith taught us. However, there are additional practices that would have been guaranteed to get a student into trouble with the teacher even though they will definitely help managers improve official writing today.

Be Friendly and Personal

Feel free to use personal pronouns in letters and memos. Employees use "I," "you," and "we" when in day-to-day conversations, so why not use them when in written communications as well? However, many people were taught to avoid personal pronouns, and this warning has residual effect for some of them. For clarity and directness, however, "I" is far preferable to archaic affectations such as "the undersigned," or "the author."

Most letters and memos should sound like spoken conversation. Achieving that conversational tone will make correspondence more direct, friendly, and personal.

Use Direct, Active Language

Ask direct questions when the situation warrants it, and avoid questions like, "Let me know whether or not you will attend." It is much more direct to ask, "Will you attend?"

The use of the passive voice is pervasive in government writing. Instead, use the active voice whenever possible. Avoid sentences like, "The contract was signed by your representative." It is much cleaner to say, "Your representative signed the contract."

Use Contractions

Use *don't, wouldn't, can't, shouldn't,* and so on, even though the use of contractions may formerly have been taught as being taboo. Contractions contribute to the natural, conversational tone that written communications should be working to achieve. Even so, many writers of official correspondence squeeze the contractions out of the writing without realizing what they are doing. The result is a formalistic style, stilted and stuffy, that serves only to create more distance between writer and reader.

Write Short Sentences

No one else is William Faulkner, who could get away with writing an opening sentence of some 180-plus words. No official memorandum or letter should even vaguely resemble the Great American Novel.

It is difficult to lay down any firm guidelines for sentence length. But consider that any sentence much more than 20 words long is edging into questionable territory. Some teachers of business writing have suggested 20 words as maximum sentence length. Others suggest that 14 or 15 words be considered maximum. Regardless, it is safe to say that the longer the sentence, the more opportunities there are for misunderstanding.

Forget Old Taboos About Prepositions and Conjunctions

It is likely that most writing students were repeatedly and sternly warned against committing two terrible "no-nos": ending a sentence with a preposition and starting a sentence with a conjunction.

A story is told about Winston Churchill and the rule concerning prepositions. When reminded that it was improper to end a sentence with a preposition, Churchill replied, "This is something up with which I shall not put." An extreme example, for sure, but it cleanly illustrates how far out of the way the search for so-called structure can lead. Go ahead and say, "This is something I won't put up with."

A sure-fire way to lose points with any teacher is to begin sentences with conjunctions, especially "and" and "but." Fortunately this archaic prohibition has been successfully shattered by professional writers. Of course if every other sentence in the letter begins with *and,* the writer will have created a different kind of monster. However, the freedom to open a sentence in this manner can eliminate a great many long sentences and a great deal of needless repetition.

Say What You Want to Say and Stop

Avoid starting a letter by repeating what was said in the letter being answered. Also, avoid opening with standard stuffing such as "In response to yours of the. . . ."

Simply say what you are trying to say. If the point of the letter is to tell a potential supplier that the bid was not accepted, do not spend two paragraphs describing the evaluation process and building the rationale for the "no" delivered in paragraph three. State the answer in the opening paragraph, preferably in the first sentence. Then go on to explain the reasons why, if necessary.

Having delivered a clear message and explained it as necessary, do not spend another paragraph or two winding down by repeating what has already been said. Simply say it and stop. Also, watch out for standard closing lines that mean little or nothing. It may be quite all right to say something like, "Call me if you need more information"—if you really mean it. It is thoughtful and it shows that you are interested, but avoid phrases such as, "We trust this arrangement meets with your complete satisfaction." For one thing, this tells the reader you expect satisfaction to result—you are not just asking. Anyway, if the reader is not completely satisfied you are likely to hear about it.

Consider something else that appears in the last example: the use of the collective "we." Few words are more likely to make a letter impersonal to a reader who is made to feel that the communication is coming from a crowd. "We" has its place, for instance, when writing to someone outside the correctional agency and speaking for the organization. However, rather than being agency-to-person or agency-to-agency, most writing will be person-to-person. As long as the thoughts are your own and yours is the only hand pushing the pen, use *I.*

■ Sample Letter

To illustrate the application of some of the guidelines offered in this chapter, a sample letter is presented in before and after versions. The letter in its original form (Exhibit 21-1) was received by the records section of a correctional facility.

Several comments on Exhibit 21-1 include the following:

- The writer used "gentlemen" as a salutation and many correctional personnel are not male; with very little effort the writer could have come much closer than *gentlemen.*

April 15, 1996

State Prison
Main Street
Someplace, New York

ATTN: Records Department
Re: John Doe

Gentlemen:

Please be advised that we are attorneys for the above named former inmate at your facility, John Doe. We are enclosing herewith authorization executed by Mr. Doe for which please forward to this office a copy of the prison record of the said John Doe.

Thanking you, I am,

Very truly yours,
Rayborn & Rayborn
Attorneys-at-Law

Exhibit 21-1 Sample letter before editing.

- "Please be advised . . ." is one of those bits of stuffing spoken of earlier. It went out of date years ago.
- "We are enclosing herewith . . ." More pomposity. The preferred form is "enclosed is," or simply "here is."
- ". . . the above named . . ." Above-named is right; within-named, also. In fact, the unfortunate Mr. Doe is named three times in one brief letter.
- "thanking you, I am . . ." This archaic form—still used by some letter writers—is made all the sillier in this case by the presence of two names in the signature block following "I am" (and the letter was received unsigned).

Now take a look at Exhibit 21-2, the same letter after reasonable editing. This letter is simple, straightforward, and to the point. And this is but one of several acceptable ways the letter could be rewritten.

Why all the fuss? One might argue that the message got across anyway, so why the criticism and the editing? True, the reader may have gotten the message. The sample letter is extremely short, so the potential for serious misunderstanding may not be evident. However, the potential for misinterpretation builds rapidly with the number of words in a letter. The body of Exhibit 21-1 contains 48 words. The edited version in Exhibit 21-2 contains 30 words. The difference represents a 33 percent reduction in the number of words used to get the message across, not to mention the mistreatment of the language corrected by the revision.

It has been estimated that most official correspondence contains from 25 to 100 percent more words than are needed to get the message across effectively. Each added word presents another unwanted opportunity for misunderstanding. Also, keep in mind that if every document you received were properly written, the two-inch thick stack of paper awaiting your attention could possibly be one-fourth of an inch to a full inch thinner.

April 15, 1996

Records Department
State Prison
Main Street
Someplace, New York

Dear _____ ,

We represent John Doe, who was an inmate at your institution between August 3, 1990, and May 17, 1995.

Please send copy of Mr. Doe's prison records. Proper authorization is enclosed.

Thank you.

Sincerely,
Richard Rayborn
Rayborn & Rayborn
Attorneys-at-Law

Exhibit 21-2 Sample letter after editing.

■ Other Writing

So far, this discussion has centered primarily on guidelines for writing letters and memos, since these make up the bulk of most supervisors' writing chores. However, a manager may occasionally find it necessary or desirable to tackle larger writing tasks such as informational or analytical reports, training presentations, speeches, or journal articles.

Many elements of the personal, direct style preferred for correspondence are applicable to other writing. For example, some speeches or training presentations can, and should, be handled with the same personal touch. However, some additional rules apply in writing more structured material such as formal reports, and still more rules apply when writing for publication in magazines or journals.

If faced with writing a major report, get a manual or handbook on the subject and do some studying. Be especially aware of the need to use one of the commonly recommended formats—one that calls for a tight summary of objectives, conclusions, and recommendations early in the report. Also, remember that the first step in preparing to write a report (or a letter, memo, or any other piece of writing) is to get a clear image of the audience.

If you are serious about improving your writing, start with the few suggestions presented in this chapter and go on to the information contained in published sources. Ideally, consider keeping a four-volume self-help library on your desk: Strunk and White's *The Elements of Style*, any other book about writing in general (preferably business writing), a reference book on report writing, and a reasonably current dictionary.

■ A Matter of Practice

Writing is a great deal of like any other skill, in that the more you work at it the better you get. If you enjoy writing, or feel that you could enjoy it, then you have a head start on the self-improvement process. But if you do not enjoy writing, if every letter, every memo, every report, every performance appraisal narrative looms before you as a painful, distasteful task, you had better examine your attitude toward writing. Is writing tough because you truly dislike it? Or is it the other way around—you dislike it primarily because it is difficult for you?

You may not have to write a great deal on your job, but chances are you write enough to make it worthwhile to try a modest self-improvement program. One thing is certain: you will never get better at writing unless you work at it.

EXERCISES

Exercise 21-1: The Matthewson Memo

To: Ted Matthewson, Warden
From: William Abernathy, Director of Corrections
Re: Use of CS Chemical Agents

Please be advised that departmental audit staff have indicated to me that your institution has continued to stock, use in training, and maintain ready for tactical deployment a large quantity of CS gas. As you know, maintaining or using any form of CS gas has been prohibited by departmental regulations for a lengthy period of time because it has been irrefutably established that this type of chemical agent, when used in closed areas such as cell houses and other buildings, can cause serious medical harm and in some cases where used indiscriminately could aggravate existing medical conditions even to the point of causing the death of an inmate. Enclosed is a copy of the controlling policy for your information.

Please consider this memo a formal directive for the discontinuation of all storage, training use, or tactical deployment of said CS chemical agents. You are required to make immediate arrangements to acquire sufficient supplies of CN-based chemical agents and to replace all CS gas supplies in your institution with the equivalent CN agents, also adjusting all emergency plans and other institutional inventories and regulations to reflect this change and the different usage requirements of CN versus CS.

If you have any questions regarding the application or interpretation of this directive, you certainly may feel free to contact me at your convenience.

Instructions

1. Shorten the Matthewson memo by reading through it once and eliminating excess words where possible.

2. Take the results of step one and rewrite the memo, rearranging passages and condensing thoughts as necessary.

3. Determine the percentage of reduction in number of words of your finished memo as compared with the original.

4. How many words are there in the longest sentence of the original? In your rewritten memo?

Exercise 21-2: The Copy Machine Letter

Write a letter to Mr. Nathan Perkins, district manager of Repro, Inc., describing the trouble you are having with your copy machine. Include the following information. (Not all points are of equal significance, and they are not presented in any logical order.)

1. You are records manager for the State Penitentiary.

2. The facility has one Repro CM-400 machine.

3. You have had the machine for 10 months.

4. You have discovered that the local address for Repro service is a manufacturer's representative and that the nearest real service agency is 125 miles away.

5. The machine has required outside service five times.

6. For the last three breakdowns your machine was down for 3 days, 2 days, and 4 days, respectively, awaiting service.

7. You bought the machine without a service contract.

8. There is administrative pressure on you to replace the Repro machine with a better-known machine.

9. The machine gives off a strong odor when operating.

10. The Repro CM-400 was the cheapest machine available.

11. The sales representative promised prompt service.

Additional Considerations

1. Your letter should probably contain complaints (poor performance, poor service, and so on) and threats (to replace, and so on). Before writing, decide which kind of message (complaint or threat) will be the dominant message.

2. You have admitted to yourself that your own organization caused a great deal of the problem when it "bought cheap" (lowest price, no service contract). Carefully consider to what extent (if at all) you might admit this in your letter.

ENDNOTES

1. United States Department of the Interior. *Gobbledy Gook Has Gotta Go.* Washington, DC: Government Printing Office, 1966.

22

How to Arrange and Conduct Effective Meetings

Chapter Objectives

- Characterize various types of meetings by the purposes for which they may be held.
- Consider the necessity for meetings but consider as well how to keep their numbers to an essential minimum.
- Provide guidelines for determining the need for a meeting and for preparing to conduct an effective meeting.
- Offer guidelines for leading a meeting in such a way as to obtain maximum benefits from the process while consuming the least possible amount of the valuable time of those attending.
- Place meetings in general in perspective as an often misused but potentially effective management tool.

Meetings: Where you go to learn how to do better the things you already know how to do anyway, but don't have time to do, because of too many meetings.

Anonymous

I have discovered that all human evil comes from this, man's being unable to sit still in a room.

Blaise Pascal

The Value of A Useful Meeting

The warden held a mini-staff meeting every day at the end of the day; he called it the "closeout." The deputy wardens, chief correctional supervisor, and executive assistant were required to attend. The agenda was simple—let everyone else in the room know anything of significance that had happened in each person's respective area that day, and what was expected to develop for the following day.

Usually the meeting took no longer than 15 minutes. Things covered included such items as intelligence information from the chief correctional supervisor, a coming audit team as reported by one of the deputies, and a notable media inquiry fielded by the executive assistant. The warden shared new information received from headquarters—everything from new policies being contemplated to personnel changes.

The value of this daily meeting—short as it was—was immense. It had considerable impact in consistent decision-making at all levels of the facility. Everyone knew what page everyone else was on, so that when the exec or the captain had to cover for a deputy for a few weeks, nothing radical was changed. It provided an opportunity to get the reaction and input of others as issues emerged and developed. It allowed an opportunity to ventilate after a day in the pressure cooker, and to be affirmed and supported. It gave a sense of closure for each day, and a sense that the coming day was not some unguided missile headed straight at the administration.

■ Meetings Are Here to Stay

Anyone who has spent more than a few weeks in management has had some unfortunate experiences with meetings. Most managers have complained about meetings, especially in regard to their number, frequency, and value. Like them or not, however, meetings are important, and frequently essential, in the operation of an organization.

Managers may often consider meetings costly, frustrating, and wasteful. Indeed, it is easy to complain about meetings, as though some indication of chronic discontent with such proceedings is expected. However, managers have to accept meetings as a regular part of work—indeed, in all reality a characteristic of organizational life.

Generally, the higher one climbs in the management structure the more time spent in meetings. At all managerial levels, however, the situations and pressures that foster convening or attending meetings are many. Put simply, it is not possible to do an effective supervisory job without occasionally (and for some supervisors, frequently) dealing with people in gatherings larger than the simple one-on-one encounter.

A great deal of grumbling about meetings is justified. Many are largely a waste of time because they lack clear purpose or are poorly led. Some meetings are a total waste of time and resources—they should not be held at all.

Meetings are costly relative to other ways of doing business. They are not necessarily costly in the sense of out-of-pocket expenditures. Their cost is reckoned largely in terms of the expenditure of unrecoverable time. The true cost of most questionable or unnecessary meetings would be best measured in terms of lost productivity. What would these people have been doing had they not been involved in a meeting?

In defense of meetings, it is necessary to point out that meetings often represent the best available technique for arriving at joint conclusions and determining joint actions. It is often possible to accomplish within minutes results that would require hours, days, or weeks by other means.

Meetings are also essential to consultative and participative leadership styles. Joint decisions and actions take longer to arrive at than do unilateral decisions and edicts. That is because true two-way communication (including discussion and feedback) requires more time than so-called one-way communication. However, the extra time spent in meetings can represent a small price to pay for the benefits afforded by an honest, open, participative leadership style.

Why meetings? This question can have any of several answers depending on the purpose of the particular gathering. Consider the different types of meetings presented next as determined by the purpose of each.

■ Types of Meetings

Information Meetings

The information meeting is held simply for the transfer of information. It can range in size from a small, section-level session to an institution-wide "recall" of all employees who are not staffing housing units and towers. In such a meeting, supervisory personnel have something to pass along to employees or others and have elected to do this with a meeting rather than by some other means. The basic purpose of any information meeting is to transfer information to the group, and in this setting the leader may do most or all of the talking. Although there are usually questions and discussion for the sake of clarification, the transfer of information is essentially one-way communication.

Discussion Meetings

The objective of a discussion meeting is to gain agreement on something through the exchange of information, ideas, and opinions. The essence of the discussion meeting is interchange. In such a setting, the exchange of information must be established between and among all participants.

Directed Discussion

A directed discussion meeting may be appropriate when a conclusion, solution, or decision is evident. The conclusion has already been determined, yet it is not simply being relayed to the group as straight information. It is the leader's objective to gain the participants' acceptance of the solution. In effect, a directed discussion is a "sales pitch."

Problem-Solving Discussion

This type of meeting is held when a problem exists and a solution or decision must be determined by the group. The answer determined by joint action may well turn out to be based on the ideas of a single participant. But at the outset the only thing that is apparent is that there is a problem with which several parties could reasonably be concerned.

Exploratory Discussion

The purpose of an exploratory discussion meeting is to gain information on which a manager or others may eventually base a decision. The objective is not to develop a specific solution or recommendation but rather to generate and develop ideas and information for others (including upper management) who must make the decision.

A Special Case: The Staff Meeting

The staff meeting may be an information meeting, a discussion meeting, or both. A staff meeting is usually held for the purpose of communication among the members of a group. (The "closeout" meeting related at the beginning of this chapter is an example of a very short version of a highly selective staff meeting.) Staff members may report on the status of their activities, and thus each may be required to effect the one-way transfer of information to others. This meeting form is also used to solve problems, sell ideas, and explore issues, and, depending on the business at hand, it may take on any or all of the three forms of the discussion meeting.

■ Meeting Preparation

Defining the Problem

To enable a group to begin dealing with a situation it is necessary to establish the nature of that situation. The first step, then, in preparing for a meeting is identification of the issue to be dealt with. Before the invited participants attend the meeting they must understand what they are going to discuss when they get there, so the convener must be able to supply a concise statement of the problem or other reason for the meeting.

Determining If There Is a Need for a Meeting

Having defined the problem, do not automatically assume that a meeting is inevitable. Other means of organizational communication are available, and all are valid and may be preferred under certain circumstances. In determining the need for a meeting, consider the following:

- How many people are involved? If very few, perhaps the task can be accomplished by telephone, letters, or memos. Useful technological advances in this area include teleconferences carried out over the phone or computer media.

- Will a meeting save time? Often a problem can be solved by a memorandum or report, but if an exchange of ideas is needed then a meeting may avoid a seemingly endless series of other contacts.
- Should everyone get the same story? Perhaps the issue is complex or technically involved, calling for different levels of participation by various categories of personnel. Perhaps there are policy issues to be considered, suggesting that certain matters be taken up at a policy-making level before they can be dealt with generally.

Deciding What Should Be Accomplished

Before convening a meeting, the convener should have a clear idea of what is to be achieved. As a definition of the problem provides a starting point, so the meeting itself provides the target. As a minimum, and before calling people together, is should be possible to say whether the meeting should result in a solution to a specific problem, the group's acceptance of an idea, some significant decision-making information, or other such results.

Selecting the Meeting Type

Based on advance determination of what should be accomplished, the convener should then decide what type of meeting to have. There are some broad differences between information meetings and discussion meetings, but there are also significant differences in emphasis among the subtypes of discussion meeting. Start with a clear understanding as to whether the discussion should be directed, or should be oriented toward problem solving or exploration. That will assist in controlling the meeting and keeping the discussion headed toward the desired result.

Selecting the Participants

Again based on the best determination of what is to be accomplished, the convener next needs to decide who should attend the meeting. This choice should be based on an assessment of who has the knowledge to deal with the problem and the authority to make decisions and commit resources to solve the problem. The objective in selecting people to attend the meeting should be to secure the broadest possible coverage of the problem without "overloading" the meeting.

Distributing Advance Information

All those invited should receive, along with the meeting notification, all information that would be helpful in preparing for the meeting. If a statement of the problem is all that can be provided, then let them know this is all the information that is available. If there is any background information in the form of letters, memos, or reports, send it to them so they can consider all aspects of the problem in advance of the meeting. Convey the expectations for the meeting. At the very least, those invited should know both the problem and the objective.

Notifying and Reminding the Participants

For all but the most informal meetings (or those which are repetitive in nature), written notification should be given. This should include time, place, preparations to make or materials to bring, the statement of the problem, and the meeting's objectives. Ideally, the convener should provide written notification a week or more in advance and plan on telephoning reminders a day or two before the meeting actually takes place. Generally, the more important the topic of the meeting and the busier the people invited to attend, the more advance notice that should be provided.

It is also a good idea to clear the date and time in advance with the key people attending. Otherwise it may be necessary to notify everyone of a change in arrangements although some have already cleared their calendars to attend. Remember that as convener, you are not the only busy person in the institution; it may frequently be necessary to adjust your schedule to accommodate the availability of others.

Arranging for Proper Facilities

This step should seem self-evident, but too often a dozen people find they are ready to meet but have no place to sit down. If a sizeable meeting room is needed, make sure it is reserved for the preferred date and time before notification goes out. It should also go without saying (but nevertheless must be said, judging from the number of times things go wrong) that there must be sufficient space for all persons involved, reasonably comfortable surroundings, and reasonable freedom from interruptions. Also, advance arrangements should be made for needed equipment such as chalkboards, projectors, and other aids.

Preparing an Agenda

For all but the simplest of meetings, an agenda should be used to guide the proceedings. If the meeting promises to be long and involved, the agenda should be worked out sufficiently in advance so it can be supplied to all attendees with the meeting notification. Whether or not it is supplied in advance, however, the meeting leader needs an agenda. It may consist of only a few broad points jotted in the corner of a note pad. But the process of writing it forces a rethinking of the purpose in calling the meeting, and some advance consideration of how to move from problem to objective.

Although the convener will not think of every necessary step or essential question ahead of time, an agenda will at least provide a reminder of certain important points. When the meeting is underway the agenda may well expand, as issues are raised and previously unknown information surfaces. This is all well and good, as long as all agenda additions or digressions contribute directly to moving the group from the problem toward the meeting's objective.

■ Leading a Meeting

Start on Time

It is prudent in some cases to allow some flexibility in how closely to adhere to a rigid starting time. If the meeting is a one-time affair involving a number of people who are organizationally scattered (perhaps including some who are management superiors), bending the rules a few minutes on starting time may make sense. In reality, the group is likely to wait more than a few minutes for a tardy person to show, especially if that person happens to be the next-level supervisor.

It is wise to remember that a one-shot meeting is not a regular part of someone's schedule or pattern of behavior. Flexibility is in order, when it does not stretch too far. It may be wise to build in some modest amount of slack in scheduling. For instance, a session might be scheduled for 1:30 P.M. although the convener knows full well it will not start until 1:45 P.M. Try not to overdo this practice. It suggests disrespect for those who do show up on time and deference to the latecomers.

On the other hand, in the case of a regularly scheduled meeting (for instance, a monthly staff meeting held at 3:00 P.M. on the third Thursday of each month), try to begin precisely on time. The more chronic latecomers are accommodated, the more likely these people are to remain chronic latecomers. Also, chronic tardiness can often be an indication of other problems, such as hostility, disrespect, lack of interest, or perhaps inflated ego.

Making it a habit of starting on time can go a long way toward curing chronic tardiness. If it is 3:00 P.M. and only half of the staff are present, start the meeting even though it is likely the remainder will be trickling in over the next several minutes. Out of respect for those who show up on time, do not repeat what has already been said for the sake of the latecomers. Rather, let the late arrivers know they have to wait until after the meeting to find out what they missed from those who were present. Make it plain, also, that the content of the early part of the meeting is not always filler or "warm up" material that people can afford to miss. Make it a habit to start regularly scheduled meetings precisely on time, and most chronic latecomers will change their ways as they get the message and become accustomed to your pattern of behavior.

State the Purpose of the Meeting

First tell the group why they are there and what they need to accomplish. Also, give them the best estimate of the amount of time the meeting should require. Ending time can be fully as troublesome as starting time in some situations. A meeting can be a form of escape for some people who lead busy, hectic working lives. Some may tend to prolong the session with irrelevancies if progress is not well controlled. (One effective way to get a 1-hour meeting concluded in 1 hour is to schedule it to start an hour before lunch or an hour before quitting time.) In short, the first item of business should be to advise those attending why they are there, what they are expected to accomplish, and approximately how long it should take.

Encourage Discussion

Do not allow the meeting to move in such narrow lines that valuable input is lost. Ask for clarification of comments that are offered. Consider requesting opinions and asking direct questions, particularly of the few "silent ones" who frequently populate meetings. Remember, if meetings are structured wisely then everyone who is there is there for a good reason. It is part of the job of the meeting leader to do everything possible to get those people talking who ordinarily tend to remain quiet.

Exercise Control: A List of Don'ts

Of Yourself

- Don't let your ego get in the way simply because you are the meeting leader and thus automatically "in control" of the proceedings.
- Don't lecture or otherwise dominate the proceedings. Remember, the setting is a meeting, not a speech or a class.
- Don't direct the others by telling them what to do or what they should say or conclude. This would amount to one-way communication, which is only marginally appropriate even when the purpose of the meeting is purely informational.
- Don't argue with participants, but always be open to discussion.
- Don't attempt to be funny. What may be funny to one person may not be to another. The best laughs generated at a meeting are those that arise naturally from the discussion.

Of the Group

- Don't allow lengthy tangential digressions to pull you away from the subject of the meeting. Granted, many legitimate problems are identified through tangential discussions. But legitimate or not, if they do not relate to the problem at hand they are diluting the effectiveness of the meeting. Should a legitimate problem arise, make note of it but sideline it for action at another time or at another meeting and proceed with the subject at hand.
- Don't allow monopolizers and ego-trippers to take over. Certain talkative people may have significant contributions to make. But their constant presence center stage serves to narrow the discussion and discourage marginally vocal contributors from opening up at all. Overall, effectiveness as a meeting leader will largely be determined by how effectively he or she controls the discussion of the group.

Summarize Periodically

Agreement in a discussion meeting is usually not reached in a single, progressive series of exchanges. Rather, agreement accrues as discussion points are sifted, sorted, and merged, and a solution or recommendation begins to take form. Capture this by periodically summarizing what has been said, giving the group, in plain words, a recounting of where the group is and where it seems to be headed. If they can agree with this summary of progress (essentially, their thoughts encapsulated and restated in the leader's words) then the meeting is on the right track.

End with a Specific Plan

When the meeting is over, the convener should be able to deliver a final summary stating what has been decided and who is going to do what and by when. Far too many meetings are frustrating affairs that may feel productive while underway, but afterward leave participants with a sense of incompleteness.

No one should leave a meeting without a full understanding of the decisions made, the actions to be taken, the people responsible for implementation, and the timetable for implementation. If the subject is sufficiently complex, it may be advantageous to call for understanding by going "around the table," asking everyone for their interpretations of what has been decided. This will disclose how they see their roles, if any, in the implementation of the decision. In any event, do not let the group leave without a clear understanding of what has been decided and what happens next.

Follow Up

As far as the manager's authority over the problem extends, it is up to him or her to follow up, assuring that what has been decided gets accomplished. It is also up to the manager to see that minutes of the meeting (should they be necessary) are prepared and distributed. Provide later assurance to all participants that what they decided has in fact been accomplished. Schedule a follow-up meeting should one be necessary.

■ Use or Abuse?

As with any other management tool or technique, meetings can be overused, underused, or used ineffectively. They can be expected to do far too much—they are certainly no substitute for effective individual decision-making. Or they can be denied the opportunity to serve their purpose appropriately, due to too little time for the meeting itself, the wrong type of meeting being used, the wrong mix of participants, and other hampering elements.

Whether meetings are used or abused is largely up to the supervisor. Meetings are often an unwieldy way of doing business, and as such it is easy for them to become wasteful and ineffective. However, the properly conducted meeting remains one of the most effective available ways of accomplishing certain tasks. Whether meetings are effective or ineffective revolves on the issue of control. If a manager fails to control meetings, the meetings proceed to control the manager.

EXERCISES

Exercise 22-1: The Conference

This exercise concerns a conference involving five persons:

1. Sue Turner, executive assistant to the warden. She functions in a staff capacity and has no supervision responsibility except for her secretary, Betty. One of her assignments is a long-term project intended to determine some of the reasons for correctional officer turnover. She called the meeting approximately 10 days ago. Two of the other persons she notified by telephone, and two she invited in person.

2. Martin Calabro, personnel officer

3. Marv England, chief of security

4. Paul Stanfield, training officer

5. Mary Hanson, union president

The meeting was scheduled for 1:00 P.M. in Turner' office. Turner returned from lunch at 1:08 to find Calabro and Hanson already there. At 1:12 England entered and Turner said, "I'd like to get started, but where's Stanfield?"

"I don't know," somebody answered.

Turner dialed a number and received no answer. She then dialed the switchboard and asked for a page. A moment later a call came in and Turner spoke briefly with Stanfield.

Turning from the telephone Turner said, "He forgot. He'll be here in a minute."

"Sue, I wish you had a larger office or a better place to meet," said England. "I don't know how we're going to fit another person in here."

"I know it's small," Turner answered, "but both conference rooms are tied up and I couldn't find another place. Say, holler out to Betty and tell her to find another chair—we're going to need it."

Mary Hanson said, "Sue, can you open your window a little? It's already stuffy in here."

Sue responded by opening the window a few inches. Then the training officer entered, squeezing into the office with the chair that had just been located. It was 1:18 P.M.

Turner said, "I guess we can get started now." She shuffled through a stack of papers before her and said, "I've got a copy, if I can, ah—oh, here it is—of a recent turnover survey done by the personnel officers in each institution in the department of corrections." Looking at the personnel officer she asked, "I assume you have this?"

"Yes, I have it. There's a copy in my office, but I didn't know I needed to bring it with me."

Turner said, "Well, I think what we can get from this thing is—"

The chief of security interrupted, "Sue, wouldn't it be better if we all could see it? Then you could go down it point-by-point."

Sue said, "I guess you're right. I just have this single copy." She turned toward the door and hollered, "Betty, can you come here a minute?"

Betty entered and Sue instructed her to run four copies of the survey at once.

Turning back to her pile of papers Sue said to the group in general, "The last time we got together there were a number of things we decided to look for. I don't remember just what we assigned to whom, but I've got it here somewhere."

For a half-minute or so Sue leafed through the papers before her. Then she turned to the file drawer of her desk and began to go through folders.

While Sue was looking, Marv England turned to Calabro and said, "Say, what have you been doing about advertising for that vacant firearms instructor position? We notified you 3 weeks ago, and Eleanor is leaving in another week and we still haven't had any candidates to interview."

Martin Calabro responded. His tone sparked a defensive reaction and a lively discussion began.

Turner located the paper she was seeking and Betty returned with the requested photocopies. Turner distributed the copies and fixed her attention on the chief of security and personnel officer as he waited for an opening in the discussion, which was now something between a conversation and an argument. At approximately 1:32 they managed to return to the subject of officer turnover.

"Now, about this survey of the other institutions," Turner began.

Mary Hanson said, "What about the survey? I thought you wanted to start with the things we agreed to do the last time we were together."

"Who cares," said Calabro. "Let's just get started."

The chief of security looked at his watch and said, "We'd better hurry up and get started and finished. I have a staff meeting at 2:00."

The meeting settled down to a discussion of the survey and the preliminary data each person had gathered since the previous meeting. At exactly 2 minutes before 2:00, the chief of security excused himself to attend his meeting. At 2:08 Stanfield was called over the paging system; he left Turner' office and did not return.

At 2:12 Turner said she felt they had tentatively decided on their next step but required some input from the two parties who had already left. She then started to excuse the other two participants with the suggestion that they get together again after 2 weeks.

At that point Turner' telephone rang. She answered it herself, her usual practice, and talked for some 4 to 5 minutes before returning her attention to the two persons left in the office. She said, "I guess that's about it for this time around. I'll get back to you and set a time for the next meeting."

When the last of the participants had left, Turner' secretary came into her office and asked if they were finished with the extra chair. Turner indicated they were, and as Betty removed the chair Turner thought gloomily of how difficult it was to get anything done in this institution, and wondered if the warden would be satisfied with her progress on this project.

Instructions

1. Perform a detailed critique of "The Conference." Make a list of points you consider errors and omissions in the way this meeting was arranged and conducted.

2. After each item on your list indicate what positive steps could have been taken to avoid each of the errors and omissions.

Exercise 22-2: Your Word Against His

This is a variant on the situation presented in Exercise 12-1. You are at a meeting chaired by your department head. Also present are another department head and four other supervisors. The subject of the meeting is the manner in which the facility's supervisors are to conduct themselves during the present union organizing campaign.

Your department head makes a statement concerning one way in which supervisors should behave. You are surprised to hear this because earlier that same day you read a legal opinion that described this particular action as probably illegal.

You interrupt with, "Pardon me, but I don't believe it can really be done that way. I'm certain it would leave us open to an unfair labor practice charge."

Obviously irritated with the interruption, your department head responds sharply, "This isn't open to discussion. You're wrong."

You open your mouth to speak again, but you are cut short by an angry glance.

You are certain that the boss is wrong; he had inadvertently turned around a pair of words and described a "cannot-do" as a "can-do." Unfortunately you are in a conference room full of people and the document that could prove your point is in your office.

Questions

1. Recognizing that you are but an attendee at the meeting and that your immediate superior is running the meeting, what should you—or what can you—do to assure that the other participants do not act on critically incorrect information immediately after the meeting?

2. What fundamental requirements of effective meeting leadership appear to have been ignored in this meeting?

Budgeting: Annual Task and Year-Long Implications

23

Chapter Objectives

- Introduce the basic concepts of budgeting and establish the importance of budget preparation to the individual manager.
- Describe the advantages of participative approaches to budgeting.
- Define operating budgets, capital budgets, and cash budgets as key elements of what is actually a *financial plan*.
- Describe the process of developing an institution's annual budget from the budgets of individual departments.
- Learn how to read a simple budget report.
- Illustrate the fundamentals of the process of monitoring expenditures against budget allocations and relate this to the management function *controlling*.
- Provide basic information on staffing considerations.

A budget is a means of telling your money where to go—instead of wondering where it went.

Anonymous

A billion here, a billion there, and pretty soon you're talking about real money.

Everett McKinley Dirksen

■ Introducing the Budget

A budget is a financial plan that serves as an estimate of future operations and, to some extent, as a means of control over those operations. It is a quantitative expression of the agency's or the component's expressed operating intentions,

Clever Budgeting

Budgeting is not simply numbers on paper—it is managing resources within reality. As such, practical budgeting often involves tradeoffs, some of which operate at a very simple level, and require ingenuity to pull off.

The unit manager and the captain agreed to a "deal" that gave the unit manager an extra officer in the unit during the evening hours, in exchange for a piece of the unit's sanitation budget, which had been targeted for a new floor buffer. The captain used the funds so acquired to buy new Lexan shields for the disturbance control squad. This reduced the unit officers to using a beat-up buffer for the small areas of the unit.

Sometimes a buffer was borrowed from another unit. When a borrowed buffer was not available then, ingeniously, the officer took a metal desk, turned it upside down on top of a wool blanket, and had inmates push it up and down the corridors to polish the floors. This continued until the next budget year, when a new buffer was again in the budget.

and it translates these intentions into numbers. And as intimated by the late Senator Dirksen in the opening quote, a budget can get out of hand if not managed properly.

A budget can be several things to the institution and to the individual supervisor. Used as a control mechanism, it can be a cost-containment tool that helps keep expenditures in line with available resources. It can also be a basis for performance evaluation, since a budget provides a quantitative indication of how well a manager utilizes the resources under his or her control. Further, a budget can be a means for directing efforts toward productivity improvement. Comparing performance against budget reveals how well resources are utilized, and can provide the basis from which to work for better utilization.

There are likely to be considerable differences in the ways supervisors approach this chapter. In some correctional agencies, individual supervisors are actively involved in the budgeting process. In others, however, the budget remains a mysterious collection of numbers assembled by the headquarters staff or the agency's business office with no supervisory involvement whatsoever. Even more variation is introduced into the budgeting process when private corrections is included in the discussion, because the entire budgeting process is tied to a revenue stream and accounting system far different than that of public corrections.

Anyone who has been directly and regularly involved in the preparation of departmental budgets will find little in this chapter that they have not encountered in practice. However, the reader with minimal or no budget preparation experience is likely to benefit from this elementary view of budgeting.

This chapter provides an overview of the budget and the budgeting process. It offers some specifics in a manner intended for consideration by correctional managers who do not have backgrounds in accounting and finance (certainly the vast majority of supervisors, both in and out of an institution). However, because of this wide variety in budget formats and budgeting processes, structured budget examples will not be provided in this chapter, which addresses instead the necessary fundamentals of budgeting for department-head-level managers.

Participative Budgeting

In most states, the legislature is heavily involved in the budgeting process. In some, the individual warden may even have to appear before legislative committees to represent the needs of the institution for the coming year. Often, this is participative budgeting in name only, as the funding parameters are set in advance and the agency representative has only a nominal impact on the final budget allocation.

In some agencies the budget is still prepared "upstairs" in the headquarters office and handed down for wardens and their subordinates to carry out. This practice leaves a great deal to be desired, as it requires the supervisor to implement a plan without having participated in its development. The supervisor may remain ignorant of many of the whys and wherefores of the budget and thus be in a position of weakness when it comes to translating the financial plan into action at the departmental level.

Other institutions bring their supervisors into the budgeting process. Sometimes this is done by the responsible managers presenting structured requests that are forwarded to the agency's headquarters as a consolidated package for the whole facility. This more team-like approach to budgeting, requiring active participation of managers at all levels, calls for close coordination of many diverse inputs and activities. It takes considerably more time and effort to assemble a budget in this manner than it does simply to allow the administration and the business office to get their heads together and issue a budget. The results of the team approach are well worth the extra work, though, because it provides several distinct advantages:

- The resulting budget is usually a far more realistic and workable plan than any that could be developed by some other means.
- The involvement of individual supervisors in the process breeds their commitment. They are more likely to believe in the budget and strive to make it work because they took part in its development.
- The interdepartmental and intradepartmental activities, and interpersonal contacts, pursued during the process tend to strengthen supervisors in their jobs.
- A spirit and attitude of teamwork is created in getting managers at all levels to work together toward a common goal.
- The encouragement of more realistic planning serves to sharpen the focus of management's efforts and produce more appropriate results a greater part of the time.

Involving the individual supervisor in the creation of the budget helps to define clearly his or her authority and the concomitant responsibility for the operation of a department. Simply put, you created your budget, so you are responsible for operating within it. In the illustration at the beginning of the chapter, getting the extra correctional office coverage and still finding a way to have shiny floors meant digging down and getting a little creative.

At an operational level, there never was a time in an institution that the sanitation officer didn't need new bug sprayers and more noxious chemicals, the laundry supervisor didn't need a bigger budget for socks, and the captain didn't want more money for CN gas and bullets. The annual budget cycle gave each of them and every other department head a chance to fight for scarce dollars, and to do the kind of horse-trading that would make a seasoned garage-saler blush.

Although the responsibility for draft budget preparation may lie with the supervisor, there is no reason for involvement not to extend below that level. As noted later in this chapter, there are certain budget-related activities (such as the accumulation of operating statistics) in which some employees could participate. Employees who understand the nature of the budget and the reasons behind its preparation are more likely to share actively in the objectives of the department.

Accounting Concepts

Before discussing how a budget is structured, explanations of a pair of simple but important concepts are in order: the fiscal year and fixed versus flexible budgets.

The fiscal year is simply the institution's or agency's 12-month accounting year. This may be any consecutive 12-month period established for accounting purposes (for instance, May 1 this year to April 30 of next year). This means that for accounting purposes the institution's "year" begins on May 1. However, a fiscal year is just as likely to be coincident with the calendar year beginning January 1. The governments of many states and the federal government use something other than the calendar year as their fiscal year. Private correctional firms are similarly free to adopt a fiscal year other than the calendar year.

Within the accounting year, it is common practice to keep track of payroll and certain other expenses on a basis of 1, 2, or 4 weeks and to accumulate this information on the basis of the "accounting period." Thus in a 4-week cycle there are 13 such accounting periods in the year. However, other important facts and figures are accumulated by month either because it is necessary to do so or because this is clearly the best data-collection period available. Accountants and others speak of the "monthly closing," the act of determining the financial results or status of operations for a given month.

Actually, this represents one of the biggest headaches encountered in budgeting—differences in the length of the periods for which some data are accumulated. Payroll data are almost always kept in 2- or 4-week periods. Many other expenses and various operating statistics are accumulated by month. In developing a budget (and especially in examining the results of operations after the fact) it is frequently necessary to manipulate some of the figures by "adding in" or "backing out" certain numbers at either end of a given period so there is complete information for the period of interest. This kind of action does not represent fraud or tampering with records. It is a necessary means of accommodating these different time parameters.

The idea of fixed versus flexible budgeting refers to the structural character of the budget relative to activity throughout the year. A fixed budget (by far the simpler of the two) assumes a stable level of operations throughout the year. It spreads all budgeted costs evenly across the 12 months. A flexible budget, however, recognizes that certain costs can vary as the level of operations varies and attempts to account for this variation in the budget. Utilities will fluctuate with the season. Inmate population levels and associated costs will rise and fall in response to factors outside the correctional manager's control. A flexible budget may treat certain costs as fixed because they are in fact fixed or because this is simply the best way to handle them. But it will also attempt to identify and allow for costs that vary as departmental activity varies.

■ The Total Budget

There are several parts to the institution's overall budget. These often are segmented out as the operating, capital, and cash budgets. Some require the supervisor's involvement, and some ordinarily do not. The parts and subparts of the total budget described in the following paragraphs are each properly identified as budgets in their own right.

The Operating Budget

The operating budget usually consists of three parts.

Statistical Budget

The statistical budget is made up of projections of activity for the coming budget year. It usually is based on a combination of past activity, current trends, and some limited knowledge of future conditions and circumstances. This is the institution's or the agency's best estimate of work activity for the coming 12-month period. These estimates are projections of statistics such as expected commitments or releases, inmate-days expected, number of meals to be served, pounds of laundry to be processed, and so on. These may be prepared as gross, fixed estimates of a year's activity, or they may be projected by month, based on certain knowledge suggesting variations in activity.

In getting ready to prepare the budget for the institution, the appropriate data may not be maintained in the individual department. In such a case the accounting section will generally compile and summarize the most recent historical data and projections available and supply this information to the various departments. However, it is often necessary for the departments to refine these figures into detailed estimates of activities for the coming budget year. This refinement is accomplished the same way the accounting department makes its projections—based on past activity in the department and the responsible manager's particular knowledge of future operations. The assistance of the agency's research department is particularly valuable in dealing with population projections that drive many of the costs involved in budgets for departments like food service, the laundry, and medical services.

Expense Budget

The expense budget attempts to account for the actual cost of operations (personnel and all other costs) for each department individually and for the institution as a whole. The departmental expense budget is the major area of supervisory involvement in the budgeting process. A simple step-by-step illustration of the preparation of one segment of a departmental expense budget is presented later in this chapter.

Revenue Budget

While public correctional agencies know the amount appropriated for the various budget areas, in the private sector a revenue component to the budget process is important. Usually prepared by the organization's accounting department, the revenue budget projects income likely to be received by the institution during the budget year, based on contractual information or inmate-day revenue projections.

Keeping in mind that revenue ultimately must cover all costs of operation, the revenue budget must attempt to consider the impact of numerous factors beyond simply earning income from keeping beds filled. A revenue budget for a private sector facility may, for instance, attempt to reflect the impact of another state sending inmates to the facility. It could include the effects of a program being eliminated or curtailed. It may reflect changes in staffing levels that have a direct bearing on cost. It certainly will contain implications from decisions and recommendations of the contracting agency, and the effects of changes in reimbursement regulation or contract renegotiations.

The Capital Budget

A capital budget is prepared to account for potential expenditures for major fixed and movable equipment. Fixed equipment (a building, a boiler, or a new roof) and major movable equipment (computers, copy machines, package X-ray machines) represent those costs that must be "capitalized." These costs apply to operations for considerably longer than just the coming budget period, so they must be spread out over a number of periods.

Sometimes the boundary between capital purchases and items that can be allocated to operations in the budget period is hazy. However, the agency or state revenue department usually has its own guidelines for determining what should be called a capital purchase. Sometimes these limits are set by statute. That guide may be expressed as some specific amount of money, some useful-life criterion, or a combination of these. For example, the institution may have decided that any purchased item having a useful life longer than 1 year and costing more than $200 must be capitalized. Thus a $250 rifle would be capitalized but a $150 gas mask would not, although both will clearly last longer than 1 year.

As a supervisor, make it a point to learn the agency's guidelines for identifying capital assets. That information usually is provided in the capital budget request or it could be secured through the departmental expense budget, depending on where it fits by agency definition.

In capital budgeting, requests for purchases often outrun the money available. Requests may come from all directions. The security staff would like new shotguns for the towers. The legislature may pass a law requiring a certain vocational training program with attendant equipment costs. Supervisors and department heads would like to have newer and better computer equipment (doesn't everyone need an agency-provided personal digital assistant?). A higher level of government may tell the institution something is needed to conform to some code, regulation, or requirement (as recently has been the case with the Americans with Disability Act).

Compiling a capital budget may also involve the development of short-range and long-range capital plans as the institution attempts to determine best what new and replacement equipment will be required for future operations. This involves looking at all equipment and listing each piece along with its age, original cost, and estimated replacement date. When looking at capital equipment within the department, list all items according to projected replacement date and then go on to add ideas of what new equipment may be required in the foreseeable future. Automated information systems in many correctional agencies assist in this and other budget-related processes.

Capital budget requests usually receive close scrutiny to sort out competing interests of the various departments. The safety manager says three new floor buffers are needed to properly maintain the new tile that was just installed in all the corridors. The food service manager wants a walk-in freezer that will help cut food costs. The chief of security needs the new high-volume chemical agent dispersal machine that every other institution in the system already has. All must be balanced in the process. As a result, a request for a substantial capital expenditure must usually be accompanied by the following:

- A realistic assessment of the urgency or priority of the purchase.
- Detailed projections of all costs involved in the acquisition.
- A full description of the project and the rationale (justification).
- A statement as to whether the requested item represents an addition, replacement, or improvement.
- The impact, in the case of a revenue-producing department, such as correctional industries, of the proposed acquisition on revenues.

Most capital budget requests are ordinarily subject to several levels of review and approval. Supervisors may be assigned a certain dollar limit for capital purchases for a department, but this amount will usually be low when reckoned against the total cost of desired capital purchases.

However, it is just as likely (perhaps even more so) that the line manager personally will have no capital budget approval authority with respect to his or her department. The immediate supervisor, or perhaps the warden, associate warden, or business manager, may retain the authority to approve capital purchases up to a given amount. Some very large capital expenditures, however, may be authorized only by the headquarters office. Large-scale computer purchases, for instance, likely will be controlled at the headquarters level in order to achieve economies of purchasing scale across the entire system, and to ensure compatibility for all computer software or hardware used in the agency. In the private sector, the warden may be given authority for procurement up to certain levels. Procurements above that level must be approved at the company's headquarters in order to authorize the amortization of those costs over a period of time.

The Cash Budget

The cash budget is prepared by the business office or accounting department and is usually done last in the budgeting process. It consists of estimates of the institution's cash needs as compared with allocated funding or (in the case of the private sector) projections of cash receipts over the term covered by the budget.

The pattern of cash-in versus cash-out that is examined in the cash budget is extremely important to the institution because of the need to remain financially solvent. Most governmental agencies have to comply with what is called an "anti-deficiency" law that requires that the agency spend no more than its allocated funding. This statutory requirement is a very compelling reason to carefully control all spending. For private corrections, of course, the impetus is to make sure there is a healthy bottom line for stockholders.

Personnel

One way to handle the personnel portion of the budget is to assume uniform distribution of worked hours at so many per day. Coincident with this, one must assume distribution of nonworked time on a basis suggested by past practice and knowledge of some future events (such as vacation scheduling). One could conceivably do a more accurate job of spreading the personnel budget with some knowledge of a pattern of fluctuation in department activity. This may be the case, for instance, in facilities dealing with the courts, when certain times and/or days of the weeks reflect notable variations in commitment activity. Many prisons have a fixed relief factor that covers this type of additional coverage need.

Supplies and Repairs

The month-by-month expenditures in these areas will vary monthly, with workload. Also, a month with 30 days will absorb a slightly smaller portion of such expense than will a month with 31 days.

Fixed Expenses

The fixed expenses for the year are spread evenly by simply assigning one-twelfth of that total to each month of the budget year.

Staff Training

The funds for training are spread evenly across the budget year. Granted, this amount is not likely to be consumed in the same pattern in which it is budgeted, but as long as the training money is spent throughout the year with that budget limit in mind the results will be satisfactory. This could be budgeted more accurately if information was available about expected patterns of attendance at programs. (For example, a department may always send two people to the same annual conference in the same month each year.) But even if those programs are not finalized at the time the budget is compiled the results should prove satisfactory as long as the year's expenditures remain within the total. Note also that this amount of money is completely manageable by the supervisor since it is usually at the supervisor's discretion that employees are permitted to attend outside programs.

The Realities of an Expense Budget

A departmental budget is far from precise. Predicting equipment repairs as a function of usage is a questionable practice. One might just as readily spread repair costs evenly across the 12 months. However, manageable cost items, such as staff education, are better highlighted by the budget and placed so they can be more readily seen, the better to control them in relation to other expenses. Some expenses are partly manageable, such as personnel. There are any number of management actions a manager can take when actual costs and activity begin to move one way or the other. A manager may, for example, wish to permit or even encourage the heaviest use of vacation time during periods when workload falls off, if there is such a period for the agency or institution.

By highlighting all cost elements in the department, the budget affords an ongoing awareness of supply expenses and other variable costs. Should the cost of supplies suddenly exceed the projected amount, the manager is alerted to the need for investigation and possible action. Without a budget, certain costs could run unexplainably high for perhaps months without the manager becoming aware of the situation.

A budget, remember, is a plan. As such it attempts to look into a time that is not yet here, and it is necessarily based on estimates and projections. In no way is it ever to be regarded as a preordained, concrete picture of future events. Rarely will a manager budget as precisely as desired, and rarely will the results be exactly as predicted. In this respect the budget is a great deal like any other management tool. It cannot do the management job by itself, but chances are a better management job will be possible with it than without it.

■ The Budgeting Process

Responsibilities

The responsibility for preparing the budget rests in part with several individuals and groups.

In the public sector, the head of the agency ultimately is responsible for everything in the budget. He or she thus retains the authority for final approval of the budget, subject to whatever legislative authorization system the state has in place. Although the director of corrections will not likely be involved in the details of budget preparation, he or she will probably rule on major proposed capital expenditures item by item, and review and approve the principal parts of the budget, again, subject to legislative review. In the private sector, overall budget authorization and profit targets lies with the company's headquarters staff and the warden has considerable discretion in spending.

At the individual institution level, the warden must assure that the budget being submitted to the agency head is consistent with the goals of the institution. It also must be realistic and workable in the light of current knowledge of expected population levels, programs, and expenses.

Ordinarily, the business manager has an active role in preparing the cash and revenue budgets and assembling all pieces of the operating budget into a total budget for the institution. Like the warden, the business manager must assure that budgeting guidelines are followed and that the resulting plan appears to be workable and realistic in terms of what is known about the coming budget year.

All managers and department heads, at least under a participative budgeting approach, are responsible for assembling the expense and capital budgets for their own departments. In accounting or budgeting terms, the smallest organizational unit for which a budget is prepared is usually identified as a "cost center." This is an organizational unit for which costs are identified and collected, and it is usually provided with expense and capital budgets prepared by the supervisor. In the case of middle- or upper-level managers who may have charge of several cost centers, the task also includes assembling individual cost-center budgets into a budget for their total area of responsibility.

At both the institutional and headquarters levels, budget committees may be established each year to facilitate the preparation of the budget. The institutional committee will usually consist of the warden, the business manager, the associate warden(s), and key department heads. The budget committee establishes and distributes the guidelines for budget preparation. It decides how particular problems should be solved and certain issues dealt with as they occur. It also keeps all stages of budget preparation activity moving toward completion.

The Budget Coordination Meeting

A budget coordination meeting is a device that some organizations use as a "kickoff" for the year's budget-building activity. If the target for forwarding the institution budget to headquarters is at the end of October, the budget coordination meeting may take place in June or July. Convened and conducted by members of the budget committee, the coordination meeting should include all managers who have an active role in budget preparation. At this meeting, current and expected cost-related operational factors are discussed. These would include per diem food and other expenses, salary fund usage levels, and so on. At this time, policies affecting budgeting are reviewed and any changes are noted as well. Planned organizational changes likely to affect budgeting factors are reviewed and clarified, and trends of likely future activity are discussed. In short, any known or suspected factor that may have a bearing on budget preparation is reviewed.

The use of a budget manual can assist in this process. The manual is a book of informational and instructional documents assembled by the budget committee for the current year's budget preparation activity. Given to each person who must prepare and present a budget for one or more cost centers, this manual contains all information, instructions, and forms necessary for budget preparation. It describes the full scope and purpose of the budgeting program; outlines all procedures for preparation, review, and revision of budgets; and defines the duties and authority and responsibility of all persons involved in the process.

The Budget Calendar

The preparation of the budget for a sizeable correctional facility is an exercise in timing. Budgets for dozens of individual cost centers must be put together, and these must be assembled into budgets for larger organizational units and eventually into a budget for the whole institution. If two or three pieces are missing, the total budget will be delayed. Also, the cash budget cannot be properly prepared until the operating and capital budgets are complete.

The budget calendar provides deadlines for all steps in the budget preparation process. The calendar will let a supervisor know when the department's draft budget must be submitted and when it is likely to be returned following initial review. It also will establish when the revision must be submitted, and how much time must be allowed for the approval process.

A great deal of the stress that can be encountered in the annual budgeting exercise is due to timing issues. Managers often tend to get things done at the deadline or perhaps a little late, and their planning rarely anticipates all contingencies. As a result, it is sometimes necessary to repeat the whole process (perhaps several times) on the way to a realistic, workable budget for the entire institution.

Review and Coordination

When all of the individual departments' draft budgets are submitted, the business office staff ordinarily is responsible for preparing the total budget for review and approval. The budget may go through several drafts before the budget committee believes it has a document appropriate for submission to the warden, and ultimately to the agency head.

This cycle is due to the fact that while many supervisors enter the budgeting process with the best of intentions, the department-level results may be unsatisfactory the first time around. Also, the key people assembling the final budget simply do not know exactly where the totals are going to fall until all the pieces are put together for the first time. Back-and-forth activity ensues, as efforts are made to shift resources in more appropriate directions and allow formulation of a budget that is realistic under the circumstances.

Adoption

While procedures vary from one agency to another, ordinarily the budget is forwarded to the agency head for review and approval. The summary information provided generally includes:

- A narrative description of the budget and some of its key elements and the reasons behind its preparation in this manner.
- Relevant appropriation information (or a condensed income statement in the case of private facilities) for the budget period.
- A summary of capital expenditure requests.
- A cash analysis covering the budget period.
- The key factors used in forecasting.
- Estimates of the impact of this budget on the institution's services and fiscal status.

The agency head will either approve or reject the budget in whole or in part. If adjustments are needed because of conflicting priorities between facilities or due to other agency needs, then necessary elements of the process are repeated until the agency head approves a budget for the institution for the coming year.

■ Using a Budget Report

A budget should be a live, working plan. For it to serve in this capacity requires timely and accurate reporting of operating results. Even though public correctional facilities do not have to deal with the problem of variable cash flow related to per diem payments, they still must be mindful of the status of their appropriation account as the year progresses.

Tracking and reporting the results of operations is generally the responsibility of the accounting or business office staff. Timeliness is essential. It stands to reason that the more recent the feedback received on results, the more valuable the budget is for highlighting the need for management action. The information received each month includes a comparison of operating results for the period with the budget projections for the period. This allows a manager to make adjustments for those aspects of operations that are wholly or partly under direct supervisory control. For example, an extremely high increase in the cost of vegetables for a sustained period of time early in the budget year can throw the food service budget into a crisis late in the year if adjustments are not made as soon as possible.

Reporting is often done in a way that highlights exceptions. The reporting system is built on recognition of the likely presence of natural variations between operating results and budget projections. It will specifically flag items that appear to be out of line beyond normally expected variations.

Reading a Budget Report

In the "actual" columns of a budget report, the occasional minus sign after a dollar amount indicates a credit to an account. However, minus signs in the "variance" columns have a clearly different meaning. A minus sign after a variance dollar amount indicates a favorable variance, a condition of being under budget by that amount. Thus a minus sign after a variance percentage also indicates a favorable variance—the account stands at that percent under budget.

A caution is in order regarding the use of minus signs in budget reports. Their use is not always consistent from organization to organization. But minus signs carry a generally negative connotation for many people that is often the reverse of their meaning in a budget report. However, in many organizations' budget reports, the minus sign attached to a variance is "good" in that it means "under"—a condition of being under the budgeted amount.

In analyzing a budget report look for the obvious, such as accounts for which there are charges but no budget. Often charges appear against a "zero" budget (that is, an account for which the department has budgeted no expenditures) because people in the department make errors filling out requisitions and other forms. This also may occur when people filling out forms do not provide sufficient detail, leaving others to place the charges in specific accounts.

The supervisor should assure that all the department's employees who incur expenditures and charges supply full charging information, which usually includes both department number and specific account number. This will reduce the likelihood of budget anomalies due to charging errors.

Occasionally, charges to an account for which there is no budget may be legitimate. Suppose the account in question appears to be the most logical place for a particular charge but there happens to be no budget. This should serve as a flag to consider that account as a legitimate line item for the coming budget year (if it appears as though such charges may occur again).

The supervisor is often provided with guidelines for analyzing the budget report and answering for variances. Such guidelines are usually expressed in terms of a variance threshold beyond which answers are expected. For example, the guidelines for the manager of a department could include strictures such as:

- For personnel costs, explanation or justification must be provided for variances beyond plus or minus 2 percent or plus or minus $1,000 for total departmental labor costs for the month.
- For nonpersonnel expense, explanation or justification must be provided for variances beyond plus or minus 10 percent or plus or minus $50 for any specific account for the month.

Answering to variances under budget is almost as important as answering to variances over budget. Being under budget for a period does not necessarily mean that money is being saved, or that a favorable condition is emerging. It can often indicate that expenditures are occurring in a pattern inconsistent with the budget allocation. It also can indicate that certain necessary or desirable expenditures have been overlooked. For instance, a low figure in a salary account (due to vacant positions) might be more than offset by higher expenses in an overtime category elsewhere in the budget.

Under a thorough financial control system supervisors are usually expected to answer to their immediate supervisor for budget variances. To do this effectively, in addition to being provided with a monthly report of expenditures versus budget, the supervisor usually receives detailed backup about the individual charges made to each account.

In working with budget reports, a manager should make it a habit to question all details that are not fully understood. Learning month by month where all the department's budget dollars are going will eventually enable at least the partial alteration of patterns of expenditures in a way that may improve unfavorable trends as they appear to develop.

■ Control: Awareness Plus Action

Preparing a budget and working with it throughout the year heightens a manager's awareness of how the department's resources are used in fulfilling the department's responsibilities. Reports of actual results versus budget projections provide all-important information on which to base corrective action when needed.

For instance, when productivity problems occur and the relationship between worked hours and the volume of work begins to change unfavorably, positive steps can be taken. These might include effective use of overtime, review of scheduling of vacations and other time off, and routine scheduling of personnel for day-to-day coverage.

This kind of tracking also would be valuable in cases where use of consumables increases unexpectedly. In one facility, the inmate population started (for reasons too complicated to explain here) to flush socks and underwear down the toilets rather than turn them in to the clothing exchange. This came to light in two ways—the staff in the sewage treatment plant began to complain about the clogging problems in their equipment, but also as the laundry exchange department's expenditures for replacements began to go up.

In reviewing the budget and feedback on actual operations from the perspective of a complete organizational unit (whether a small department, a correctional facility, or even an entire agency), consider what is obtained from the process. A manager who knows the position of the organization relative to its goals also knows where to look for improvement.

■ Staffing and the Budget

This final section is included in the budgeting chapter because staffing has an inevitable impact on the budget. Positions equal dollars, and personnel costs are usually the single greatest item in any budget. Consequently, controlling salary dollars is essential to controlling the budget. Yet few management texts include information on the fundamentals of staffing itself.[1]

First, a distinction should be made regarding the difference between staffing and scheduling. Staffing levels (personnel ceilings) are assigned to agencies and their components in what are often called Full Time Equivalents (FTEs). An FTE represents funding for one full-time employee's position for a single fiscal year. Scheduling, on the other hand, is the process of making actual assignments to posts within an agency.

Positions and posts are not the same. A position is essentially represented by a person, who corresponds to a single FTE. In contrast, a post is a discrete assignment in an organization, which may be filled only 8 hours a day, 5 days a week by one person; 24 hours a day, 365 days a year by five or six persons, or some other combination of days and daily hours. This distinction is important when discussing the actual number of employees (positions) needed to cover all institutional jobs (posts)— a topic to be covered next in the discussion of the shift relief factor.

Technically correct roster formulation and analysis are increasingly important to contemporary correctional agencies. In an era of scarce financial resources, optimal utilization of available institutional manpower is a must. This is particularly true with regard to the security department, which constitutes the largest block of staff in virtually all correctional facilities. In addition, critical analyses of roster structure and manpower utilization can be valuable adjuncts to agency administrators in seeking and justifying additional resources from legislatures and other funding bodies.

While there may be a variety of common funding and entitlement factors within a system, the actual utilization of allocated manpower must be done on an institution-by-institution basis. Those determinations take into account the physical plant, programs, and local practices, as well as any labor-management provisions in effect. Nevertheless, there are a number of common principles that can be applied in virtually every correctional setting in order to effectively deploy available security personnel. These include the following:

- Centralized Roster Management
- Relief Factor Analysis
- Position Tracking System
- Recapitulation of Manpower Utilization

Centralized Roster Management

Roster management usually is centralized through the use of a coherent, coordinated management scheme. A master roster yields daily rosters that show the actual array of manpower available to each shift supervisor. These daily rosters provide the basis for an accurate assignment system, but also are important as documentation of manpower utilization for timekeeping purposes.

Without a centralized system, it is conceivable that one or more shifts in an institution could be operating with excess manpower and the others with a shortage, with no method of compensating other than by resorting to overtime. A tightly managed, centralized roster system is essential to overcome the handicaps resulting from fragmented assignment patterns.

Roster development often encompasses the use of "relief brackets," which organize permanent assignments in concert with their relief posts, in order to efficiently deploy all available manpower. In this system, a block of seven posts are aligned together; two specific posts are allocated to provide relief functions for five other posts.

For instance, a post designated MR-1 (Morning Relief 1) would relieve one of the five main posts in the block on Tuesday and Wednesday, the second post on Thursday and Friday, and on the fifth working day (Saturday) MR-1 would relieve the third post in the block (covering the first of that officer's two days off). The MR-1 officer would then have Sunday and Monday off. Paired with MR-1 would be MR-2, which has Monday and Tuesday off. The MR-2 officer would begin the week by relieving the two posts with Wednesday/Thursday and Friday/Saturday days off and complete the week on Sunday by relieving the third officer in the block for his or her other day off.

Also, comprehensive rosters enable staff to structure ideal days off for each post. For instance, a rear gate post, if used at all on weekends, would be a relatively undemanding assignment, and thus ideal for a relief officer. The regular day off for such a post should be Saturday and Sunday. Similarly, it would make little sense to have the regular visiting room officer off on the heaviest visiting days, and so the days off on that post should be selected accordingly.

Annual leave and training often are coordinated in this system to maintain necessary staff coverage at all times. The institution's goal is to have a consistent number of staff on leave throughout the year, except for periods set aside for any institutional training for the entire staff. Also, with a regular relief and sick/annual relief structure, staff know the post requirements when working reliefs, rather than moving around from post to post on a frequent basis. By carefully arranging these factors, personnel are arrayed throughout the institution optimally, not for staff convenience, but for the efficiency of the agency.

A master roster contains the necessary elements of an institution's manpower array. Relief blocks also are set up so that supervisory staff (and staff in positions of similar level of responsibility) relieve each other in a predictable fashion. All relief staff are not on duty on the same days, and days off are not concentrated on weekends, although posts with logical days off have had those days assigned.

Daily rosters represent the changing days off for all staff in the security department for the week in question, and show not only who will staff every post each day, but also what each employee's status is on every day. A roster in this form can be annotated by each shift supervisor as to any on-the-spot changes, and then tabulated on a daily basis for recapitulating total manpower use for that day. The rosters then form the basis for not only the recapitulation system, but also for the employee time and attendance records. As another feature, they also provide a prime document that can be used in the event an internal or outside investigation ever requires the location of a given staff member on a particular date, or what staff member was working a given post at a certain time.

Relief Factor Analysis

Before any realistic assessment can be made of necessary workforce levels for a correctional agency, a shift relief factor must be computed. This factor is usually expressed as a multiplier, and is applied to the number of posts in a given facility, in order to determine the actual number of staff that must be on board to cover all posts. **Table 23-1** depicts the calculation of a sample relief factor.

Table 23-1	Shift Relief Factor Calculation Summary	
Step	**Procedure**	**Value**
1.	Regular days off per employee per year (52 weeks $\times$ 2 days per week)	104.0
2.	Remaining work days (365 − Step 1)	261.0
3.	Average vacation days per employee per year	16.4
4.	Holidays per employee per year	10.0
5.	Average sick days per employee per year	13.2
6.	Average LWOP per employee per year (includes Workman's Compensation)	20.7
7.	Average training/employee/year	5.0
8.	New employee training beyond initial training	0.0
9.	Percent of employees employed one year or less	10%
10.	Other days off per employee per year (Union meetings, litigation, military leave, etc.)	2.0
11.	Total days off per employee per year (Add Steps 3–7, 10, then add Step 8 $\times$ Step 9)	67.3
12.	Number of actual work days per employee per year (Step 11 − Step 2)	193.7
13.	5-day Post Relief Factor (Step 2/Step 12)	1.34
14.	6-day Post Relief Factor (Step 13 $\times$ 6/5)	1.61
15.	7-day Post Relief Factor (Step 13 $\times$ 7/5)	1.88

The relief factor, when multiplied against each post and position in the institution, compensates for around-the-clock operation, days off, vacation days, holidays, sick days, training, and other leave days such as funerals, injury, and perhaps discipline time. When completed, it yields the actual number of staff required to maintain all posts filled as established. The data used to derive shift relief factor computations ordinarily are provided by the personnel staff of the agency, and will often apply throughout the whole system. They must reflect actual usage patterns, as well as the statutory entitlements for holidays and other factors impacting post coverage. Because these underlying factors differ from agency to agency, the relief factor differs as well.

As a final comment in this area, many agencies have a workforce utilization committee that tracks and manages all workforce utilization issues. This group is responsible for maintaining the integrity and accuracy of the roster management system.

Position Tracking System

One general problem in the analysis of many existing rosters stems from the difficulty in following positions from their original allocation through to their present utilization. Without such a system, a position could be requested under justification for a critical security need, and then reprogrammed into a less important area at a later date, without administrative review at a high level. Most agencies have such a system, in concert with approved master rosters, to ensure that proper accountability and utilization are enforced regarding all allocated positions.

In this system, every allocated position is assigned a distinctive number and its use and incumbent are associated in a formal structure. Whenever the position becomes vacant, proper approval must be obtained from agency headquarters before filling it. Similarly, whenever it becomes necessary to convert a position from one use to another, approval must be given only after a written rationale is reviewed and filed with that position's other documentation for future review.

Recapitulation of Manpower Utilization

Most roster systems also have a method for recapitulation of all manpower services provided to the institution by the security department. This facilitates accounting for every hour of paid staff time, and allows reconciliation of those numbers with the positions and salary dollars allocated to the agency.

The typical system used involves a relatively simple tabulation of the totals of all utilization categories daily, usually on the daily roster. Staff then periodically total all categories for reporting purposes. Except when an institution is over its authorized complement, this recapitulation figure should equal the authorized complement. Whatever system is adopted, it should be adapted to current needs in the institution, but also be sufficiently flexible to encompass program, staffing, and population changes.

ENDNOTES

1. This section on staffing is adapted from unpublished consultant materials compiled by author Phillips, James D. Henderson, and Jerry A. O'Brien.

Quality and Productivity: Sides of the Same Coin

24

Quality never costs as much money as it saves.

Anonymous

We are not at our best perched at the summit; we are climbers, at our best when the way is steep.

John W. Gardner

The Total Quality Movement: "Excellence" All Over Again?

This chapter is important because no organization can provide high-quality services to its customers without an internal focus on quality itself. As correctional

FROM THE INSIDE — Trying Out Total Quality Movement

The head of a major prison industry operation decided that a total quality movement was what the organization needed. He went to a fancy retreat where he learned all the principles involved, and he came back fired up to make things happen.

He generated a great deal of "publicity" in the industrial side of the prison system for this enterprise. Among his first steps was to delegate responsibility for the program to one of his deputies. This individual was very senior in the organization, and almost eligible to retire.

The program was kicked off with the usual commitments by the top man to back the program personally, 100 percent, for as long as it took, with whatever resources it took, and so on. Managers in the field yawned—they had seen it all before, in various other configurations. And they had seen those efforts die natural (that is, bureaucratically natural) deaths.

They were right. The top man got diverted to other interests. The deputy retired. The program faltered. No one talks about it anymore. Whether any lasting benefit was achieved, no one will ever know.

managers, our customers are the public, our employees, and inmates. So a discussion of this precept is central to how managers do their jobs.

This is not to say that managers should launch major, structured quality initiatives on their own. Such activities clearly require higher management approval and an immense amount of interdepartmental coordination. But every good manager is concerned about the quality performance of his or her department, and the principles discussed in this chapter can be useful in achieving that important goal. And when a manager's organization launches such a program, knowing these fundamentals will be immensely useful.

The issue of quality became a fashionable business term in recent years, in the way that "excellence" was management's trend word of the 1980s. Although the total quality movement and the excellence movement of the recent past had somewhat different origins, the apparent effects of the quality movement have been very much the same as the visible results of the excellence movement. A basically sound and well-intentioned philosophy is being adopted, adapted, promoted, and implemented with mixed results.

The excellence movement was sparked by one very good, highly readable, phenomenally successful nonfiction book, *In Search of Excellence*.[1] This book was so successful that being concerned with excellence in the conduct of enterprise of all kinds suddenly became fashionable. It also quickly became apparent that almost anything labeled with that newly popularized word would be immediately noticed by a great many people, some of whom would buy those products. Overall the fundamental message of *In Search of Excellence* was copied, reworked, repackaged, and supposedly expanded upon by countless authors, consultants, trainers, and others. Countless business organizations of all sizes began to claim excellence as a goal that they had attained, or were on their way to attaining.

Many of the organizations that attempted to adopt excellence as a guiding philosophy of operations ran into the same problem that has stopped many otherwise effective organizations in their tracks. That is how to instill a specific philosophy in other people so that it will eventually drive them to behave in a desired manner.

Between the philosophy (which may initially be accepted by a few members of management at or near the top of the organization) and the actual practice (which involves many employees living out the philosophy) there lies a matter of process. There has to be some process available to transfer the philosophy from the few to the many.

A great many people never see past the process and never truly adopt the philosophy. They simply go through the motions, appearing to do what they perceive top management wants them to do. Invariably, when a philosophy is proceduralized—that is, when a process is superimposed upon something as ethereal as a concept, idea, or belief—something is lost. And those who simply adopt the process as part of the job without buying into the philosophy will not truly reflect the philosophy in their behavior.

When a philosophy of management is over-proceduralized, over-promoted, over-publicized, and over-praised, it becomes a fad. It becomes fashionable for its own sake. In this manner excellence went down the same path traveled years earlier by management by objectives (MBO). There is reason to wonder, then, whether the quality movement may be just the latest management fad, destined to go the way of MBO and excellence. Many employees, including many experienced managers, seem to think so, which makes the effective implementation of any new program along these lines somewhat more difficult.

Quality Control, Assurance, and Management

For years, many manufacturing and service industries had what was referred to as quality control. Quality control ordinarily concentrated on finding defects, rejecting defective products, and providing information with which to alter processes so they would produce fewer defects.

Along those lines, some correctional facilities and headquarters offices conducted periodic audits of each institution's department's functioning. However, many agencies had very little of what is called quality assurance in other service organizations. In this context, quality assurance consists largely of scrutinizing records to detect and tally departures from some established standard. Procedural and staffing problems thus could be identified, and managers had a starting point for reducing the same kinds of errors in the future.

In addition to correcting the processes that produced the errors, both quality control and quality assurance were often responsible for instituting more frequent quality checkpoints during the work process itself. In this manner, errors might be caught earlier. However, the most important similarity between quality control and quality assurance was that both focused primarily on finding errors after the fact. Both were, and still are, retrospective processes.

Recently (using philosophical grounding and methods exported from the United States to Japan decades earlier and later brought back as "new, revolutionary management techniques") the emphasis on quality has shifted from "catch errors before they go out the door" to "avoid making the errors in the first place." Thus, this forms the basis of the quality movement, as embodied today in labels such as *total quality management* (TQM) and *continuous quality improvement*.

Old Friends in New Clothes?

Many of the tools and techniques utilized under the TQM umbrella should look familiar to some people who have been in the workforce for a few years. Many of the "current" tools and techniques have been around for quite a while, some for decades. They are presently being resurrected and revitalized, and in some instances renamed.

For example, a number of TQM-implementation case histories mention the acronym TOPS, standing for team-oriented problem solving. As the name suggests, workers who have active concerns with various aspects of particular problems approach problem solving as a team, with a common goal and purpose. These problem-solving teams espoused under TQM look, sound, and function the same as the quality circles promoted during the brief popularity of "Japanese management."

Other TOPS look-alikes, such as self-directed work teams and team-oriented process improvement, are essentially quality circles by another name. These particular labels are but two of several similar designations that have emerged as representing a significant part of the path to continuous quality improvement.

Quality circles were themselves nothing new when they were so named. In years past many work organizations utilized what were called work simplification project teams. These were, in function and intent, essentially identical to quality circles and the problem-solving teams of TQM. Written about in the 1950s and earlier, these teams began to find a place in some service organizations as early as 1956.[2]

Many of the specific tools used by today's TQM problem solvers go back 50, 60, 70 years or more. Serious managers who believe in the potential of total quality management are finding that these techniques are fully as valuable today as they have ever been.

The Common Driving Force

It is not really important how many previously popular techniques are returned to the spotlight or how many genuinely new wrinkles are added. There remains one ingredient that is as essential to TQM as it has been to any other approach by any other name. That crucial ingredient is top management commitment.

This should come as no surprise. Top management commitment to new ideas and approaches has been a prerequisite to complete success for as long as organized enterprise has existed. Without it, most organized endeavors are destined to, at best, generate results that fall short of intentions. At worst, they fail altogether and cause harmful results or leave residual damage.

One cannot imagine any rational top manager openly avowing opposition to the principle of quality improvement. Ask any top manager whose organization has espoused TQM if he or she is truly committed to it. For that matter, ask any top manager at all if quality, period, is a personal commitment. Predictably, each will state unwavering commitment. Many such endeavors fail because of insufficient top management commitment, but since almost all managers will voice commitment there is but one conclusion to be drawn. Top management commitment is a matter of degree. And it is the degree of commitment that is critical.

None of today's total quality programs will work as intended unless top management is actually involved and actively promoting the concept. Superficial, lip-service-only commitment at the top results in similar weak commitment at lower organizational levels. Beware the skyrocket commitment of the top manager who gets all fired up over TQM, distributes information to everyone, creates a TQM steering committee, advisory committee, or similar body, and chairs the first meeting or two or three—but then starts missing meetings because of "pressing business" and transfers the guiding role to subordinates. Those, of course, were (to the old hands in the system) the first signs of likely failure for the quality program described in the beginning of this chapter.

A total quality program also will not work if managers, especially first-line supervisors, will not let go and truly delegate to employees. This means not simply giving employees the responsibility for doing different tasks or determining more efficient methods. It means also giving them the authority to make the decisions necessary to implement their own decisions. Further, letting go also means accepting what the employees decide and living with it.

Letting go to the extent just described is difficult for the majority of managers. A great many managers, far more than would be able to see it in themselves, possess a recognizable streak of authoritarianism. Upon reflection, the reasons for a fairly strong presence of residual authoritarianism are understandable. Modern management (true, open, participative management) is a phenomenon of the recent two or three decades. While steady, its spread has been gradual. There remain many areas of organized activity in which employees have yet to experience any management style other than straightforward "bossism."

Managers learn about management mostly from other managers, and especially from those organizational superiors who, for good or ill, were by virtue of position role models for those newer to management. At one time virtually all correctional management (and in most other organizations as well) was authoritarian. Even now, management that is at least partly authoritarian predominates. Most management role models thus convey at least a modicum of authoritarianism.

Subtle proof of the existence of the authoritarian streak can be experienced by the manager who might ponder his or her reaction to being pushed abruptly into a fully participative management situation. The manager may feel that participative management equates to weakness, and that delegating decision-making authority to subordinates is somehow abrogating one's responsibility.

It remains clear, however, that changes in management style and approach may have to occur in order for a quality management program to be successful. In most instances the manager will need to shift from being the supervisor (from planning, telling, instructing, and controlling) to being the leader of a team and being a counselor, teacher, coach, and facilitator.

Management's commitment, then, has to involve two things: First, a total commitment to participative management and employee empowerment, but also a commitment to intra- and interdepartmental teamwork and improved communication throughout the organization.

Convincing Staff TQM Applies to Corrections

Many staff think that the correctional setting is so different as to invalidate the core concepts of TQM. It is easy to say that output cannot be measured in relation to inmates in any meaningful way, and thus the program is useless in correctional settings. Neutralizing this view is a key starting point for any TQM program in corrections.

The easiest examples to use in refuting it are seen in jobs such as secretary or clerk. These positions involve performance of tasks that are easily observable, quantifiable to a certain degree, and thus easy to monitor. The quality of typing as measured by typographical errors, or the quality of filing as measured by the rate of misfiled documents, can be clear indicators of the performance level of the employees involved. And certainly there is no reason why staff in those categories cannot be included in a participatory scheme that solicits and incorporates their views on how their jobs can be done more effectively.

But even the performance of correctional officers can be brought into such a framework. There is nothing preventing supervisory staff from tracking the number of cell searches conducted each day, the manner in which pat searches are conducted as inmates go in and out of housing units, or the contents of safety and sanitation reports on those units. These are real-world measures that can be used to assess correctional officer performance. And there is nothing to stop a TQM-oriented manager from structuring a participatory scheme that includes the input of those same officers. To a certain extent, the unit management system mentioned in Chapter 4 is a solid starting point for just this kind of interaction and participation in the quality process.

But in any case, there is no validity to the claim that the principles of TQM cannot be put into place in correctional organization. The old saying, "If you can't find time to do it right the first time, how are you going to find time to do it over?" is appropriate here. It is equally applicable when talking about a poorly written memo about an inmate, a poorly typed report on the facility's budget, a poorly conducted cell search, or a poorly repaired plumbing leak in the powerhouse.

Will Total Quality Management Prevail?

Total quality management has every chance of working where previous and perhaps partial efforts undertaken under other names have failed. Activity undertaken in the name of quality improvement is presently widespread elsewhere in the workplace. Consequently, the impression that "everyone is doing it" places considerable pressure on the supposed few who have yet to undertake such an effort. With so much happening in the name of TQM one will find that, in some organizations, top management commitment is real and lasting. Also, some organizations' genuine successes will inspire others to try following a similar path.

Within corrections specifically, it may be difficult to avoid considering the adoption of quality management because of growing pressure to control costs, which can be accomplished by improving quality. This can only be done when TQM is accepted, not as a "project" or one-time fix, but rather as a permanent philosophy of operations.

More Than Just a Job

For total quality management to work, the majority of employees in the organization need to experience sufficient commitment to their employing institution to see it as more than just a place to work. This can be a tall order indeed in these times of staff shortages, increasing numbers of inmates, shrinking funding, and growing pressure from the courts.

Today, correctional managers are wrestling with increased population levels and limited fiscal resources, notwithstanding massive construction programs that have barely been able to keep pace with commitment activity in most states. It seems clear that the tools and techniques of total quality management offer a workable means of reconciling concerns about quality and cost, and to some extent helping corrections deal with explosive growth. Once again it comes down to a matter of commitment. If employees at all levels are sincerely committed to the organization, its mission, and its patients, there is no reasonable way that TQM can not work.

For reasons suggested throughout these pages, some attempts to implement total quality management will fail. For some people, quality will gather dust in the old-management-approach graveyard along with all the other techniques that were tried, half-heartedly applied, and discarded. Some attempts, however, will succeed. And correctional organizations that are successful in implementing TQM in the years to come will find themselves among the stronger, more adaptable operations. Such successes will translate operationally into a safer, more humane environment for staff and inmates, and a greater level of public safety overall.

■ Productivity in Corrections

Can You Have It and Hold Managers Accountable for It?

For public correctional employees it is reasonable to ask why one should be concerned about this issue, especially in the face of government budget systems that only indirectly encourage concern for productivity. Of course, for private sector corrections the answer is obvious—the bottom line. However, for both private and public corrections, productivity deserves attention because:

- It is simply good management to want to apply available resources to best effect.
- It is possible, in an organization as varied and complex as a modern correctional institution, to redistribute resources from areas of savings to other essentials.
- The inflationary spiral affects all business and financial activities, and when resource inputs grow at a rate faster than system outputs the costs to the consumer increase.
- There is increasing external pressure to hold costs down while continuing to provide quality programs to inmates.

Further, concern about productivity needs to be ongoing. One of the common errors of productivity's heyday was to regard productivity improvement as a one-time effort that could sweep through an organization and make it more efficient for all time. Not so. Some of the larger, more obvious corrections are one-time changes, but the need for attention to productivity is ongoing. Left unto itself an activity will be subject to creeping inefficiencies that eventually grow to be a significant problem. Left unto themselves most human activities will suffer diminished productivity; regular attention is needed to keep them tuned and running efficiently.

The issue of accountability is an important one in connection with this topic. If public managers are not held accountable for their actions, an important segment of our system of governance will fail. In addition to the fiscal issues resulting from such failure, with corrections there are vital public safety implications as well. While in the private sector the accountability is to stockholders, the public safety ramifications are no less.

A Return to Favor

Concern for productivity in the correctional setting is not an isolated event. It is going on throughout our society as trade imbalances, outsourcing, off-shore operations, and other elements modify the traditional balances in our economy. Fortunately, in the first decade of the new century there are clear indications that productivity in the American workplace is on the rise.

With corrections accounting for a growing portion of government budgets nationwide, there is no denying the importance of productivity in the correctional setting. States, in particular, have to make difficult choices about whether to fund schools, bridges, or corrections. Today's government fiscal crisis means that there inevitably will be calls for a close examination of correctional operations for potential improvements in productivity in order to reduce costs.

Some of this can be achieved through technology. The use of perimeter detection systems and roving vehicular patrols can reduce the personnel costs for perimeter security, for instance. Other automated processes can speed staff in their duties–digitalizing photographs and fingerprints, for example. But other pressures continue.

Starting in the mid-1990s there have been calls at the federal level to curtail correctional programs perceived as amenities, although these initiatives were for punitive reasons as much as for alleged cost-cutting. Paradoxically, "three strikes" legislation was being enacted in many jurisdictions, which had the effect of driving correctional populations (and costs) up even further. If an institution is expected to do more with the same or fewer resources, it follows that responsible management will seek all reasonable ways of improving productivity. That makes TQM something that every correctional manager should consider.

■ Sides of the Same Coin

The total quality management movement has its concern for productivity at a time when a variety of other factors are also pointing toward the need for improved productivity. Indeed, the quality issue is inseparable from productivity. One has direct implications for the other, and there is a direct relationship between them.

In the future, however, concern for productivity may not be limited to just the methods-improvement, cost-cutting, staff-reduction activities frequently associated with management engineering. Although work measurement may play a large part in productivity improvement, the view of productivity and the scope of future effort will likely be more global than past efforts.

In any activity productivity is represented by a relationship between input and output; that is:

$$\text{Output/Input} = \text{Productivity}$$

Any process or activity has associated with it a level of quality, with quality broadly described as the relative acceptability of the output. Any change in productivity involves changes between and among all of these factors. Thus productivity may be said to be improved if:

- Output is increased while input is held constant or decreased and quality is held constant (doing more with the same or less).
- Input is decreased while output is held constant or increased while quality is held constant (doing the same or more with less).
- Quality is improved while input and output are held constant or reduced (doing better with the same or less).

It follows, then, that productivity is reduced if the opposite of any of the foregoing changes occurs, such as if output decreases while input is held constant.

Productivity may have received more or less attention at various times. But the concept of productivity and the need for constant attention to it have always been with us. The overall objective of productivity improvement efforts will remain the enhancement of the accomplishment of work by doing things in less time or with less effort or at lower cost while maintaining or improving quality.

The principal factors influencing productivity, and thus quality, are:

- Capital investment
- Technological change
- Economies of scale
- Work methods, procedures, and systems
- Knowledge and skill of the workforce
- The willingness of the workforce to excel at what they do, and in all instances to do the right things in the best possible way

Given the direction from which most of the forces urging organizational improvement are coming, for the first decade of the 21st century, it is reasonable to expect to hear at least as much about quality as about productivity. These concepts remain two sides of the same coin, and the supervisor's tools that have traditionally been associated with productivity improvement (see Chapter 26) are also the tools of total quality management.

EXERCISES

Exercise 24-1: In Search Of—?

Select one actual application of a "new management approach" for analysis and study. This could be an application of management by objectives (MBO), some formal program or process emanating from the "excellence" movement, a quality circles program, an organized methods improvement program, or any other change-oriented mechanism intended to improve organizational success. The ideal choice would be an application in which you are personally involved, but lacking such involvement any familiar application is an option.

Questions

For the application you chose to study, answer the following questions:

1. Expressed in no more than one or two sentences, what was the guiding philosophy of the program?

2. In your view, was the program's guiding philosophy successfully translated into practice? How was this done or not done?

3. Do you believe that overall the program succeeded or failed? In either case, identify what you believe to be the three most important reasons behind the outcome.

4. What do you consider to be the single most important lesson learned from the organization's experience with the program?

Exercise 24-2: The "Elevator" Speech

You are a supervisor in a correctional facility that has just announced the introduction of a total quality management process that will eventually involve all employees in all departments.

Write a short, informal speech, perhaps no longer than one or two paragraphs, consisting of a half-dozen sentences in total, that you can use to explain concisely to someone who asks, "What's this quality stuff all about and how can you use it in an institution?" (Since time is frequently limited in circumstances in which people may communicate almost literally while passing, it is helpful to have a clear, well-thought-out response to such questions ready at all times. This kind of response is often referred to as an elevator speech because you can deliver it in full between elevator stops.)

ENDNOTES

1. Peters, Thomas J. and Robert H. Waterman, Jr. *In Search of Excellence.* New York: Harper & Row, 1982.

2. Maynard, H. B., ed. *Industrial Engineering Handbook,* 2nd ed. New York: McGraw-Hill, 1963, Section 10, pp. 10–183 through 10–191; Smalley, Harold E. and John R. Freeman. *Hospital Industrial Engineering.* New York: Reinhold, 1966, pp. 68–69.

Teams, Team Building, and Teamwork

25

Chapter Objectives

- Differentiate between types of teams: ad hoc or special-purpose teams versus departmental teams.
- Review the composition and structure of ad hoc or special-purpose teams.
- Review potential legal problems associated with employee teams and suggest how legal entanglements can be minimized or avoided.
- Review the problems encountered by teams functioning in a company environment in which recognition and reward systems are based on individual performance.
- Introduce the concept of team building within the context of the departmental group.
- Introduce the stages of team building, from original formation to continuing maintenance.
- Identify individual employee motivation as the primary driver of team performance.
- Identify the leadership qualities necessary for successful team building, and provide guidance for the team-building department manager.

Man's greatest discovery is not fire, nor the wheel, nor the combustion engine nor atomic energy, nor anything in the material world. Man's greatest discovery is teamwork by agreement.

B. Brewster Jennings

Nature's stern discipline enjoins mutual help at least as often as warfare.

Theodosius Dobzhansky

A True Team Effort

It was the early 1970s and civil disturbances in the nation's capitol were expected. Federal prisons on the east coast were put on notice that they might have to provide staff for temporary detention centers in the vicinity of Washington, DC. This in fact happened, and on short notice several facilities had to essentially put together a complete staff complement for a medium-security institution. This was done using personnel from a half dozen locations—none of whom had ever worked together before.

Correctional officers, caseworkers, medical personnel, laundry staff, secretaries, bus crews, receiving and discharge personnel, records specialists—all the staff components of a 500-bed institution were involved. They had to be chosen, notified, moved to the DC area from their home institutions, housed in temporary quarters, fed, and assembled on-site at the prison. A supervisory structure had to be developed, staffing patterns developed, rosters compiled, supplies acquired, key rings developed and distributed, and a thousand other things necessary to the operation of a prison had to be done. Within 48 hours of notification, a massive and effective team exercise was carried out, just to be ready to receive the first busload of Pentagon demonstration arrestees. A large, complex team had been formed.

■ Types of Teams

There is no doubt that corrections is a team enterprise—from the need for people to work together to process in a busload of inmates, to a classification team, to a group assembled to develop a solution to a laundry problem in segregation, to the crisis situation where trained response teams save lives. Regrettably, human beings are not inherently team players. In fact, the male of the species is particularly inclined toward individual action (often competitive) that can hamper unstructured group efforts. That is why this chapter is so important in the development of correctional managers, because managers inherently form and often lead teams in their workplace.

There are two general types of teams to be encountered in the workforce, and it is not always immediately clear what type is being referred to when one speaks of team building. It is therefore appropriate to begin by differentiating between team types to establish the direction for this chapter.

The first type of team, often referred to as a project team or an employee team is the focus of the first major section of this chapter. It is characterized by the group that is assembled for a specific purpose, perhaps including people from a number of different departments or disciplines. This kind of team may be ad hoc, assembled for a specific one-time task, or it may be ongoing in that it consists of permanent or rotating membership and handles a certain kind of business or problem on a regular basis. These are the teams of team-oriented problem solving in total quality management. These are the teams that at one time may have been referred to as quality circles. And these are the teams of the form taken by an organization's safety committee or product evaluation committee and other such groups.

The other type (represented in the extreme by the lead example of this chapter) is the departmental team, which is simply a group of employees and a supervisor. The word "simply" is used because the team composition is indeed simple. Most people readily understand the relationship between the supervisor and the direct-reporting group. Most also understand that everyone in the group has a job to do, and together this accomplishes the work of the group. Such a group can of course continue to operate as a number of individuals doing their jobs for a common purpose, but when these individuals are united into a true team the potential of the group is expanded dramatically.

■ The Project or Employee Team

Team Organization

The organization of a project or employee team can be relatively simple. A decision is made to pull together a number of people with the appropriate knowledge, experience, perspective, or whatever to address a particular problem. They may, as part of the project, provide input on particular issues, undertake productivity improvement projects, or address certain ongoing concerns such as safety, quality, education, and so forth.

Team organization is addressed in Chapter 25, where the composition of both the ad hoc and ongoing team are described. These kinds of teams most often consist of both managers and rank-and-file employees. And because there are nonmanagerial employees on such teams (often constituting the majority of team membership) there are potential legal problems depending on the kinds of problems or issues addressed.

Legally Maximizing the Use of Employee Teams

Effective people management is of course important to the success of any form of team. However, fully as important for many of these kinds of teams is sensitivity to the potential legal pitfalls of team operation. Whether through total quality management or simply out of a sincere desire to solicit employee participation, a number of organizations have discovered, much to their dismay, that a team can easily stray into questionable legal territory.

This section is going to lay out some guidelines for addressing this issue. However, it cannot possibly address all of the contractual and legal issues that may impact the subject of employee teams. The labor-management environment of each agency is different, and it differs significantly from public to private sector.

In these days of increasing employee participation and involvement, department managers are increasingly likely to hear suggestions similar to the three that follow:

> "Let's pull an employee team together and find out what the workforce thinks about this issue. After all, there's nobody who knows the

ins and outs of most of our jobs better than the people who do them every day."

"We have no real idea what the workforce thinks about our benefits package. If we pull together a group of representative employees we might learn something that will help do a better job of allocating limited benefits resources."

"Let's give this problem to a hand-picked group of employees to see what they can come up with. There's no danger, because all they can do is recommend. As management, we're free to accept or reject what they suggest."

In fact, each of these suggestions contains some fairly conventional management wisdom fostered by experience and supported by common sense. It can be summarized as follows:

In most instances nobody does know a given job better than the person who does it every day.

It only makes sense to try to account for employees' needs and desires in designing a benefits program.

In each case, management is free to accept or reject employee suggestions (as long as they remain aware that the final responsibility always resides with management).

Participative management and employee involvement have been talked about for decades and have been practiced in an increasing number of work organizations. This has been particularly noticeable since the human relations approach to management began to make inroads into the authoritarian atmosphere of the past. The recent popularity of TQM programs spurred even greater emphasis on participative management. It is tending to bring more employees further into self-determination on the job. More and more is being done with the involvement of employees by way of employee teams.

There are, however, areas of employee involvement in which teams can be an intrusion on the prerogatives of labor unions. If not carefully formulated and managed, there is a constant risk that a given employee team could be judged an illegal labor organization under the National Labor Relations Act (NLRA). The three suggestions advanced above, all well-intended ideas for employee involvement, could all readily lead to groups that could be considered as infringing on the rights of collective bargaining organizations. Whether and how potential infringement principles affect a particular agency in the public sector is a matter for agency legal counsel to determine. However, anyone managing in the private corrections sector should be particularly mindful of the information presented next and also seek legal guidance if none is already contained in company policy.

Old Problem, Different Emphasis

The problem of potential conflict has actually existed since the NLRA became law in 1935, but it was brought into sharp focus by a particular case. During 1989, Electromation, an Elkhart, Indiana, manufacturer of electrical equipment, established several employee committees. One was created to investigate bonuses, another to look at premium pay, one to study absenteeism, a group to examine employer-employee communications, and one to deal with a no-smoking policy. Management defined the subjects, set the number of members for each committee, appointed managers to all of the committees, and paid workers to participate.

When Electromation's five employee committees were challenged by the affected union, the National Labor Relations Board (NLRB) agreed with the challenge. The NLRB said these five employee representation committees were essentially employer-dominated labor organizations that discussed wages and other terms of employment. The NLRB said that the company was not simply dealing with quality, productivity, or efficiency. Instead, it ruled that the company was creating the impression among employees that their differences with management were being resolved bilaterally.[1]

Following the Electromation case and citing that ruling, the NLRB ordered the DuPont Company to disband seven labor-management committees, six created for safety issues and one for recreation. The NLRB said these committees were dominated by management and were dealing with issues that should have been bargained with their union. As key elements of the committees, the company retained veto power, established and controlled agendas, decided membership and dictated the structure and purpose of each committee, and discussed what could be called "conditions of work." The committees' activities supposedly produced changes and benefits that the union had sought but failed to obtain through collective bargaining.

In a third case, similar to the DuPont situation, the Polaroid Corporation was forced to disband its long-standing employees' committee after the Labor Department determined that it was actually a "labor organization" and its "election procedure was undemocratic."

Taken together, these cases illustrate the problems that have to be avoided in forming such committees. They can be useful management tools, but only if properly used. Again, agency legal counsel will be aware of the law in this area and whether these cases are applicable to any management strategies being contemplated or already in use.

What Are the Problem Areas?

Generalizing from the experiences of Electromation, DuPont, and Polaroid, and expanding on pre-Electromation information, an employee team or committee might be considered an employer-dominated illegal labor organization for any of a number of reasons.

First (and probably foremost) among these reasons is whether the group is seen as dealing with wages, hours, benefits, grievances, or other terms and conditions of employment. These are, of course, among the issues most frequently subject to collective bargaining and are seen (at least by the NLRB) as the exclusive province of unions.

Second, an employee team or committee also might be seen as an illegal labor organization if committee suggestions or recommendations result in management decisions. This is applicable if the group itself does not have the power to make the decisions, and if employees are elected to the group as representatives of larger bodies of employees.

Third, the team may also be improper or illegal if employees see the group as a means of resolving their concerns with management, and if meetings appear to involve "negotiation" between employees and members of management.

Experiences since the Electromation case was decided seem to suggest that a violation will most likely be found if an employee team is actually set up during a union campaign (this can attract an unfair labor practice charge), participation in a committee is made mandatory, and the employer picks the members or controls the method of their selection.

Not the Last Word

It seems to be a matter of individual opinion as to whether "quality" committees or committees established to serve as communications devices to improve quality are labor organizations under the law. Probably the worst-case consequences of sanctioning a committee that is actually an illegal labor organization occur when the committee itself votes to affiliate with an outside union. If, by dealing with a committee and accepting some of its ideas and recommendations the employer has already recognized the committee as representing employees, that representation cannot be withdrawn just because the committee decided to affiliate with a union. The employer may be legally forced to recognize the union and deal with it without an election by employees.

It is doubtful, however, that a non-union employer would likely be subject to NLRB charges concerning employer-dominated labor organizations unless active union organizing was occurring.

Avoiding "Committee Paralysis"

Because of this legal climate, there may be a tendency for some organizations to shy away altogether from teams or committees that include rank-and-file employees. Some justify the de-emphasis of teams by pointing out that attacking employee committees has become an active tactic of unions that are either in place or seeking acceptance.

However, the Electromation decision did not open up any truly new issues; it simply surfaced some that had existed for years. Long before that decision it was recognized that, "The more effective the committee is, the more likely the National Labor Relations Board is to find that the committee is an illegally dominated and supported 'company union' in violation of the Taft-Hartley Act."

Regardless, these cases and their related difficulties should not be allowed to deter completely the use of employee teams. The active use of employee participation and input via teams or committees lies at or near the heart of every total quality initiative. If management believes what is said to employees about the value of their input, about "empowerment" and about "owning your job," then management had best make maximum use of participative processes including teams.

Shortcomings

As popular as it is in corrections to blame the courts for problems, litigation should not receive all the blame for diminishing the effectiveness of such teams. Completely aside from labor law implications, employee teams or committees are not without a variety of other problems. A sufficient number of problems surface from time to time to suggest that individuals and the organization itself can stifle a committee or dramatically limit its effectiveness.

Some team members (especially managers serving on teams with non-managers or others of perceived lesser rank) are unwilling to set aside position and power for the sake of the team. Also, unequal levels of knowledge and ability among team members can lead some to dominate and others to become overwhelmed or "lost in the crowd."

Some extremely important and potentially highly disruptive effects on teams lie in company reward and compensation systems that continue to focus on individual effort, not on team performance. This has been a frequently encountered barrier to successful total quality implementations as organizations have tried to alter how they do business without first (or at least concurrently) changing the systems by which they do business.

Performance appraisals that do not account for team performance also present barriers. In fact, an organization's performance evaluation process is one of the major business support systems that has to change dramatically for successful total quality implementation. It is not nearly enough to simply change the language of the appraisal. It is necessary to change employees' concept of evaluation from a focus on the individual ("I had my evaluation today") to emphasis on the team.

In addition to the foregoing, lack of top management commitment to the process is a sure means of destroying effectiveness. It should go without saying that the top management that fails to "walk the talk" will be recognized as insincere. Members will quickly come to see their teams as do-nothing or rubber-stamp bodies if the participative process has no visible effect on top management.

Some problems with teams lie in the labels in use for these bodies, labels such as self-directed, autonomous, self-managed, and the like. These names are misleading in that they convey the belief that these groups are independent and free to act as they choose. No effective teams in business really provide their own direction. Each team should be directed by its specific charge or mission and by the goals of the organization. As such, all teams are actually interdependent with other organizational elements. Effective teams require clear direction, comprehensive guidelines, and open, nonthreatening leadership. Efforts to encourage employee participation with self-directed work groups will not succeed without serious attention given to the development of strong, appropriate leadership for participative management.

What to Avoid

Considering the applicable litigation issues, plus common-sense organizational concerns for participative activities, it is possible to enumerate five potential obstacles to avoid in forming and working with employee teams:

1. Never allow an employee team to deal with terms and conditions of employment, such as wages, hours, benefits, grievances, and such. Even consideration of "working conditions" in general should be avoided. As a member of management who might be part of a team, do not deal with other team members (specifically nonmanagers) concerning terms and conditions of employment. If a team's activities take it from a legitimate topic into the realm of terms and conditions of employment, its direction should be altered.

2. Do not solicit complaints, grievances, or suggestions about terms and conditions of employment from teams. If such issues arise on their own, refer them

to the proper point in the organization (for instance, refer health insurance issues to the benefits manager).

3. Do not let team meetings degenerate into gripe sessions in which members simply complain about aspects of their employment.

4. Do not mandate employee participation, ask employees to represent other employees, or sanction employee elections to choose representatives.

5. Do not allow an employee team or committee to exist and function without a clear, understandable mission or charge and without fully and plainly delineated limits on its authority and responsibility. In other words, the group must know exactly what it can and cannot do.

For Effective Employee Teams

Short of actually establishing teams or committees to wrestle with certain issues, there are a number of steps that can be taken to encourage employee participation. It is possible, and frequently advisable, to consider bringing together loosely defined groups of managers and employees to simply brainstorm ideas, gather information, and help define problems. This is a generally safe strategy, as long as no proposals are offered or recommendations made. It also is proper to assemble an employee group to share information and observations with management, again as long as no proposals or recommendations are made.

Beyond one-time or limited, informal gatherings, it is important to keep the following seven points in mind regarding actual teams or committees:

1. When establishing a team or committee, identify it up front as not intended as an employee channel to management.

2. Have a clear mission or charge in place before soliciting team membership, and have the team's functions and limits identified before any team activity begins.

3. Keep the team focused on work-improvement topics only. This requires clear guidelines and plenty of continuing vigilance. It is difficult to talk about quality, efficiency, productivity, and such without conditions of employment becoming involved. Be constantly aware of the potential need to redefine the team's boundaries periodically.

4. Staff teams with volunteers, or use rotating membership selected by some means that is not management dominated.

5. If a team is empowered to make a final management decision (that is, the team decides in place of management, not just recommends to management), it can be seen as acting as management. This is acceptable. In fact, it has been suggested that the ultimate protection against being ruled by an illegal labor organization exists when the team can make final decisions in its own right.

6. If an issue is sufficiently narrowly defined that all persons affected by it can be included in a single group, a "committee of the whole" including everyone is usually legally safe. In such an instance nobody can be seen as "representing" anyone else.

7. For standing committees or long-lived teams, maintain a majority membership of managers. A committee or team composed of a majority of managers stands

less chance of being judged illegal. However, a significant drawback of such teams should be obvious. A team composed mostly of managers is far less likely to be seen as a legitimate vehicle for employee participation.

Avoid creating teams or committees that tend to develop a continuing existence. Instead, consider establishing specific problem-solving or work improvement ad hoc groups, each with a specific, well-defined charge and a specific problem to solve. Disband each group after its goal has been attained. Such ad hoc groups can much more safely consist of a majority of nonmanagers than can "permanent" teams or committees. However, for teams composed largely of rank-and-file employees it is legally safest to have management representatives serve as observers or facilitators, without the power to vote on proposals or dominate or control the group.

For collaborative group problem solving and participative decision making in general, it is always appropriate to bring into the group those people who have the skills needed for dealing with the group's charge. It is necessary, however, to recognize that they have skills pertinent to the problem at hand and will likely have greater influence on group decisions. Also recognize that teams or committees become unwieldy as they grow in size. Small groups are generally better; active participation in tasks seems to decrease with increases in group size. In fact team participants tend to rate small groups as more satisfactory, positive, and effective than larger groups.

Team Up for Success

As to the critical factors concerning the establishment of employee teams, it may help to remember that the highest legal risk involved in establishing groups is probably encountered while active union organizing is occurring. A modest amount of risk exists if teams are established in the presence of existing union contracts. The least risk occurs when there is neither organizing nor an active union presence.

Common sense advice would suggest that employee teams or committees should not be established while organizing is occurring. It could be extremely helpful to have had some productive employee participation in place beforehand, but after the organizers arrive is no time to jump into employee teams. Even if such teams are legally established and operated, they will foster incorrect perceptions concerning their creation. Rather, with or without a union in place (and as long as active union organizing is not occurring) employee teams can be employed legally and productively.

Employee participation may well be the key to continuing increases in quality, efficiency, and productivity. Employee participation is essential. As noted earlier, nobody knows the inner workings of a job better than the person who does it day in and day out. Also, there are few if any problem solutions that are not enhanced by multiple viewpoints and inputs. A team brings to the problem the power of the group. To cite a highly pertinent quotation from an anonymous source: "I use not only all the brains I have, but all I can borrow."

The Departmental Team

Team building also can refer to building and maintaining an effective, productive departmental team—a working group dedicated as a body to continuing superior performance in specific operational areas. As far as the correctional manager is concerned, ongoing project teams experience their periods of inactivity and ad hoc teams come and go. However, the departmental team is (or should be) a constant presence, and represents most of the context for this book.

A departmental team starts with a collection of individual performers whom the manager must organize and guide in a common direction for a common purpose. If the individuals and the manager are conscientious, well motivated, and honestly share a common purpose, this collection of individuals can be reasonably effective. However, a collection of individuals working as individuals cannot achieve the level of performance possible from a well-functioning departmental team even if they are each doing their jobs well. Unless they are able to tap the potential of true teamwork, their output will rarely amount to more than the sum of their individual efforts.

Team Building and Its Purposes

Team building is an organized, systematic process of unifying a group of employees with common objectives into an effective and efficiently functioning work unit. The ongoing challenge of team building is to encourage a diverse collection of people to think and behave as a single entity rather than as a collection of individuals. What is sought in a properly functioning team is a synergistic effect—a total effect and resulting output that is greater than the sum of individual contributions. The synergistic effect is literally the team effect in which the whole is decidedly greater than the sum of the individual parts.

Like it or not, the trend of today (indeed the continuing wave of the future) in corrections is the need to accomplish more with less. As a result, it is not surprising to find that a great deal of today's effort toward improving productivity is directed at improving the performance of work groups. True team building can:
- Foster increased productivity while maintaining or improving quality.
- Improve work climate, enhance work relationships, and increase employee satisfaction.

Team building is, by its very nature, a continuing process. It is essential to understand that it is a long-term strategy, an activity that is never complete. A supervisor who manages the activities of a few people has a choice. He or she can provide the employees with what they need to do their jobs, insert new employees into the gaps inevitably created by turnover, and accept the work that each employee turns out as long as it meets or exceeds minimum standards. Or the supervisor can take the longer view, focusing on people and their needs and capabilities and striving to build and maintain a functioning unit rather than trying to influence individual productivity.

■ Recognizing Employee Potential

The supervisor who would endeavor to build and maintain an effective team must first and foremost ascribe to an optimistic view of employees overall. This has to be a genuine belief in people and their capabilities—a Theory Y approach to the management of people as opposed to a Theory X philosophy (see Chapter 11, "Leadership: Style and Substance").

The Theory X manager will ordinarily not trust employees to do a proper job unless they are watched, pushed, and regulated. He or she is a "boss"; people are viewed primarily as producers of output. Accompanying a lack of trust, one often finds lack of respect as well, and the manager who neither trusts nor respects the employees will in turn be neither trusted nor respected by the members of the work group. Yet mutual trust and respect are fundamental to the creation of a fully effective team. The manager who wants to build and maintain an effectively functioning team must believe the majority of employees want to apply more of their abilities to their work. The accompanying view is that when they are not allowed to do so, they find their work less satisfying than it could be.

Thus team building requires the following:
- Shared power and authority
- Increased employee participation

These requirements lead back to what has been said about effective people-centered management throughout the first half of this book. A fundamental of successful team building is the leader's generally positive view of employees and their potential.

Toward Strengthened Motivation

The majority of workers will put forth a sustained effort to do a good job when (and only when) they want to do a good job. Some workers will do a good job in the short term out of fear or apprehension. But quality that is forced or coerced is not sustainable and is usually delivered with resentment. How, then, does the manager inspire a group of employees to the extent that they are willing to do their best consistently?

Recall from the material concerning motivation that it is not possible for the manager to motivate employees directly. All true motivation is self-motivation. So the best the manager can do is foster conditions under which the employees will become self-motivated. Among the actions the manager can take to improve the motivational climate in the department are the following:

- Get serious about proper delegation (empowerment), not using the process to shift work around or shed unpleasant tasks, but rather to provide some opportunity for employee learning and growth.
- Consider decentralization of certain tasks when appropriate, giving people at what might be scattered locations complete control over work that might otherwise flow through two or more employees.
- Look into the possibility of job enhancement, appreciating that a mix of more tasks and more varied responsibility can provide greater interest and thus improved job satisfaction.

- Try cross-training when possible, with a goal of eventually having all employees of a given job grade able to do all jobs pertinent to that grade. This can greatly increase the flexibility of the work group while increasing interest and challenge for individuals.
- Remain conscious of every opportunity for increased participation by employees and take their contributions seriously. This will strengthen mutual trust and respect in addition to opening up some productive avenues the manager might never think of alone.
- Allow employees who have proven themselves responsible to take on special assignments, especially those over which they can exercise complete or nearly complete control.
- Constantly utilize enhanced feedback on performance. Recall that one of the job-related factors that means the most to employees is the full appreciation of work done. When praise is deserved it should be immediate, sincere, and specific. When criticism is unavoidable it should be immediate, diplomatic, and constructive.

Will absolutely every employee respond positively when so treated? No, of course not. There will always be some who cannot be reached in this manner. There will be others whose preference, whether they would admit it or not, is to "check their brains at the door" and put in their 8 hours in the same old way. However, enough people can be reached to make a difference. Enough can be reached so the conscientious manager can build a team and create an environment that puts the resistors on the spot and causes most of them to move in one of two directions: (1) buying in and moving with the team or (2) removing themselves from the group.

■ The Stages of Team Building

Although a continuous process, team building occurs over the following separately identifiable stages:

- Formation
- Disequilibrium
- Role definition
- Maturity
- Maintenance

Formation

It is of course not often that a departmental team has to be formed from the ground up with all new participants. One exception certainly would be the activation of a new institution or a new housing unit under the unit management system. There also can be significant disruptions in the balance and composition of a team, to an extent that the resulting configuration is essentially a new team. One such event is turnover in team leadership, especially when the new supervisor or manager is a complete unknown. Another is a new departmental configuration such as that resulting, for example, from an interorganizational merger that throws two or more groups together under common leadership that is new to many of them. These and similar changes in composition can throw the team into the formation stage, as the participants work to reestablish common ground and build what amounts to a new team.

In the formation stage, individual commitment is tentative. A number of people are likely to adopt a "wait-and-see" attitude, especially where new leadership might be involved. A new leader (even one who is philosophically similar to the previous leader) will have his or her own approach to different facets of the job. New relationships must be established, and until this has been accomplished a great deal of team communication will be cautious and guarded. Trust and confidence will not be extended in full until the new leader or, in the instance of a merger, the new group has acquired a level of comfort with this new combination of personalities.

Disequilibrium

In this second stage of team building most of the people in the group are having their effects as individuals according to their own needs, desires, and capabilities. They may be doing their jobs well, but they are doing them as individuals. Many do not yet see themselves as part of a team. With each other they tend to be competitive more than cooperative. Some fear loss of credit to others and others may strive to be the ones who are "on the good side" of the manager. Some will view the portion of the work they do as their exclusive "territory," and they will become protective of that domain.

Some groups never go appreciably beyond this stage, forever remaining in a state of disequilibrium. The work can and does get done in this stage, at least at a minimally acceptable level, but it is the aggregated work of individuals, not the output of a functioning team. When a group is in a state of disequilibrium there can be no synergistic effect. In this state, at best, the whole is no more than the sum of the parts.

Whether a group moves beyond disequilibrium is dependent on the skill, determination, and desire of its leadership. The purely production-centered manager may have difficulty moving a group beyond this stage; complete and successful team building requires a people-centered focus.

Role Definition

As mutual trust and respect build among the members of the group, and as individual roles are clarified, then the interrelationships of these roles are established. Cooperation begins to build and a team identity begins to emerge. Distrust is rare, and mutual respect is high. Communication becomes more open as participants begin to behave less like competitors and find a level of comfort in relationships with other members of the group. Individuals understand their place in the larger picture, and they have essentially taken on the objectives of the group as their own.

Relatively stable departmental groups, especially those consisting of people who have worked in harmony for a considerable time, will be at least at this cooperative stage. What is needed most for a group to move beyond this stage is the appropriate leadership of a manager who sees himself or herself as part of the team as well as the catalyst that moves the team. This will be the kind of leader who can honestly equate personal success with team success.

Maturity

Maturity, of course, describes the stable, functioning team. The mature team will exhibit high levels of cooperation, mutual trust and support, and productivity. The mature team will experience normal or below-normal turnover. Any turnover that does occur will be driven not by dissatisfaction but by other, more positive forces such as career advancement or forces beyond departmental control such as disability or death. For the most part, the members of the mature team will experience pride in team accomplishments and exhibit high levels of job satisfaction.

Maintenance

In team building, maturity is not a fixed destination. Rather, it is a state of existence that must be maintained or it will deteriorate. Nothing remains stable. System inputs change in character, the environment changes, subtle and not-so-subtle changes in organizational mission and direction occur. In the face of these forces, a team that does not continuously adapt will fall behind. Because nothing remains stable, team maintenance requires a regular infusion of new ideas, novel approaches, and challenges to maintain team effectiveness. The mature team (if it is to remain fully effective) must change as the circumstances surrounding it and the demands made on it change.

Another important dimension in which a team must be conscientiously maintained is the manner in which new team members are assimilated. All too often new employees are thrown into a job with inadequate orientation and incomplete preparation. When this occurs, one of two results is likely to follow. The first possibility is that the new employee may feel adrift and uncomfortable to the extent of resigning shortly after placement. A second possibility is that the new employee catches on in bits and pieces, fits and starts, by guesswork, by asking coworkers, and so on, and develops as an individual performer, not as a team player.

Team maintenance places a high priority on bringing in new members appropriately. The group must provide education in the agency's mission and objectives and the departmental team's role in pursuing them. The group also must offer a thorough orientation to the functioning of the team. Direct mentoring (perhaps teaming a new employee with an experienced person for as long as it takes to achieve comfortable assimilation) can be a useful strategy, as well as regular coaching. And finally, encouragement is needed as the cement in a process that brings a new person completely up to the status of a fully functioning team member.

■ The Power of the Team: The Individual

A great deal has been said up to this point about the departmental team as a functioning unit. This should not be allowed to convey an incorrect impression of the individual's place as part of a team. People are not absorbed into some form of "group mind" and robbed of their individuality. On the contrary, individual identity and individual motivation must be perpetuated as the heart of an effective work team. Motivation always remains individual; recall the assertion that all motivation is in fact self-motivation. Also, it is reasonable to claim that individual motivation is the point at which all human productivity begins. As to what motivates (or more correctly, what causes most people to be self-motivated), the sources of motivation come in the form of opportunity. These include the opportunity to achieve, to learn, to do interesting and challenging work, to do meaningful work, to assume responsibility, and to become actively involved in deciding how the work is done (for more detail refer to Chapter 13 on motivation).

What teamwork actually accomplishes is focusing the efforts of individuals on common goals and objectives, and the process of reaching them. Teamwork itself is not a goal. Rather, it is simply a means of harnessing the individual efforts and contributions of the people who make up the team. To maintain team effectiveness it is essential that individuals work together in pursuit of team objectives, with the aim of accomplishing together something larger than any of them could accomplish alone. A team objective should consist of the same elements as an organizational objective or an individual objective. But individual, team, or otherwise, the statement of an objective is not complete unless it includes an expression of what is to be done, how much is to be done, and when it must be done.

For example, a possible objective of a case management section of an institution or probation office might read: Reduce the number of pending classification case (what) from the present 48 to 30 (how much) by September 1 of this year (when).

■ Team Building and Leadership Style

The department manager's leadership style is the greatest single determinant of successful team building. The development and maintenance of a mature, functioning team requires an open, participative, people-centered approach to management. The difference between success and failure at team building is as fundamental as the difference between participative management (genuine leading) and authoritarian management ("bossing"). The manager who would build an effective team cannot set himself or herself above the group, but must rather be an integral part of the group. Consider the following differences between the old-fashioned authoritarian (called "boss" for this comparison) and modern participative manager ("leader" in this comparison):

- The boss wields authority, while the leader cultivates good will. The leader possesses authority but does not openly flaunt it.
- The boss tells people how things should be done, while the leader, insofar as possible, shows how things are done.
- The boss pushes people, while the leader coaches, counsels, and encourages.
- The boss has an "I" orientation (my department, my staff, my accomplishments, etc.) but the leader has a "we" orientation (our department, our group, etc.). The appropriate focus for a team's leader is always "we," "us," or "ours," and never "I," "me," or "mine."

- The boss, in effect, simply says "Go," and watches as the group moves toward a goal. The leader in effect, says "Let's go," and moves toward the goal as part of the group.

■ Guidance for the Team Builder

This heading could generate an entire educational program on its own, which would incorporate material from a dozen or so chapters from this book. The paragraphs that follow may duplicate a few points made throughout those other chapters. But in briefest possible form they are meant to provide a picture of the approach to be taken by the department manager who truly wishes to develop one of those exceptionally effective teams of which far too few are seen.

What is suggested for the team-building supervisor is a people-centered leadership approach, based on mutual trust and respect of all concerned. Certainly the employees will always see the manager as the manager; this is not an attempt to immerse the manager in a totally egalitarian situation. The employees know that the manager has the authority of the position, regardless of how directly or indirectly that authority is exercised. Also, the employees know that the manager is in charge. However, it is how this responsibility is fulfilled that makes the difference.

A truly effective leader sets himself or herself above the staff in only one dimension—the acceptance of responsibility. The manager is not paid more than the staff because he or she is smarter, better educated, or in some way superior to the others. The manager is paid more for bearing the burden of responsibility. To enhance the likelihood of building and maintaining an exceptionally effective team, the department manager must do the following:

- Be continually aware of what employees want from their employment (refer to "Demands on the Organization" in Chapter 13), and take these expressions of need seriously. Of particular importance in building an effective team is the acceptance of each person as a full-fledged member of the group, recognizing each as a contributing partner and not simply a servant of the system.
- Be seen as identifying more readily with the department than with that higher-level collective known as "management." The manager whose focus is clearly upward is likely to be viewed by the staff as an outsider, one of "them" (management). In contrast, the manager viewed by the staff as an insider, a member of their group, is far more likely to inspire the cooperation needed to build an effective team.
- Be readily visible and available to the department's employees. Visibility and availability reinforce the manager's commitment to the group and provide assurance that support and assistance are always close at hand. The manager who is frequently gone or consumed by meetings, special assignments, external conferences, and such will not be viewed by the staff as a true team member. An absentee manager cannot be the catalytic agent the manager must be in building an effective team.

- Be serious about practicing proper delegation (empowerment), and give all capable individuals the opportunity to gain experience that might open new opportunities to them.
- Offer to all staff the opportunity to learn and grow, and work to develop those employees who seriously take advantage of this opportunity.
- Practice consistently the use of deadlines and follow-up in assigning work. Any task worth doing at all deserves a specific deadline (not "when you get a chance"). Never let a deadline pass unanswered without following up.
- Practice true participative management. Be ever aware that the department's employees can be a nearly bottomless well of improvement opportunity. After all, no one knows the detailed ins and outs of a task like the person who performs it every day.
- Encourage team participation in decision-making. The successful team-building manager will recognize that giving employees a voice in decision making does not indicate weakness or abrogation of responsibility; rather, doing so indicates just the opposite—strength and sufficient confidence in the employees to bear the ultimate responsibility for the decisions made jointly.

Attitude and Commitment

The previous discussion has been directed toward securing employee commitment to the process of creating and maintaining a fully functioning departmental team. There is a direct and undeniable relationship between employee commitment and an employee's sense of being part of a team. Employee commitment cannot be mandated or manipulated into existence. It must be inspired by the manager's own commitment and continually reinforced by the manager's attitude and behavior.

Attitude (especially the manager's attitude) can be an extremely powerful force.

Attitude—whether positive or negative—is invariably contagious. And the leader's attitude can affect the entire group for good or for ill. Sometimes the most difficult task a manager can undertake is to convey a positive attitude in the face of difficulty and often downright hardship. But without a positive attitude concerning what can be accomplished, there will be little chance of building a fully functioning, productive team. In the face of constant pressure to accomplish more with less, one of the few ways of doing so is through the improved productivity possible from an efficient and effective departmental team.

EXERCISES

Exercise 25-1: Can You Build an Effective Team from the Enemy Camp?

The setting is a privately managed detention center for illegal aliens. Business manager Helen Williams was hired from outside the institution. She believed strongly in the power of a unified departmental team. In her previous position she had inherited a disorganized collection of individuals who openly competed among themselves for high evaluation scores and the manager's favor. Over time, she was able to make them into a cooperative team that became a model for the institution.

Helen accepted her new position even though she strongly suspected it was something of a "hot seat." She was the fifth person in that position in 3 years. Although Helen did not know the reasons behind the short stays of her predecessors, after a month she decided that the atmosphere in the department was decidedly unhealthy. Her staff appeared divided into two distinct rival camps that she began to think of as "Camp A" and "Camp B." (Actually, she had heard members of each group refer to the other group as "the enemy camp," but she did not use "enemy" in her descriptions of these groups.)

From her first day on the job it was apparent to Helen that many of the problems in the department stemmed from poor intradepartmental communication and lack of cooperation among staff members. She was surprised to learn that her immediate predecessor never held department meetings but instead met occasionally with groups of two or three to deal with specific problems.

Helen instituted the practice of holding a weekly 30-minute meeting for all employees in the department. She made it plain that everyone was expected to attend. After 4 months of staff meetings it seemed to Helen that the atmosphere of rivalry between the camps had scarcely diminished at all. To her, it seemed evident that the group was still divided on many matters. It also seemed to Helen that Camp A was slightly more supportive of her than was Camp B, but this seemed to make Camp B all the more difficult for her to deal with.

Helen's initial assessment of the department was that productivity was low relative to the number of people available to do the work. In the inmate billing section, for example, the average time lapse between transfer/release and billing was the longest Helen had ever seen. She had come to the job vowing to increase the department's productivity, but at the 6-month mark in her employment she had to admit that little had changed. Camp A and Camp B still went their own ways and at times seemed to be doing their best to undermine each other.

Instructions

- Consider what actions Helen might take to improve the level of cooperation and productivity in her department and to point her employees in the direction of becoming a functioning departmental team.
- Discuss how and why these two "camps" may have formed. What impact might the supervisory turnover in the department have had on this development?
- Such groups can have friendship bonds as well as work bonds. How will Helen deal differently with these two dimensions of the problem?

- Are there additional ways to use meetings and team-building exercises to break down the barriers between the two groups?
- What can Helen do to avoid the formation of new "camps" in the department?

Exercise 25-2: The Silent Majority

As the newly placed manager of the receiving and discharge department, it didn't take long for you to discover that morale in the department had been low for quite some time. As you worked to become acquainted with employees by meeting with each of them alone, you were rapidly inundated with complaints and other evidence of discontent. Most complaints involved problems with custodial department (staffing) and the business office (availability of inmate release funds), with a few problems from committing law enforcement agencies thrown in. There also were complaints about other department members and some thinly veiled charges about fellow employees who "carry tales to the administration."

In listening to the problems you detected a number of common themes. You decided that most misunderstandings could be cleared up by airing the gripes openly with the entire group. You then scheduled a staff meeting and asked all employees to prepare to air their complaints (except those involving specific other staff members) at the meeting. Most employees thought this was a good idea, and several assured you they would speak up. You were encouraged by what you heard; it seemed as though most employees were of a similar mind, indicating something of a team outlook. However, your first staff meeting was extremely brief; when offered the opportunity to air their complaints, nobody spoke.

The results were the same at the next staff meeting 2 weeks later, although in the intervening period you were bombarded with complaints from individuals. This experience left you frustrated because many of the complaints you heard were problems of the group rather than problems of individuals.

Instructions

- Discuss what can be done to get this group to open up about what is bothering them.
- What steps should be considered to get this department on track toward becoming an effective team? Identify at least two significant actions and explain why they should be considered.
- How should the specific problem of the employees who supposedly "carry tales to administration" be handled?

ENDNOTES

1. "Special Report: Employee Representation Committees at Electromation Are Illegal: NLRB," *Management Policies and Personnel Law* (January 15, 1993): 1–2.
2. Swan, J., Jr. "The Most Effective Employee Committees Are Probably Illegal," *Personnel* 63, no. 11 (1984): 91.

Methods Improvement: Making Work—and Life—Easier

26

Chapter Objectives

- Convey the belief that there is usually "room for improvement" in the way most tasks are performed.
- Outline a simple but logical approach to the improvement of work methods.
- Establish the individual worker, the person who does the job every day, as the greatest potential source of methods improvement knowledge.
- Introduce some of the more common tools and techniques of methods improvement, describe their uses, and provide guidelines for their application.
- Outline an organized approach to methods improvement that is applicable institution-wide.
- Provide practical examples of methods improvement activity.
- Identify the role of the supervisor in encouraging a "methods-minded attitude" on the part of the department's employees.

There's a way to do it better—find it.

Thomas A. Edison

Progress, man's distinctive mark alone.

Robert Browning

Technology and Inmate Health Care

For years, inmates needing medical care received it in the institution. Only in dire emergencies were they taken out to community medical facilities. Routine matters were handled with the limited staff resources available in the institution infirmary.

Over time, it became somewhat more acceptable to escort offenders to local clinics, doctor's offices, and hospitals for treatment. But there still were security concerns, particularly for inmates with backgrounds that included dangerous behavior. The need existed for delivering quality health services to inmates without unduly jeopardizing public or staff safety.

Today, technology allows a great many procedures to be carried out in the institution, which heretofore had required outside trips. Taking an electrocardiogram can tell the institution physician a lot about an inmate's heart function, but it still is a limited diagnostic measure. One of the first expansions in technology usage began with the use of fax machines. The paper printout of the cardiogram could be faxed to a cardiologist to obtain an expert opinion on the readout, even if a doctor was not available in the institution. This change set the stage for further remote examination and diagnostic options. Digitalized inmate X-rays now can be sent to a radiologist for reading. Blood, urine, and other testing can be done in a time-efficient manner using express delivery methods. Closed circuit television systems can be used to expand the expertise available for in-facility exams. Institution physicians can carry a personal digital assistant that has in its memory an immense amount of pharmaceutical data and treatment information. In short, technology has improved the delivery of inmate health care in important ways not thought possible just a few short decades ago. These changes not only enhance public safety, but deliver higher quality health services to offenders at a lower cost to the public.

■ Edison-Plus

You will notice there is no hedging in the opening statement by Thomas Edison, no qualifier that might keep the door to improvement closed. Edison did not say there is sometimes, often, or even usually a better way. In fact, there often are several better ways.

Some of the pioneers of industrial engineering (the branch of engineering dealing with work methods) insisted on the need to seek something they called the one best way. In recent years, however, the notion of a single best way to do something has come to be regarded as something of a theoretical ideal. Now, rather, it is recognized that there are usually several better ways of accomplishing a given task and that a manager must seek one of these better ways that reasonably meets needs and fulfills objectives.

For the purposes of this chapter, <u>methods improvement</u> is an organized approach to determining how to accomplish a task with less effort, in less time, or at a lower cost, while maintaining or improving the quality of the outcome. This process may be referred to in different organizations by various other labels. Among these terms are those of methods engineering, job improvement, and work simplification. Regardless of label, however, the intent remains the same: to alter the performance of a task in such a way that the results represent a desired improvement over the way it was formerly accomplished.

Lest the reader think this is a topic not very applicable to corrections, reflect a moment on the number of correctional functions that in some way are repetitive, systematic, and/or resemble an assembly line. Food service operations are replete with all three characteristics, to the extent that some institutions that rely on pre-plated meals for serving in housing units actually move trays down a conveyor belt as portions are added. The receiving and discharge process is (at times of mass movement) an assembly line of bodies and property being processed. A records office involves the movement of paper and the systematic application of intellectual activity to it. Virtually all office operations are routinized to a certain extent. Every one of these typical institutional functions, and many more, are susceptible to examination and improvement by the methods discussed in this chapter.

Methods improvement is understandably a subject that is not near and dear to a great many supervisors. Rarely is the need for change in a specific work procedure so pressing that it takes priority over the day-to-day, hour-to-hour operating problems of a department. However, methods improvement is another of those management activities that puts practitioners in something of a bind. It rarely is critical and it is usually freely postponable, but unless it is made part of today's effort managers will never realize its near-future benefits. Much like delegation, it is one of those activities managers promise to get serious about "after the rush is over." Somehow, however, they never quite get around to it.

A modest but active methods improvement program can be a valuable part of any supervisor's pattern of management. There are economic benefits to be gained. Think about what happens when a job is managed with the goal of accomplishment in less time, with less effort, or with fewer material resources. Tangibly, departmental efficiency is increased by improving the relationship between output and input. Intangibly, there are positive returns in employee attitude and workplace morale.

■ Room for Improvement

Methods improvement may exist as a formal program in the institution or agency, coordinated by a specific department. It also may exist at the department level (at the discretion of the supervisor) as an informal, ongoing effort to "tune up" and otherwise improve work procedures. At the very least, methods improvement should exist in the department as an <u>attitude</u>. Such a "methods-minded" attitude—a practiced belief in the possibility of continuing improvement—will itself encourage employees to utilize time and other resources more effectively.

Methods improvement applies largely to tasks, to problems of things (such as procedures, processes, forms, equipment, and physical layout) as opposed to problems of people. Generally people recognize that the overall task of getting things done through people puts the supervisor in a position of being largely a solver of people problems. However, work procedures and other essentially non-human factors are nevertheless shaped or influenced by the people who do the work. A certain procedure may lose efficiency over months or years because of the changing environment in which the job is performed. But efficiency may also be lost because people bring their own habits, attitudes, and preferences onto the job. Those elements can influence the way they work in a very real sense. Dig deeply enough into causes, and a manager will often find that many apparent nonpeople problems have human origins.

Each person coming to work in the department is a unique individual, and as such exerts a unique influence on the workplace. Thus a work procedure (especially one that was perhaps not particularly well defined in the first place) becomes at least in part a reflection of the worker. Run several consecutive individuals through a specific task and the result is a composite "procedure" that incorporates the subtle and perhaps not so subtle influences of these several people. Since organizations train many employees on the job using the example of the person actually doing the job, they pass along not only essential job knowledge but also the influences of other workers.

Often, the procedure for performing a given task exists in the department in three distinctly different ways. It can exist in one form in the mind of the supervisor who (perhaps having done the job at one time) thinks of a specific, and quite likely outdated, collection of steps. The procedure can also exist in the mind of the worker in the form of all the steps this person actually performs to accomplish the task. It may also exist in written form in the department's procedure manual or as a set of post orders on a post somewhere in the facility. This latter document will almost certainly be different from either the supervisor's concept or the worker's practice. And how often are policies referred to authoritatively when people believe that the same old task is being done in the same old way?

In some larger agencies, the concern for methods improvement techniques is left to a specific function such as a management engineering specialist in the headquarters office, for example. However, the power to improve procedures and make work easier need not be limited to some special staff. A simple philosophy of methods improvement might suggest the following:

- There are few, if any, existing tasks that cannot be improved.
- The people who actually do the work are potentially valuable contributors to improvement.
- The power of participative management can bring out the best in each employee.
- From a healthy participative management point of view, methods improvement may be simply described as applied common sense.

With variations in terminology and occasional differences in emphasis (owing to the agency environment in which it is applied) the process described in the remainder of this chapter is largely the process that drives work improvement within a TQM program. This process often is referred to as team oriented problem solving (TOPS) or as the mission of "self-directed work teams." It has also been called "team-oriented process improvement" and "work simplification," the latter label identifying an organized approach to work improvement dating back to 1932.[1]

The overall process has also been known by the most broadly generic label of "methods improvement." Regardless of label, however, two critical characteristics run through all of these approaches. At the heart of total quality management, these are:

1. A multidisciplinary approach is taken to problem solving, involving all departments that have a relationship to the problem.

2. Solutions are generated—and implemented—at the level of the people who actually do the work.

■ The Methods Improvement Approach

The generalized multistep approach to methods improvement presented next will be followed by a review of some of the simpler tools and techniques of methods analysis. Successful methods improvement ordinarily begins with concentration on a single, specific task. A supervisor has far too much to do to attempt to improve everything at once. A shotgun approach will lead to the start of many projects but the likely completion of few or none. Simply select one task or isolate one problem—one that is causing trouble or one that some members of the department feel could be done more efficiently—and go to work.

Select a Task or Isolate a Problem

You must of course become aware of a problem before it can be solved. The need for methods improvement often becomes apparent on a number of fronts when typical problem areas in the department arise and are examined. The significant signs that often suggest a particular task deserves analysis and improvement are:

- An activity is costing you more in staff time or dollars than you know or believe it should cost.
- Bottlenecks occur as work backs up at various stages of a process, or there are chronic backlogs of incomplete work.
- Confusion is evident as employees appear uncertain as to what a task involves, what happens next, how it happens, and why.
- Poor morale prevails in the work group.
- Labor-management problems begin to surface.
- Inappropriate staff conversations suggesting idleness and other nonproductive activities appear to occur in excess.

In locating and defining problems that may highlight the need for improvement, it is important to distinguish between symptoms of problems and true problems.

For example, what may appear to be a bottleneck caused by poor work methods in a jail's records office might be correctable through a small amount of overtime, a bit of extra effort, or a rearrangement of task assignments. However, suppose the true cause happens to lie in ineffective scheduling practices (not enough people on duty on certain days of the week when court call is heavy).

The initial (overtime) fix is not likely to work for long in such a case. When the area addressed head-on is really only a symptom of the problem, the difficulty will probably return within a short time or resurface in a somewhat altered form. Thus it is necessary to look beyond the apparent immediate problem and identify the force or forces that lie behind a manifestation of difficulty.

Gather the Facts

The next step is to learn thoroughly how the task is done now—everything one could possibly learn about the way the job is being performed. The total process involved in the task under consideration must be described through narratives, sketches, flowcharts, or whatever other techniques are available. Sketching out "the system" often increases understanding of the relationships among system components. A manager can often observe a logical order or reasonable interrelationship among the pieces once they are all captured on paper. Recording the present method in full also allows others who may be able to provide insight into the problem to examine and compare it with their understanding of the situation.

In getting the facts (in applying the information-gathering tools and techniques of methods improvement) one may gather information from:
- Written policies, procedures, and post orders, if they exist.
- Architectural and layout drawings, if involving physical layout or space.
- Work schedules.
- Samples of pertinent forms and records.
- Accounting data, payroll data, purchasing information, and other financial records.
- Statistical reports and other records of operating results.

Rest assured, however, that information collected from any of the foregoing sources will often be insufficient. Additional effort usually must be exerted in the one major area of concern that cannot be determined from existing information—how the task is currently being performed.

Reduce the Problem to Sub-Problems as Necessary

Often a single apparent methods improvement problem may seem too large or complex to tackle as a whole. Such a problem can often be broken down into a number of smaller, separately identifiable problems that can be solved independently.

Suppose, for example, that there is a significant, regularly recurring backlog of classification transcription work. One central problem seems to be behind the issue—caseworkers often neither provide clear, understandable dictation nor do they consistently complete it within the organization's required 15-day period.

Suppose also that the various aspects of this apparently sizeable difficulty have been distilled down to three prime areas. Many case managers are not doing their dictation within the required period, so they are not following the necessary procedure. Many case managers have complained about dictating equipment that is unavailable, or operates poorly or not at all. There also have been numerous complaints about high noise levels in the unit offices where the dictation takes place.

Thus the problem may be divided into several sub-problems, each of which may be addressed separately, at least at first: (1) the recording procedure, (2) the state of the dictating equipment, and (3) the physical environment of the dictation area. Once the problems have been separately addressed and tentative solutions developed, the sub-solutions may be combined and integrated into a solution to the total problem.

Challenge Everything

All of the information gathered up to this point (and most especially the record of how the job is now being done) must be subjected to searching study. Take nothing for granted. At every step along the way, ask: What is actually being done? Where is it done? By whom is it done? How is it done? Always, for every step, ask: Why?

Examine every step and challenge every detail in an effort to:

- Eliminate activities whenever possible. Just because something is being done is no reason it must be done. Ask of each step: If this were not done at all, what would be the result?
- Combine steps whenever possible.
- Improve the manner of performing various steps.
- Change sequence of activities, location where work is performed, assignments of people involved, and any other factors that might have a bearing on efficient task performance.

The process of challenging every last detail lies at the heart of methods improvement. At this point methods improvement (while appearing to many to be a logical "scientific approach") must become a wide-open, no-holds-barred creative process. It is here that participation and teamwork have far-reaching implications. A number of factors come together to challenge every detail of the problem thoroughly. This stage carries problem solvers toward these key elements:

- Brainstorming and other creative idea-generating techniques.
- Optimism, other "up" attitudes, and genuine belief in the necessity for improvement; a healthy positive attitude toward the need for constructive change.
- Willingness to experiment, innovate, fail occasionally as part of the learning process, and try the untried.
- Knowledge of the tools and techniques of information gathering and analysis.
- The complete involvement of all parties who have a stake in the problem, regardless of their level in the organization.

Develop an Improved Method

Having learned thoroughly about the task and uncovered its weaknesses and strengths, the results of this analysis are available to develop a proposed new method. However, the process is not as simple as replacing the old method with the new method with no further preparation. Rather, it is necessary to:

- Check out the proposed method thoroughly using flowcharts and other analytical techniques. Subject it to the same kind of detailed examination that you applied to the present method, to provide reasonable assurance that the difficulties leading to the consideration of this task have been corrected.

■ Analyze the impact of the proposed method in terms of human involvement, since the effective application of the new method will fall to people who actually do the work each day.

The best methods-engineered procedure ever to hit the department is bound for eventual failure if it includes elements that the worker (the critical human factor) considers demeaning or dissatisfying. When combining the "perfect" solution (perfect in a technical sense) with the human element, one sometimes needs to back off from technical perfection to accommodate the needs of employees. For example, take the problem of production efficiency in a large agency's highly automated personnel department. At first glance, one might suggest that a single personnel assistant sit at a keyboard for 8 hours a day and feed the system raw data on employee sick and annual leave usage. However, it is likely that such a rigid assignment will lead to the physical discomforts of backache and eyestrain, and repetition and boredom. This can lead to dissatisfaction, which in turn often leads to reduced efficiency and diminished productivity. It is far better to sacrifice some theoretical efficiency up front, perhaps by rearranging the work of two people so they alternately divide their time between the necessary keying activities and other, nonkeyboard activities. Or, have a representative of each department come to the personnel area, train them in the program, and have them key their own data.

If at all possible, make some trial runs of experimental applications to test proposed methods. In this way you can also test variations in methods to determine which might work better in practice or be more palatable to the employees who do the work. In the above example, start with a small department and see how hard it is to train someone else in the computer program before trying to do a large, more critical department.

In general, assess the practicability of the proposed method in the light of any possible constraints. It is one thing to idealize a method. It is quite another matter to make that method work in the face of constraints of cost, time, and quality considerations. One can spend only so much, the task must be accomplished within a given length of time, and quality must ordinarily be maintained at present levels or perhaps even improved. The manager must discern a workable solution within these limits. (See the discussion on constraints in Chapter 19.)

Implement and Follow Up

Action is a must. The best work method ever devised is useless as long as it remains solely on paper.

Be prepared to accept a certain amount of "tuning up" and "debugging." Steps that appear to work well on paper often require modification in actual practice. Also, conditions are constantly changing, and needs and circumstances present at the time of implementation may be different from those existing when the analysis was started. Just as military leaders will say that no battle plan survives contact with the enemy, rarely is a plan of any consequence implemented

100 percent as planned in every detail. As a result, it is necessary to stay on top of implementation until all the hitches have been worked out of the method.

Be ever aware of potential human relations problems and of the value of employee participation in methods improvement. Often managers will find that the extent of participation that may have preceded implementation will have already determined the success or failure of a proposed new method. These human problems are often the classic problems of resistance to change. As for employee participation, there is usually an identifiable relationship between employee involvement and resistance to change. The more intimately involved employees are in determining the form and substance of a change, the more likely they are to accept that change and thus be less resistant.

When it comes to implementing new methods or procedures or in general instituting any change, supervisors have three available avenues of approach: tell the employees what to do, convince them of what must be done, or, involve them in determining what must be done. Employee involvement in determining new and revised methods is advisable for a number of important reasons:

- The employee who is involved will thoroughly know the details of the new method as it is created.
- The involved employee will "own a piece" of the new method and will thus be more likely to accept it.

The employee is one of the most valuable sources, if not the single most valuable source, of information on how the task is performed now and how it might be performed better. The supervisor might have a better perspective on departmental operations. He or she may be able to stand back and observe what pieces do not fit together well, and see what needs to be improved. However, when it comes to specific tasks it is the employee performing these tasks day in and day out who knows how to do them most effectively and efficiently. Employees are a source of information that should not be allowed to go untapped in methods improvement.

Close follow-up on implementation is essential, and for many tasks it must be maintained for a considerable length of time. Time is necessary for new habits to form and replace old habits completely, and for new methods to dominate employees' actions fully and negate any fond remembrances of "the way we used to do it." Also, any method installed today is immediately subject to unforeseen problems and "creeping changes" that can affect performance. Thus supervisors must follow up a new method until it has completely taken its place among the department's normal processes. Even then the new method (as every departmental procedure) should be subjected to a periodic audit of effectiveness and appropriateness.

■ The Tools and Techniques of Methods Improvement

A number of analytical tools and techniques are available to a person pursuing methods improvement. This chapter will introduce very few of them, but they will be easy to use and broadly applicable to many situations.

One of the most valuable tools available for examining an activity is the flow process chart, used for tracking and recording the steps of an activity and assessing the nature of each step. Flow process charts appear in many forms, but most make use of the five basic activity symbols shown in **Figure 26-1**.

For proper analysis, all work activity should be divided into at least these five basic categories:

- An *operation* is any activity that advances the completion of the task. It is the actual doing of work, whether it is writing on a form, dialing a telephone, writing a report, searching an inmate, making a photocopy, tightening a bolt, or whatever.
- A *transportation* represents any movement encountered in the process. It may be moving an inmate from here to there, delivering a memo from you to your supervisor, carrying a record from here to the copy machine, or any other activity that has movement as its primary characteristic. Note that transportation involves only physical movement. Except for the relocation of some person or thing from one place to another, nothing actually happens to advance the completion of the work.

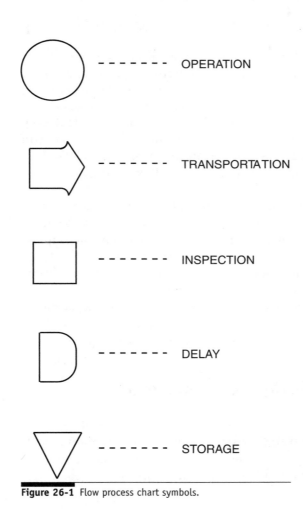

Figure 26-1 Flow process chart symbols.

For instance, the chart that just arrived on person B's desk in the records office is just as incomplete as when it left person A's hands in the chief of security's office.

- An *inspection* is an activity in which the primary emphasis is on verification of the results of prior operations. For example, a housing unit supervisor may periodically check areas that have recently been cleaned. The supervisor in the receiving and discharge area may check a set of admission forms to assure they are appropriately completed. A business office supervisor may assign an employee to the task of verifying payments before they are mailed. Inspections, like transportations, do not advance the completion of the work. Rather, they are used to determine whether necessary work has been properly performed.

- A *delay* is an unplanned interruption in the flow of necessary steps. A pencil breaks and must be replaced. A computer locks up and must be rebooted. Losing some work. A beverage spills, necessitating cleanup and replacement. The person who must sign a hand-carried requisition is on the telephone. All of these are examples of delays that might be encountered in flow process analysis.

- A *storage* is indicated whenever a step results in an anticipated interruption of the process, or it may represent the end of the process. More in-depth analysis would likely call for distinguishing between temporary and permanent storage. For instance, that completed invoice just dropped in the mailroom out-basket is in temporary storage; the process will be resumed after some period of time. However, although the completed inmate record just filed on the shelf may be retrieved some time in the future, it is for all practical purposes in permanent storage.

Although the activity symbols of Figure 26-1 can be used in charting activity in freehand fashion, they are ordinarily incorporated into prepared forms. A simplified version of a portion of a flow process chart form appears in **Exhibit 26-1**. If this form is used, it is necessary only to describe each activity, preferably in no more than two or three words; indicate the appropriate symbol for that activity; and, as you wish, connect the symbols to indicate sequential flow of activities.

An important indicator of the proper use of the flow process chart is embodied in the need for the user to note whether the chart follows the activities of a person or the flow of material. This decision must be made before beginning the analysis.

For instance, if you are observing a clerk preparing and processing a three-part form, you may reach a point in the activity where the form is taken apart and goes in three directions. At this point, the clerk either remains working with one part of the form or perhaps goes off in a fourth direction. It is necessary for the user to make an early decision on the focus of the analysis (for example, either "purchasing requisition" or "purchasing clerk") and stick with it.

Note also the space on the chart form that summarizes total numbers of work activities and total distances traveled in transportation steps. This part of the form can be used for comparing possible changes with present methods.

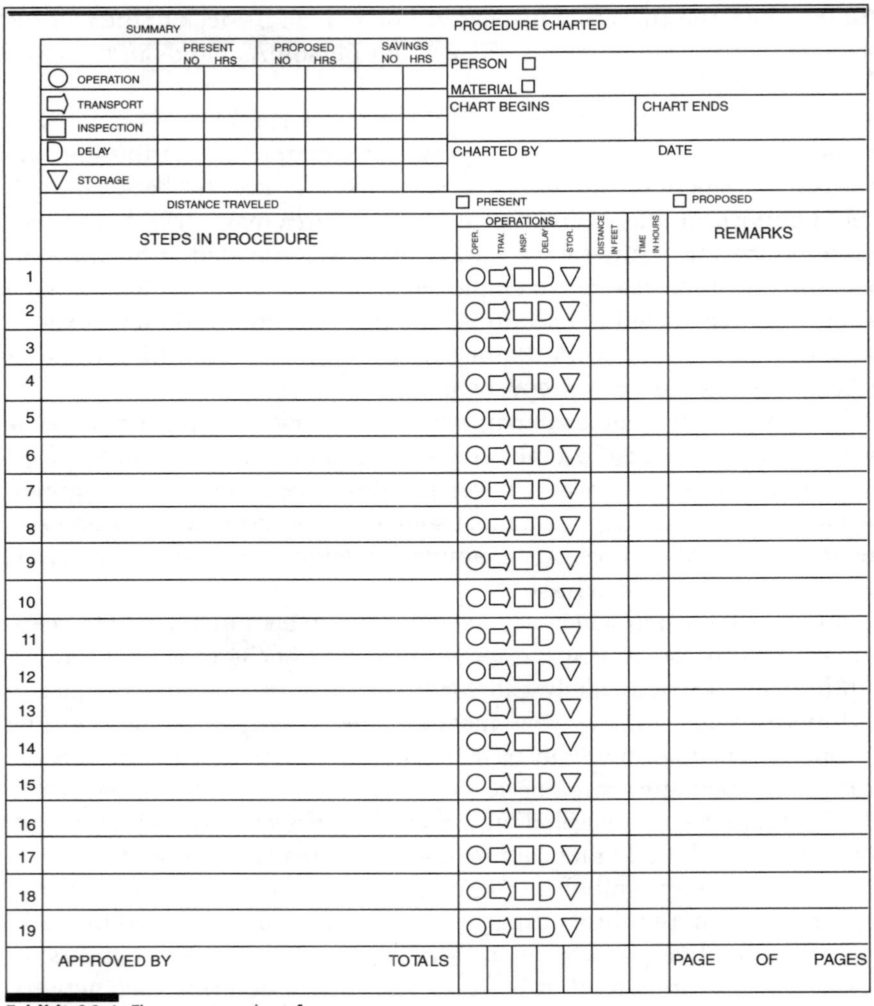

Exhibit 26-1 Flow process chart form.

Another simple but helpful tool is the flow diagram, a sample of which appears in **Figure 26-2**. A flow diagram is simply a layout drawing drawn to approximate scale, on which lines are placed to indicate paths of personnel movement or material flow. In this example, the X-ray process in a correctional infirmary is portrayed.

This can be especially helpful in determining how to rearrange equipment, furnishings, and work stations for improved effectiveness. This technique may be used to assess the approximate amount of movement necessitated by a particular layout. Any potential new layout may thus be tested on paper to see if it represents improvement over the present layout.

Numerous additional techniques are available for analyzing work activities of various kinds. Some of these are:

- **The operation process chart.** A flow process chart captures an entire sequence of all operations, transportations, delays, inspections, and storages during a process or activity. However, an operation process chart

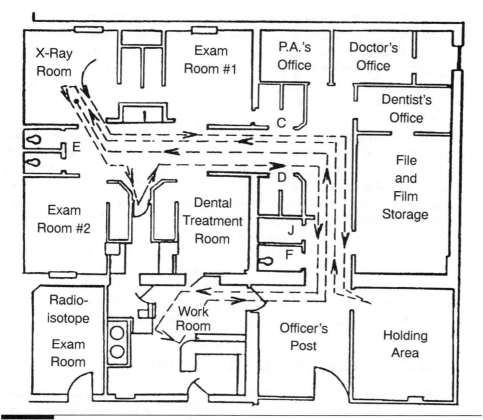

Figure 26-2 Flow diagram charting route of X-ray technician travel.

typically focuses only on operations (the circular symbol) and inspections (the square symbol). It is most conveniently used to examine all the separate small activities or suboperations that may take place within a given larger task. For example, "take fingerprints" may appear as a single operation on a flow process chart covering the activities of a staff member in receiving and discharge. If you wish to examine "take fingerprints" in considerable detail, you might focus on this activity alone. That would require creating an operation process chart that breaks fingerprinting into a number of smaller work activities (operations) and verification steps (inspections).

- **The multiple activity flow process chart.** This tool is most useful if you wish to examine, for example, a multipart form that is processed over a period of days. Using the standard charting symbols, this chart would perhaps display the multiple parts of the subject form in a vertical column at the left and a time scale of perhaps several days of the week across the horizontal dimension. Using this technique you can illustrate (usually on a single sheet of paper) what happens to each part of the multipart form on each day of the processing cycle.
- **The multiple activity chart.** Formerly referred to as a man-and-machine chart, this form is used to chart the activities of an employee in conjunction with one or more pieces of equipment. The heart of this particular

chart is a vertical time scale calibrated in either minutes or decimal parts of an hour. The activities of the worker and the utilization of the equipment are indicated parallel with this time scale. This technique makes it possible to identify times when the worker or equipment is waiting or being productively applied. It permits considering rearrangements of activity that may reduce idle (nonproductive) time for workers, equipment, or both.

■ **The gang process chart** is a flow process chart applied to the analysis of the activities of a crew of two or more persons who must work together to accomplish a given task. The steps taken by each member of the crew are charted relative to each other on a time scale. That is, what worker A is doing at any particular time appears directly beside the activity that worker B is performing at the same time.

The foregoing overview of methods improvement barely scratches the surface of management engineering technology. For the supervisor who wishes to pursue this subject in considerably more detail, helpful published references are available. Two of the best such references are *Motion and Time Study,* by Ralph M. Barnes, and the *Industrial Engineering Handbook,* edited by H. B. Maynard. These are described in more detail in the Suggested Reading section at the end of this book.

■ Example: The Information Request

Exhibit 26-2 is a flow process chart of 17 steps illustrating the processing of a single information request in a prison's records department. The activity begins when a request for information is pulled from the incoming mail. It ends when the fulfilled request is "filed," that is, placed in the inmate's file. The nature of the intervening steps should be evident from their brief descriptions.

With the complete task of answering an information request recorded on a flow process chart, one can begin to question the individual steps involved to determine whether there might be room for improvement. In this example, observation has revealed that considerable travel is involved in what would seem to be a small task—in all, about 100 feet of walking. This should raise questions as to whether any furniture or nonpermanent fixtures could be relocated to reduce personnel travel.

The delay recorded on line 8, waiting for the copy machine to warm up, should raise some questions. It may be perfectly legitimate (there are some copy machines that should be shut down between infrequent uses), but this will never come to light without some probing to determine why.

In this example it is also reasonable to consider whether certain steps could be eliminated or combined. For instance, is it possible to obtain the inmate file as in step 3, then go directly to the copy machine and perform the next several activities there and thus eliminate a pair of transportations? Or, if the copy machine warm-up should indeed be required, would it be productive for the employee to turn the copy machine on before getting the file so the machine can be used without delay when needed?

Other pertinent questions should occur in studying the chart. However, the most important question relative to this example has not yet been raised. How

SUMMARY	PRESENT NO / HRS	PROPOSED NO / HRS	SAVINGS NO / HRS	PROCEDURE CHARTED: *Information Request (Single)*
○ OPERATION	10			PERSON □ MATERIAL ☒
⤷ TRANSPORT	3			CHART BEGINS: *Get Request* CHART ENDS: *File*
□ INSPECTION	1			
D DELAY	1			CHARTED BY: *CRM* DATE: *3/9/91*
▽ STORAGE	2			

DISTANCE TRAVELED **100 ft (for 1)**   ☒ PRESENT   □ PROPOSED

	STEPS IN PROCEDURE	OPERATIONS (Oper / Tran / Insp / Delay / Stor)	DISTANCE IN FEET	TIME IN HOURS	REMARKS
1	Get, Open, & Read	operation			
2	Verify Authorization	inspection			
3	Get File	operation	15		
4	File to Desk	transport	15		
5	Locate & Remove Pages	operation			
6	To Copier	transport	20		
7	Start Copier	operation			
8	Wait for Warm-Up	delay			
9	Make Copies	operation			
10	Return to Desk	transport	20		
11	Re-Assemble File	operation			
12	Re-File File	operation	30		
13	Type Envelope	operation			
14	Assemble to Mail	operation			
15	Into Outgoing Mail	storage			
16	Enter in Log	operation			
17	File Request	storage			
18					
19					
	APPROVED BY TOTALS				PAGE OF PAGES

Exhibit 26-2 Flow process chart: information request (single).

many information requests are involved? If requests arrive at the clerk's in-basket in batches, then several additional questions become important:

- Can these requests be processed in batches, doing each step for several requests at a time?
- Can all necessary inmate files be obtained at once?
- Would it be practical to save all such activity for one time of day?
- Should all filing be done at the end of the day?
- Should physical changes be considered, such as moving a desk, to reduce transportation?

In the actual situation from which this example was taken, mail arrived in the records department once in the morning and once in the afternoon. The clerk was in the habit of handling information requests twice a day (usually within a few hours after the mail arrived) whether the "batch" consisted of a single request or several requests. A few simple changes in procedure were indicated:

- The copy machine was turned on at the beginning of the day and left on (a little investigation showed this to be the manufacturer's recommendation).

- It was decided to run the requests in batches, with little change in the steps indicated (trial runs involving a clerk doing several steps at the copy machine produced a situation that was sufficiently awkward as to be counterproductive).

- An ideal batch size of 10 to 12 requests was established, thus providing that no request went unanswered longer than two days (the decision was simply to run 10 to 12 at once or a maximum of two days' requests, whichever accumulated first).

- A simple change in layout—turning the clerk's desk around and moving it about 8 feet—reduced total travel from 100 feet to about 70 feet. (A flow chart of the resulting activity appears as **Exhibit 26-3**.)

The amount of effort put into such an analysis will depend largely on the total workload involved. In the situation just described, for example, if information requests amounted to only one or two a day (as might be the case in a very small institution) there is little to be gained by shaving a minute or so from each repetition of the task. The time employed by the supervisor or other person analyzing the operation is fully as valuable as the time of the employee. In this and other cases, hours of effort can be wasted to produce obviously minuscule returns. However, under conditions of high volume there may be significant gains to be made through a few hours' effort. Suppose, for instance, the institution was of sufficient size that processing such requests required a full-time person or more. Analysis could make a difference as to whether one person could or could not do the job alone. Assess the potential savings in time and resources as early as possible in the methods improvement. Let the best efforts be directed toward improvements having the greatest potential for savings in time or other resources.

■ Implementing Methods Improvement

A formal, institution-wide methods improvement program should ordinarily consist of three phases: philosophy, education, and application.

Philosophy

Introducing a methods improvement effort starts at the top of the organization. As with so many other enterprises, for a program to achieve long-run success, top management support should be active and visible. However, regardless of where the idea begins, all employees must see top management as 100 percent supportive of an organized effort to maintain constant cost-containment pressure through regular, positive questioning of work methods. This commitment must be reflected in a shared belief that few if any activities cannot be improved.

Granted, the dissemination of the methods improvement philosophy throughout the organization may require a great deal of promotional effort.

SUMMARY	PRESENT NO / HRS	PROPOSED NO / HRS	SAVINGS NO / HRS	PROCEDURE CHARTED — Information Request (Batch)
○ OPERATION	9			PERSON ☐
▭ TRANSPORT	3			MATERIAL ☒
▭ INSPECTION	1			CHART BEGINS — Get Requests / CHART ENDS — File
D DELAY	0			CHARTED BY — CRM / DATE — 3/9/91
▽ STORAGE	2			

DISTANCE TRAVELED 70 ft (for 10–12) ☐ PRESENT ☒ PROPOSED

#	STEPS IN PROCEDURE	OPERATIONS	DISTANCE IN FEET	TIME IN HOURS	REMARKS
1	Get, Open, Read & Sort by #	⊗ ▭ ☐ D ▽			Max. 10–12 / Batch
2	Verify Authorizations	○ ▭ ☒ D ▽			
3	Get Files	⊗ ▭ ☐ D ▽	10		
4	Files to Desk	○ ▭ ☐ D ▽	10		
5	Locate & Remove Pages	⊗ ▭ ☐ D ▽			
6	To Copier	○ ▭ ☐ D ▽	15		
7	Make Copies	⊗ ▭ ☐ D ▽			
8	Return to Desk	○ ▭ ☐ D ▽	15		
9	Re-Assemble Files	⊗ ▭ ☐ D ▽			
10	Re-File Files	⊗ ▭ ☐ D ▽	20		
11	Type Envelopes	⊗ ▭ ☐ D ▽			
12	Assemble to Mail	⊗ ▭ ☐ D ▽			
13	Into Outgoing Mail	○ ▭ ☐ D ▽			
14	Enter in Log	⊗ ▭ ☐ D ▽			
15	File Requests	○ ▭ ☐ D ▽			
16		○ ▭ ☐ D ▽			
17		○ ▭ ☐ D ▽			
18		○ ▭ ☐ D ▽			
19		○ ▭ ☐ D ▽			
	APPROVED BY TOTALS				PAGE OF PAGES

Exhibit 26-3 Flow process chart: information request (batch).

Moreover, the business of promoting the benefits of the continuous oversight of work methods is necessarily a management "selling" job that is never truly complete. However, once a program is underway and some positive results have been generated, these results will in turn help to sell the philosophy.

Through meetings and printed matter, everyone in the organization should be encouraged to retain the basic message that cost containment is here to stay. Every employee should realize that he or she can help keep the institution operating smoothly, safely, and cost-effectively. This can be done by emphasizing continued attention to making each hour of labor and each dollar's worth of material input go further, do better, and produce higher quality results.

Education

Education in the tools and techniques of methods improvement should begin some weeks after the introduction of the philosophy. This activity should bring supervisors and middle managers together with management engineering professionals, and begin to acquaint them with the ways and means of work analysis. Ideally, the educational process should include some pilot methods improvement projects taken from the actual jobs of the participants.

At this point a number of supervisors and managers should be at least partly oriented and committed to a working knowledge of methods improvement tools and techniques. Then, other selected nonsupervisory employees (especially those having skills applicable to potential projects) can be brought into the education process.

Application

Ideally, an ongoing methods improvement program should be guided and monitored by either an administrative steering committee or a methods improvement coordinator who preferably is a member of administration, a management engineer, or in rare cases both. Many agencies (and particularly individual institutions) may not have the latter specialist available, and may rely instead on a staff member who has received specialized training for this purpose.

In addition to assisting actively and coordinating necessary skills and resources, the primary task of the steering committee or coordinator is to keep the methods improvement projects moving. Methods improvement will not necessarily be particularly high on the average supervisor's priority list. There is usually more than enough to do without it. In constantly monitoring the program, the committee or coordinator serves as an ever-present reminder that keeping methods improvement alive requires periodic activity. This monitoring also makes the process easier for supervisors who are involved in active projects.

Common sense suggests that it is best initially to pursue projects that hold the greatest potential for returns, or that are most readily completed. Although this may seem like "skimming" or going after just the "easy stuff," there is a distinct strategic advantage in proceeding in this fashion. Some early, visible successes often serve as a much-needed boost to the momentum of the entire program.

Individual project teams may take two forms: the ad hoc group assembled for a single project, or the ongoing group formed to deal with a particular department or specific function.

The ideal ad hoc project team might consist of:

- The supervisor responsible for the task being studied.
- The assistant to the supervisor (if there is one).
- One or two of the persons who regularly perform the task.
- One "expert" knowledgeable in the most prominent technical specialty related to the task (for example, an accountant to deal with aspects of a billing problem in the business office, a materiel management specialist to deal with a transportation problem in the prison industry factory, a personnel professional to deal with personnel policy issues).

An ongoing project team formed to deal with a series of projects relating to a specific department or function might consist of:

- The department head or assistant.
- A representative of the administration.
- A business office representative.
- A management engineer or systems analyst.

- One or more "experts," persons from particular technical specialties that might be involved in most projects.
- One or two nonsupervisory employees of the department, including, perhaps, a representative of the bargaining unit (bearing in mind the concerns raised in Chapter 25 about legal issues of team formation).

Given the history of many methods improvement undertakings, a great deal of the foregoing may appear idealistic. In many organizations great things have been started, only to die of their own weight or gradually dwindle to nothing. Regardless of the organization's overall attitude toward methods improvement, though, the opportunity remains for individual supervisors to pursue methods improvement strictly within the confines of their own departments.

Granted, some projects can never be undertaken or successfully completed without interdepartmental cooperation. However, such projects aside, in each department are countless ways in which a conscientious supervisor and a few positively motivated employees can improve work methods and contain costs. In the average institutional department, a great deal of improvement is possible. The supervisor has only to recognize the possibilities for improvement and provide the example of leadership required to bring the employees into the improvement process.

Total Quality Management in Action

The preceding few paragraphs also have largely described the TQM process. Certainly the major phases of the process (philosophy, education, and application) are identical, as is the critical need for solid top management commitment. A great deal of the organization for TQM is the same, with a high-level "steering committee" providing guidance, and/or a methods improvement coordinator.

A true total quality management approach rightly calls upon the resources of the organization as a whole. Each internal function or activity must identify its "mission" (the essential reasons for its existence). It also must identify its "customers," (both internal and external) and pursue the goal of quality as the agency defines it.

Regardless of label or variation in application, however, the ultimate goals of TQM and methods improvement are identical. Ensure that the organization is doing the right things, and doing them right the first time.

■ The Methods-Minded Attitude

The "methods-minded" attitude referred to earlier has its foundation in each employee's belief that improvement is always possible. It presumes that left unattended for prolonged periods, work activities will experience creeping changes, which usually are not changes for the better. Methods-mindedness will often be reflected in the person who develops efficient work habits and always approaches a task in such a way as to achieve satisfactory results with minimum effort. Methods-mindedness, then, might seem to be a talent. Rather, however, it is a habit, or, more appropriately, a collection of habits. And as a habit, methods-mindedness can be learned.

The manager's attitude toward methods improvement will do the most to encourage employees to become methods-minded. A full-scale plunge into methods improvement is not necessary. Focus on one or two troublesome tasks at a time—or simply on one activity or process that clearly has room for improvement. Even simply keeping one such project open at a time, working on it as time allows, will inescapably lead to noticeable progress and stimulate the interest of some employees.

Of course, employees are once again among the keys to any supervisor's success. It may be possible to stand back at an objective distance and see problems employees cannot see. But dealing with specific procedural details reverses the situation. Nobody knows most of the inner workings of a task better than the employees who do it day in and day out. Employees can see a great many important facts that are hidden from their supervisor.

Effective methods improvement in a department requires a merger of structural and human considerations. Referring to the organization types discussed in Chapter 2, methods improvement is able to produce an effective blend of the Job Organization System and the Cooperative Motivation System by tapping the individual enthusiasm and motivation of employees in order to organize and refine the task they must perform.

EXERCISES

Exercise 26-1: The Pencil

Using the basic flow charting symbols, chart the following simple activity. Get a wooden pencil from your desk, sharpen it at a sharpener located near the door of the room, and return to your place. Also, account for the following:

Somebody else got to the sharpener first, so you had to wait.

The sharpener gave you a poor job. You found it packed with shavings that you had to clean out before you got an acceptable point.

Assuming "pencil sharpening" to be of sufficient importance to deserve a few minutes further study, sketch a flow diagram showing a user's relationship to the sharpener (use your actual surroundings). If you had to sharpen at least 10 pencils each day, what could you do to reduce the time and effort spent on this task?

Exercise 26-2: The Form

Select one printed form that either originates or is largely processed in your department. Do the following:

Utilizing the tools and techniques outlined in the chapter, record the flow of the form (and its copies, if it is a multipart form).

Further apply the tools and techniques to develop a proposed new processing method. (This will be possible in most cases.)

Assess the form itself for possible changes in the arrangement of information that you might want to recommend.

Given that you might wish to alter the processing of the form or redesign the form itself in some way, outline the steps you would take in implementing your proposed changes.

ENDNOTES

1. Maynard, H. B., ed. *Industrial Engineering Handbook,* 2nd ed. New York: McGraw-Hill, 1963, Section 1, p. 13.

Reengineering and Reduction in Force

27

Chapter Objectives

- Provide a basic understanding of the process referred to as "reengineering" (and known by various other labels).
- Relate reengineering to the concepts of methods improvement and TQM.
- Review the obstacles to reengineering that are frequently encountered.
- Examine the pros and cons of using external help (consultants) in a reengineering effort.
- Provide a perspective on reduction-in-force as an undesirable but sometimes necessary adjustment for the sake of organizational viability.
- Stress the need to attend seriously to the "survivors," those who remain following a reduction in force.
- Outline the individual supervisor's likely involvement in a staff reduction.
- Stress the essential nature of all forms of communication with the survivors of a staff reduction.

The future never just happens; it is created.

V. Clayton Sherman

Progress, therefore, is not an accident, but a necessity . . . It is a part of nature.

Herbert Spencer

FROM THE INSIDE — Managing Difficult and Dangerous Offenders

Almost everyone in the United States and many people throughout the world know about Alcatraz. As the most secure federal prison in the first half of the 20th century, "The Rock" was a symbol of the ultimate confinement sanction the U.S. government could impose. The institution was thought to be a permanent fixture of American corrections.

In penal circles, Alcatraz was the ultimate symbol of what is called the "concentration" model for managing difficult and dangerous prisoners. Take all the bad guys, put them together in one place, and concentrate your resources to make sure they are managed properly. That way, inmates and staff in other institutions are safer, and you don't need to use such intense resources elsewhere in the system.

By the early 1960s, despite its symbolism, a series of incidents and attempted escapes racked the institution. That, coupled with deterioration of the physical plant because of the salt air environment in San Francisco Bay, made it clear the institution could not continue to function as had been planned. Indeed, it was closed in 1963 and its inmates distributed around the country to other federal prisons. This was a return to the dispersal model of managing difficult and dangerous offenders.

Many Bureau of Prisons personnel had worked at the prison with every intention of completing a career there. Instead, they were told that they would be offered jobs at other federal institutions, or they would be discharged. The dreaded reduction in force was a reality as the entire prison closed for good.

By the late 1970s, it became clear to Bureau of Prisons officials that the dispersal model was no longer working as it had originally after Alcatraz closed. Increasing numbers of highly dangerous inmates, many of them associated with prison gangs, were creating serious problems throughout the system. Traditional penitentiary options seemed insufficient to safely contain and control them. After careful deliberation and planning, the Bureau returned to the concentration model. It established the U.S. Penitentiary in Marion, Illinois as its most secure institution, and the designated place of confinement for all such offenders. Nobody called it by that name, but inside of 20 years, the Bureau had twice reengineered its confinement strategy for high-security inmates.

■ Reengineering: Perception, Intent, and Reality

The term "reengineering" is not particularly common in the correctional world—it is more often used in business circles. However, just because the term is not familiar to the average correctional manager does not mean that the process it represents is not valuable in the progressive management of an agency or institution. Indeed, as the Alcatraz narrative suggests, correctional agencies do reengineer, sometimes addressing the same issue through different strategies. Consequently, this chapter deals with the concept of reengineering, how it dovetails with other management improvement practices discussed elsewhere, and how individual managers can use it to improve their areas of responsibility.

Regardless of the context in which it is applied, reengineering is a concept that actually has been used for years. It replaces a number of other "-ings," such as repositioning, rightsizing, downsizing, reorganizing, revitalizing, and modernizing. One can argue about differences in perceived meaning of all of the "-ings." For instance, "reorganizing" can be perceived as simply "moving things around." Perhaps "downsizing" is seen as no more than "cutting back." One of the oldest of the "-ings," "modernizing," at one time inspired fear of technology as it was perceived as people being displaced by machinery. But there remains a common thread holding all of the "-ings" together: change in the way business gets done.

But regardless of label, there has always been one critical thing these terms have in common—the way they have been perceived by the majority of employees. They all seem to convey the extremely unsettling message to staff that some people are going to lose their jobs. Reduction in numbers of employees has never been the intent of these "-ings." Rather, reduction in force (RIF) has frequently been a by-product of a process intended to improve operations overall. This improvement might result from decreasing costs, increasing revenues, or improving operating efficiency by getting greater results from the same or less resources.

Personnel reductions are something that public agency managers think they don't have to worry about. There is a common belief that there is an aura of protection around public employees when it comes to job security. But strictly speaking that is not really true. One of the authors lives in the state of Illinois, where (despite an increase in the system's inmate population) department of corrections employees in several locations have actually seen their facilities close in the last year. Other states have done likewise when faced with budget problems occasioned by the economic decline of the first few years of the 21st century. When such reductions occur, agencies may actually lay off newer staff. In other cases, they offer longer-tenured employees involuntary transfers or untimely retirements in lieu of being "riffed." These kinds of unfortunate impacts are not common in the public sector, but they can occur.

And of course in the private corrections sector, the loss of a contract, a change in the type of offender, new contract requirements, and other factors can very quickly change the kind of staff a prison needs. A change from a juvenile population to adults can mean the institution no longer needs the same type of vocational training instructors, for instance. In such a case, the old staff are simply told their services are no longer needed. This might be called compulsory reengineering.

A Working Definition

Reengineering may be defined as the systematic redesign of an organization's core processes, starting with desired outcomes and establishing the most efficient possible processes to achieve those outcomes. But the reader should not think of engineering in the specific sense of building a bridge or a product line in a factory (although a prison industry operation certainly might be an application of that approach).

Think of engineering more generally, as the process of identifying a desired result in mind and determining the most effective and efficient way of attaining that result. It does not matter whether the envisioned result is a physical product or other specifically desired outcome (such as maintaining security, producing classification reports, or supervising parolees). Reengineering literally means engineering something again, looking at the same product or outcome again and determining how it could be produced or attained in a better way. What may seem simple in concept, however, can be extremely difficult to achieve in practice.

An original engineering effort begins cleanly: Nothing exists in the way of a process, so there is nothing in place to channel thoughts or efforts in any particular direction. Note that at the heart of all classical methods improvement or problem-solving processes is how something is presently done.

Once a process is in place, however, especially a process that has been considered by some to be reasonable and workable, this previously traveled path would seem to be the logical starting point. Reengineering, on the other hand, ignores (or at least tries to ignore) the manner in which something is done, focusing instead on the desired outcome and independently establishing a new way to achieve that outcome. This new way might or might not look like the old way. What is important is that it be developed without undue influence from the present process and that it represents the most effective means of achieving an essential outcome. And the word *essential* is the key. Methods improvement and problem-solving processes focus on doing things the right way. The concept translates to reengineering when the focus becomes one of doing the right things in the right way.

Obstacles to Reengineering

The appropriate path to reengineering includes the following steps:

- Identification of the organization's mission.
- Identification of the outcomes necessary to fulfill the organization's mission.
- Establishment of the processes needed to achieve the necessary outcomes.

Many things conspire to frustrate attempts at innovation. Existing processes, traditional divisions of work, and assumptions and managerial expectations are prime offenders. Thus the principal barriers to reengineering continue to be:

- Current processes that distract from an essential focus on desired outcomes.
- Existing rules and regulations that place boundaries around thinking from the outset.
- Departmental boundaries that reinforce perceived limitations on the scope of a manager's responsibility and authority.
- Functional and occupational boundaries that cause a problem-solver to enter each problem with preconceived notions of who can do what and when.

- Normal resistance to change that is experienced as managers are forced into unfamiliar and uncomfortable territory.
- Individual paradigms and perspectives, through which perceptions are narrowed. Managers see no more than what they have been conditioned to see, and in some instances (not surprisingly) they are actually closed off from certain kinds of information.

Using Outside Help

Many organizations use outside consultants to assist in reengineering. Although consultants are dramatically overused and misused in many organizations, reengineering is one area of need in which consultants can be highly effective precisely because they are outsiders. The outsider's perspective is likely to be broader and more open than that of the insider. Moreover, the person from outside does not have the insider's emotional or intellectual stake in current processes. Frequently, the outside perspective is far more accurate and encompassing than that of an insider.

Another important reason for the outsider's involvement in reengineering lies in the need for a view that is as unbiased as possible of the organizational changes that appear to be needed. Employees who may become involved in a reengineering effort can hardly be expected to recommend themselves out of a job. The insider who becomes involved in a serious reengineering effort potentially affecting his or her own position is ill prepared to participate constructively.

■ Reduction in Force and Beyond

Reducing labor cost is probably the most common goal of organizational restructuring. Reduced labor cost comes about occasionally from reduced salaries, sometimes by way of reduced hours for employees, and most often in the form of reduced numbers of staff. Of these three, the latter is usually the only one available to the public sector manager. Wages are usually set in a contract or other fixed pay schedule structure. Positions are usually established as full- or part-time by stature or regulation that the individual agency cannot change. Only in the area of numbers of staff is there some (albeit limited) flexibility for agency and institutional managers. It may be useful to look into this a little further to dispel the myth that managers can't redesign flexibility into their manpower structure.

Many agencies are allocated a certain number of positions as part of their budget, above which number they cannot hire. It may be broken down into part- and full-time allocations, and almost certainly will be segmented into job categories—representing 27 doctors, 3,325 correctional officers, 215 case managers, 387 secretaries, and so on. Managerial discretion may be broad enough to allow for deliberate (and totally legal) calibration of the actual use of these hiring ceilings, in order to adjust for new priorities or unexpected circumstances.

For years, the Federal Bureau of Prisons only allocated a percentage of its total salary dollars (dictated by Congress) to field locations. The balance (often in the 3–5 percent range of total agency budget for salary expenses) was withheld for emergencies. Individual institutions (and indeed the Bureau itself) were not allocated specific funds for responding to riots, major escapes, institution evacuations (such as Hurricane Andrew required in August of 1992). Consequently, the withheld funds were used to deal with those contingencies as they arose. As the end of the fiscal year approached, hiring, equipment purchases, and other expenditures could be ramped up, while calibrated to stay within the total agency appropriation.

A variant on this strategy can be seen at the local facility level as well. Under the Bureau system, each facility is limited to a specified percentage of their salary dollar total. Additionally, each position has a certain annual salary amount associated with it. If a high-paying position (a psychologist for instance) was held vacant for all or part of a year after a vacancy occurred, then more correctional officers could be hired with the dollars saved. Such hiring might be limited to filling the 3–5 percent shortfall created by the original headquarters withholding. But it might not, so long as the dollar cap for the institution was not exceeded. Thus, a creative manager could develop a local staff deployment strategy that met local needs without exceeding legislative or regulatory bounds. This amounts to a controlled, time-limited RIF. It goes without saying that a considerable amount of thought and effort are required in structuring and implementing a RIF in a manner that will be as fair as possible to all concerned while supporting the public safety goals of the agency.

However, when this is done on a large scale, the effects can be dramatic. The managerial effort associated with this significant undertaking cannot stop simply when the staff who have been appropriately identified for separation have been released, or the position vacated for other reasons. In the vast majority of workforce cutbacks, the people who remain are far more numerous than those who leave. For those staff, the implementation of the RIF is the beginning of a completely new work situation in what will, and what in fact must, become a dramatically different organization. Although many will tend to seek a "business as usual" state of affairs following a staff reduction, they will find that this is not possible.

What Follows Reduction in Force?

A major RIF will forever alter many employees' beliefs and attitudes concerning their employment. Consider the fact that for many years correctional workers saw private sector workforce reductions occurring in their communities while feeling relatively safe against the likelihood of ever having to share the experience. For a long time they felt certain that, as an absolutely essential service, prisons would remain untouched by the severe economic concerns that plagued other employment areas. In some states, correctional workers have been forever awakened to the hard fact that their chosen profession is subject to many of the external forces from which they believed they were relatively protected. While it is not highly likely that public sector correctional managers will have to be involved in a RIF, it is not unprecedented.

The immediate responses to a RIF can include the following:

- Many employees may at first (and permanently, if positive steps are not taken) feel more like a cost of doing business than a valued organizational member.

- Employee commitment to the organization will tend to erode as perceived employment security is diminished.
- Employee morale will be automatically reduced.
- Some key staff the organization wants to retain may resign to seek employment in a field they may perceive as more stable, further impacting the morale and outlook of those who remain.
- Managers and supervisors (with their thinking still bound by former ways of doing things) may attempt to compensate for lost staff by increasing the use of overtime and temporary help.
- Managers will experience additional frustration as controls are placed on hiring and on the use of overtime and temporaries.

In the time immediately following a RIF, there also is a severe risk of cost reduction becoming universally perceived as a higher priority than people. It is true that successful cost control is an essential element of survival; the organization that cannot adapt to financial reality will not survive to employ anyone.

People, however, still remain the driving force. It is people working together who must bring the organization into line with financial reality. Yet the same organization's continued existence depends on serving the public.

If done for the right reasons and in the right way, then a revitalization of the remaining workforce will follow a RIF. An organization cannot and should never attempt to simply "lay off" a number of employees and call upon those who remain to close ranks and continue as before. All who remain have a more difficult and more responsible task looming before them. The agency's top management should endeavor to give all of the support and assistance that can reasonably be provided in making the transition to a leaner, more purposefully directed organization.

The Necessity of Reducing the Workforce

Although the scenarios have differed to some extent from one jurisdiction to another, correctional agencies across the nation have been experiencing very real fiscal pressures. This should come as no surprise to those who have been aware of the budget and public safety debates that continually take place at all levels of government. Add to this the competitive forces that private corrections has introduced and the mix is even more difficult. Communicating with the affected employees is critical to making any workforce reduction a productive act.

When a RIF is planned and before the cuts occur, the workforce must be given every opportunity to understand why this is going to happen. The more openly the employees have been treated all along and the more frankly they have been advised of the agency's real circumstances on a continuing basis, the easier it will be to communicate why. Any RIF, while preferably designed and recommended by senior management, should proceed after all other reasonable efforts to reduce costs have been explored as follows:

- All realistic short-term savings opportunities should be identified and implemented.
- Maximum effort should be expended to reduce staff through attrition before the actual reduction is done by freezing hiring in most positions and, as much as possible, transferring current employees into areas of greatest need.

- Overtime should be severely curtailed, essentially reserved for true emergencies only and approvable by only a select few.
- The use of temporary help should be curtailed (reliance on contract and temporary help can tend to increase under staff-reduction pressure if not closely monitored).
- Supply inventories should be reduced to levels conforming with the true needs indicated by reduced levels of activity.
- It must be stressed that no matter how much cost-control effort precedes a RIF, the reduction itself is never the end of the process. For the organization's continued fiscal stability and effectiveness, it becomes the job of all employees to pursue continuous cost control in concert with continuous quality improvement.

The Employees Who Remain

Stress and stress-related fear among those who remain following a RIF is natural, predictable, and essentially universal throughout the organization. A fully understandable feeling among survivors is the fear that they may be the next to go. A RIF instantly establishes two entirely different groups of employees: those who leave and those who remain. Management must recognize that the manner in which it deals with the reduction's survivors has a considerably greater bearing on the organization's future than how the terminations attendant to the RIF have been addressed.

Those who have departed are gone (probably forever, perhaps to another institution), but the survivors are there and are critical to the agency's or institution's future. It becomes necessary to unite the survivors into a forward-moving team and to motivate them to work harder in a leaner, more efficient, and yet initially a completely alien, situation. Through a concentrated and continuing communication program, the survivors of the reduction need to learn several things:

- Why they remain and what will be expected of them, why the old organization is gone forever, and how they can help shape the new organizational culture that will be emerging.
- That as the survivors of the reduction they are among the best in their occupations, and that is essentially why they are still in place.
- That in the future continually doing more with less will remain critical to organizational survival and continued employment.

Immediate and Natural Reactions to Staff Reduction

The initial issues that emerge in the wake of a RIF are people issues.

- There is the actual loss of productive employees that the agency otherwise might wish to retain.
- There is the potential loss of valuable "free-agent" employees, those professional and technical employees whose primary loyalty is to an occupation and whose movement between organizations may be governed more by labor market circumstances than by ties to a specific organization or group. Psychologists, medical and mental health specialists, and tradespersons would fall in this category, but there are many others.

- There often is an immediate drop in productivity, precisely at a time when productivity increases are needed for the sake of long-term survival. This occurs because morale has dropped and employees are preoccupied with issues of security and concern for their future.
- It also is likely that there will be increases in the use of sick time and health care benefits, on-the-job accidents, medication errors, and other lapses in quality. These are often experienced during and after a RIF, again owing to employees' concern for their employment.

During and after a reduction there may be a fully understandable tendency to cut financial corners and curtail all possible expenditures in an effort to save jobs and achieve the budget goals set under the new program. However, this may be precisely the time for the organization to be devoting money and effort to developing new ways to do things, and to ensuring the increased flexibility, adaptability, and effectiveness of the remaining workforce.

Employee Motivation Following Reduction

Under normal circumstances (without the prospect of a RIF and with each employee's reasonable expectation of continued employment), job security and wages are not particularly active motivating forces. Rather, they are potential dissatisfiers. As long as wages and job security are perceived as "reasonable," the concern for them is secondary. However, when they are disturbed (when a pay freeze is in force, for example, or when security diminishes via rumors of a RIF), these become factors in heightening employee dissatisfaction. This, in turn, negatively impacts motivation.

It becomes necessary to help the surviving employees reestablish a sense of equilibrium with their modified surroundings and achieve a relative sense of security. An employee who may come to work each day wondering "Will I be next?" will be neither effective nor productive. As long as an employee is preoccupied with personal survival, individual productivity will decline at the time its improvement is needed more than ever.

It is necessary to communicate with employees fully, completely, and repeatedly until they understand that:

- Nobody (neither the agency nor a labor union) can absolutely guarantee continued employment.
- A certain amount of stress is inevitable regardless of what management does following a RIF, but stress can be energizing as well as debilitating and can serve as a spur to improvement.
- A future emphasis on improved productivity is essential to survival as an organization and as individual employees.

Employees' aggregate job performance is the organization's best survival guarantee, and as far as individual employees are concerned, their performance is their own best job security.

The most potent motivating forces (perhaps the only true long-run motivating forces) are inherent in people's work. These forces are largely opportunities. They consist of the opportunity to learn and grow, to do interesting work, to contribute, and to feel a sense of accomplishment and worth. However, these motivators can work only when employees are able to feel relatively secure and reasonably compensated. Management needs to provide conditions under which all employees can become self-motivated and then act on that belief.

Attendant to employees' motivational needs, the organization might also consider the creation of incentive award programs. These and other flexible rewards can encourage and acknowledge innovation, commitment, and enhanced productivity. Overall, top management should at all times let employees know what is expected of them and tell them exactly how this desired behavior will be rewarded.

Changes in Supervisors' Roles

Any significant RIF includes the elimination of some positions, and the reconfiguration of positions responsible for managing the work of others. In the presence of a generally flatter management structure (overall fewer levels from top to bottom) supervisors and managers are likely to find their roles expanded. They will essentially assume new roles, roles that are more challenging and that require more direct decision making. This is a polite way of saying that in addition to line employees, under the new arrangement managers who survive the RIF also may be expected to do more with less.

The individual line supervisor will be the organization's primary conduit for communications with staff. At each management level, the supervisor is always a critical link in the movement of information up and down the chain of command, facilitating the flow between each direct-reporting employee and the rest of the organization. Since the size of direct-reporting work groups generally increases following a RIF and flattening of the organization, the influence of the individual supervisor becomes even more significant. Some of the supervisor's key concerns after a RIF are as follows:

- The need to be conscious of the employee's motivational needs and to work to control turnover both immediately and over the long term.
- The need to function as a strong advocate for the staff, to achieve the best for those who must leave as well as for those who remain.
- The need to begin preparing to work with the survivors, helping them to internalize the dramatic change well before the reduction is fully implemented.
- The need to actively encourage employee participation more than ever before, stressing involvement and drawing all possible employees into the decision-making processes. More than ever the supervisor's focus needs to be "we," never "I" or "you."
- The need to develop and utilize employee teams to the maximum possible extent.
- The need at all times to communicate, communicate, communicate, remaining in touch with employees' fears and concerns even when some of the answers have to be "We simply don't know yet, but we'll keep you informed."

Training after Staff Reductions

Supervisors and middle managers can be better prepared to meet the demands created by a RIF through enhanced training. Yet it is almost a foregone conclusion that an organization's training function is among the first to be cut during lean times. It often is seen as a frill—something that is dispensable when compared to most other functions. Training departments are usually small, and a cutback usually removes one or more full-time positions and may cut travel expenses as well.

No staff reduction rationale is more counterproductive than the across-the-board cut in which every department is expected to surrender an equal percentage of staff. Yet this rationale is frequently applied out of a misguided sense of "fairness" that suggests the pain must be equally borne by all. Some functions are simply more valuable at certain times than are others. In a manufacturing company that is reorganizing in response to a declining market share, for example, it could be folly to reduce staff in marketing and new product development since this may be precisely the time to be enhancing and reemphasizing these functions. Likewise, in many correctional agencies it can be short-sighted to reduce the resources allocated to training as part of a RIF that leaves in its wake a dramatically increased need for that very function.

Following a RIF, it is necessary to assist all remaining employees to become more flexible and adaptable. To that end, an enhanced program of training or continuing education is strongly recommended for employees at all levels.

■ Coping with Expanding Responsibilities and Ongoing Change

In a RIF situation, the manager will find that responsibilities have expanded considerably. It may even seem, at least until the combined position has been brought under control, that the workload has grown to unmanageable proportions. The span of control and number of staff under supervision may have increased. The physical area for which the manager is responsible may increase as well.

To cope under these conditions, the manager must take an approach that combines similar tasks in the most efficient possible manner. There will probably never be enough time for everything that must be done, so it will be more important than ever to be conscious of priorities to ensure that the most important tasks are the ones being addressed.

High among the skills that will become more important are:

■ Delegation, since there will be a need to get more done by others.
■ Planning, since there may be a greater area and larger staff to worry about.
■ Time management, since there will be more to do and more to keep track of.
■ Decision-making, since there will be more decisions to make. Employee relations because a larger staff invariably raises more personnel concerns.

Generally, a manager's style may have to become more open and participative than it might once have been. The primary concern should be, "How can we make this new work structure function as efficiently as possible?"

Rare as they might be, a RIF in a correctional agency involves a forced paradigm shift that affects workers in a dramatic and considerably disturbing fashion. As a RIF approaches, visible signs (such as deliberate slowdowns in filling open positions or hiring freezes) begin to emerge. A great deal of the resistance of managers to delays in filling positions will center around the contention that operational security will suffer without the one-for-one replacement of departing personnel. The paradigm at work here is extremely strong—the current way of organizing the agency or facility is the best and most efficient. It assumes that cost and effectiveness exist in a direct relationship, so if managers take cost out (in this instance remove staff), effectiveness will automatically decline. Therefore (since the administration would never willingly or knowingly tolerate reduced operational efficiency) managers believe they have to have the staff, whether the money to pay them exists or not.

Implicit in this argument is the belief that the answer lies in more money for the system or, at the very least, money diverted from other parts of the system "Cut their program, not mine" is the (not always unspoken) motto. This and other arguments do not prove to be persuasive very often with legislators who must make difficult decisions about education, health care, highway safety, and child welfare, as well as prison operations.

The strength of the old way of doing things hampers the ability to quickly find a new way. When forced into unwanted change, people often respond through a grieving process. (Actually, it is not change itself that is resisted, so much as being changed.) Loss and uncertainty prevail. Those who survive a staff reduction may even experience guilt over having done so. There is considerable trauma experienced by those who are forced to leave the organization, but the trauma may be just as great for those who remain. Those who remain have to make the painful adjustments that organizational survival requires.

The survivors of a RIF typically perceive dwindling control over their work circumstances. Therefore, every effort should be made to keep employees advised of the agency's circumstances. Using both direct management action and union channels, employee communication and involvement are critical in establishing the future attitude and disposition of the remaining workforce. To ensure maximum possible communication and involvement, management should:

- Hold regular employee meetings to answer questions and to explain where the organization is going.
- Make necessary changes rationally, whenever possible calling upon the participation of affected employees.
- Seek regular employee feedback by way of surveys and other means, so as to be continually aware of employee concerns.
- Continually explore new operational strategies and other potential means of enhancing its value to the state or other jurisdiction, as well as reinforcing its fiscal responsibility.

Many employees who remain also tend to see the dramatic changes forced upon them as marking the end of the world as they knew it. Since none who survive (neither line managers nor rank-and-file employees) can control what happens to the agency in the face of external pressures, it indeed can seem that way. But for the survivors it can also be the beginning of a challenging new world.

EXERCISES

Exercise 27-1: The Unknown

You are the unit manager of 3 West, a 64-bed maximum-security, long-term psychiatric treatment unit. Your operation is located in a major medical referral facility for a large correctional system. There is an adjacent similar-sized and designed forensic unit known as 3 East, where court-ordered psychiatric studies are performed. Unit 3 East is a mirror image of 3 West, and the two are laid out such that their control centers are back-to-back in a central core, although they are fully separate and secure from each other. Over the recent 12-month period each unit has averaged approximately 60 percent occupancy. While some vacant beds are expectable (and need to be maintained for contingency use), some questions have been raised about the efficiency of this arrangement—two identical units with similar missions and staff levels, but with less-than-optimal bed utilization.

For several months your employees asked you questions concerning rumors of an impending reorganization. Some of these issues originated in the union, but most were of undetermined origin. You did all that you could to answer their questions, but you were able to say very little of substance. Finally, after 2 months of speculation, a consolidation was announced.

The result of this change would be the loss of several line staff positions. Morning watch patrol coverage inside the unit would be provided by one officer instead of two. Only one of the control centers would be staffed, with controls for the other being remotely operated by the remaining officer. A recreation officer position in one unit would be eliminated. Instead of having two psychologists and one psychiatrist in each unit, three psychologists and one full-time and one part-time contract psychiatrist would be assigned to the combined unit operation. Most importantly to you, your counterpart in 3 East is to be reassigned to another facility, and you are to be the manager of both units.

Instructions

As you think about this situation, consider how you would address the following questions:

1. What would be the likely impact on your individual span of control as a supervisor?
2. Considering that you were always constructively occupied managing 3 West, how can you merge the responsibilities of the two units and successfully run both?
3. What kinds of tasks you once engaged in will now be performed less often than when you ran 3 West alone?
4. What skills will increase in importance because of your new role in running the combined units, and how might your management style have to change?

Exercise 27-2: The Directed Decision

The setting is a large private correctional facility with no union representation for employees. Dan Carey, chief of facility operations, was meeting with Arthur Brooks, power house supervisor, about the possible need to soon identify one or more employees from the power house for possible layoff. Layoffs were not new to the institution, having occurred twice over the previous 3 years, but they always had involved program staff, not operational personnel. Reducing the number of current power house employees will require internal staffing changes within the facility maintenance department as a whole, but the administration has deemed it necessary.

Said Dan to Brooks, "I'm new to this place. What did you do the last time there were layoffs?" Art answered, "We were given a sort of scale to use for comparing people. It gave roughly equal weight to four factors—basic job qualifications, seniority, performance evaluations, and what I'd guess you'd call conduct." "What's that?" "It involves warnings and suspensions and such for violating institution policy," Art said. "You know, like a written warning for being late too many times."

Dan Carey said, "I don't see how we're going to escape losing at least one person in this next cutback. I want you to make sure the one we lose is your maintenance guy, Fredericks." "He's been here a while," Art said, "Lots longer than most. Offhand I don't see how to make him fit the criteria." "You make him fit," Dan said. "You and I both know he's one of the biggest goof-offs in the department and our least productive worker. We can't help it that past management didn't do what they should have done." "Should have," Art said, "but didn't."

Dan shook his head vigorously. "We've got tons of work-order history that proves he's our least productive person." "But not a negative word in his file," said Art. Dan Carey and Art Brooks discussed Fredericks and his situation for a quarter hour. Their positions are summarized as follows:

Dan: "We shouldn't have to keep the worst employee of the whole lot over all the others. We have no union here so there's no contract to make us observe seniority. If I have to I can prove he's the least productive person. We may not like it, but it's a legitimate layoff. This is a lot easier and cleaner way of dealing with Fredericks than going through a whole long, drawn-out process that someone else should have followed long ago."

Art: "Laying off Fredericks is risky. Seniority has always been a major factor in determining who gets laid off here. It's an established practice, even though there's no written policy. You might be able to prove he's our least productive person, but there's no file we can open up and use to prove that he ever knew how he was doing. There's only a bunch of average evaluations. He might be the one who really deserves to go, but this isn't the way to get him out of here."

Instructions

1. Consider the summaries of Dan's and Art's positions as alternatives for addressing the potential impact on Fredericks. Which alternative are you more willing to support, and why?

2. Fredericks reports to Art Brooks who in turn reports to Dan Carey. Imagine that Art has recommended against laying off Fredericks, instead recommending the department's newest member for the reduction. Dan Carey, however, chooses to reject Art's recommendation and directs him to lay off Fredericks. If you found yourself in Art's position, how would you proceed?

ENDNOTES

1. Lumsdon, K. "Mean Streets," *Hospitals and Health Networks* 69, no. 19 (1995): 44–52.
2. Note: Portions of this chapter adapted from C. R. McConnell, "After Reduction in Force: Reinvigorating the Survivors," *The Health Care Supervisor,* 14, no. 4 (1996), pp. 1–10.

Training and Continuing Education

28

Chapter Objectives

- Establish the importance of training and continuing education as legitimate concerns of every supervisor.
- Stress the necessity for management commitment to training and continuing education.
- Describe various approaches to training and education.
- Describe the application of certain elementary in-department educational activities, such as cross-training.
- Establish the role of the supervisor in providing and guiding the department's training and continuing education program.
- Offer simple advice concerning identification of a work group's training needs.
- Identify avenues of training and continuing education available to the supervisor.

The person who sees a career as one of perpetual investment in education stands a much better chance of surviving in today's world.

Tom Peters

The ideal condition
Would be, I admit, that men should be right by instinct;
But since we are all likely to go astray,
The reasonable thing is to learn from those who can teach.

Sophocles

FROM THE INSIDE

Why You Should Pay Attention During Training Sessions

Every staff member received training in how to use the new gas masks. If you were issued a new one, you tore the end off of the hermetically sealed wrapper, removed the entire mask, reached inside and removed a cardboard protector from inside the transparent face mask, twisted the canister to be sure it was tight, and pulled the mask on, starting with the straps over the back of your head. Every staff member was told (and shown) this at least once a year in annual refresher training.

The time came when there was a disturbance in one of the cell houses. The two-story unit was sealed and response squads assembled at the upper and lower entry doors to each range. The captain, unit manager, and several other managerial officials were at the main entry, trying to persuade the inmates inside to surrender. After an appropriate amount of verbal activity, it became clear that some degree of force was going to be needed to restore order. Gas was used, and the cell house almost immediately became saturated, with clouds billowing out into the entry area.

It was at that point that the hastily assembled officials realized that everyone else involved had gas masks but them. The captain yelled down the corridor for masks to be brought from the control center. An officer ran up with a half dozen new masks, still in the sealed wrappers. The captain, half blinded and choking from the gas, grabbed one of them, yanked it from the wrapper, and pulled it on over his head.

The unit manager next to him finally got his mask on, looking over to the captain, who was standing there muttering, "Gawd, this gas is thick. Where are those squads anyway?" The unit manager reached over, pulled up the canister of the mask (lifting the face plate away from the captain's face) and removed the cardboard liner so that the captain could see.

It was a long time before the captain lived down that little bit of forgotten training.

■ Why Training and Continuing Education?

In typical usage, training is an in-house activity provided by the agency, which is intended to upgrade or refresh employee skills and often relies on in-house resources. (This is the kind of activity referred to in the story above.) Continuing education often is conducted off-site (or as an internal program presented by outside resources) and is intended to maintain or improve professional abilities, often as required by a licensing or accreditation body. For the purposes of this chapter, the term *training* will refer to both, since that is the more commonly encountered usage in day-to-day correctional work.

It can be difficult and at times nearly impossible for people working in the field of corrections to keep up with all of the changes affecting their work. Some correctional workers (subject to continually evolving techniques, procedures, and equipment) must feel as though the more they learn, the more there remains to be learned.

The rate of technological change was discussed in Chapter 20, which asserted, in part, that what managers know and how they apply that knowledge does not (and in fact cannot) remain stable. Today's knowledge is not sufficient for the needs of tomorrow. Without the presence of a learning attitude managers may be doomed to remain forever behind the times. The rate of technological change will not have equal impact on all skills and all professions. But there is sufficient reason to suggest that standing still relative to developments in one's field (however slowly these developments may seem to accrue) is to guarantee falling short of the ability to meet some future job needs. In some areas of work activity, including administration, psychology, medical, finance, and automated technology, to name only a few, it is often necessary to absorb and react to change at a rapid pace simply to remain abreast of the times. (In a shameless plug for the American Correctional Association, its biannual conferences provide an excellent means of keeping current in each of these areas by way of seminars, speeches, vendor displays, and other programs.)

To the manager, training affords a significant means of bringing change under control, of making change work for, rather than against, management. From using a new style gas mask, to learning hostage negotiation skills, to acquiring new parolee interview skills, training is critical in the correctional environment. Properly focused training provides the ability to increase knowledge, improve skills, and change attitudes as job performance needs change. In addition, it serves at least two other major functions. One of these is the provision of reinforcement through refresher training. Knowledge that lies unused or infrequently used can get rusty from lack of application. Relearning what one has learned before can revitalize knowledge and sharpen skills. The other major function of training is to increase individual capabilities and thus improve the potential effectiveness of all members of a work group. Later, this chapter will highlight the benefits of cross-training, the practice of assuring that employees become proficient in the performance of jobs other than those to which they are regularly assigned.

All of the foregoing benefits of training aside, there is also one compelling reason driving a great deal of training in some segments of the correctional setting. A great deal of it is required by accreditation and regulatory bodies. The American Correctional Association has numerous accreditation standards on training.[1] Accreditation requirements and state hospital codes applying to a prison's in-house medical facilities likewise will include training for staff.[2] With the regulatory requirements of various governmental bodies increasing steadily, there is no chance that such requirements will abate.

Thus some training takes place (and grudgingly, at that) simply because it is required. However, a great deal of essential training—and specifically the training of supervisors and managers and the cross-training of employees—remains voluntary. It must therefore be accomplished through desire and commitment.

■ Commitment

Training receives a great deal of verbal support in most organizations. Indeed, few managers will deny the supposed value of training for working people. In practice, however, training in the work organization is often little more than a bit of surface activity that produces little or no behavioral change.

It is true that training presents some problems in the functioning correctional organization. There are numerous activities of necessarily greater immediacy than training. It remains something that "we ought to do when we have time." But since it is eminently postponable it is usually put off until "after the current crunch is over." Also, some forms of training cost money, a basic resource often in short supply. It is natural that various activities are seen as competing for available funds. It is just as natural that limited funds are channeled toward points of urgent need, of uses possessing the potential for generating readily measurable returns.

Quite often (perhaps too often) training loses out in direct competition for management's time and the organization's resources. There is certainly not a sense of immediate need for most proposed training or education endeavors. Neither is the return on investment in training necessarily immediate or measurable in any direct sense.

A great deal of training is undertaken without much out-of-pocket spending, so the principal avenues of commitment become time and effort. However, it becomes difficult for an individual supervisor to devote sufficient time and effort to training if such a posture is not encouraged by higher management.

Regardless of the resources (the time available to apply and the available materials and assistance) going into a training program, there is one critical element without which no educational undertaking can long succeed, an element that cannot be mandated. This is the personal and organizational commitment to training and training as a necessary and desirable activity. This commitment must be present and visible at a level and scope sufficient to support the training effort. That is, management training for department heads must have the support of administration; supervisory development for shift commanders must have the support of the chief of security; skilled training for case managers must have the support of the chief of case management; and so on. Some measure of commitment is behind every successful training endeavor, whether personal, departmental, or organizational.

■ Many Options

Approaches to training can range from the formal and highly structured to the informal and almost totally unstructured. Training need not always be a structured program of formal classes on specific topics. It may indeed be this, but it may also be, for example:

- A simple demonstration of manual skills of work methods for a small group of employees, or perhaps one-on-one for a single employee.
- Employee orientation and guided on-the-job training in which an employee learns while doing.

- Self-study, perhaps job-related reading, correspondence courses, or out-side classes.

Thoughts about training tend to inspire visions of classes, teachers, and classrooms, and indeed a great deal of training is approached in "schoolroom" fashion. Such programs and classes, however, can vary considerably in type according to their purposes.

Informational Programs

Informational programs are intended to implant specific information for later recall and application. This is "going to school" in the fundamental sense. Students receive information that they will be expected to absorb as received, recall when necessary, and apply as appropriate in the future.

Informational programs are subject to evaluation by direct testing or examination. The pattern is familiar. Attend classes and participate in a number of assignments, then deal with a number of questions designed to test recall of the information conveyed in the program.

Skill Programs

Skill programs are intended to impart specific skills: perhaps how to operate a computer, how to use a new type of gas gun, how to write a report, or how to cook using a new piece of equipment. To some extent, the results of skill programs can also be measured by test or examination, especially if these require the learner to demonstrate the skill.

Concept Programs

This type of program deals with concepts and concept-based patterns of behavior that, when translated into action and refined through application in specific situations, result in the development of "skills" of a sort. Most management training falls under this heading. Although the term "skills" is used often, as in "supervisory skills development program," this really is something much more fundamental.

Consider as an example one topic drawn from the manager's bag of "skills"—delegation. The participants may sit through a class or a series of classes on delegation and learn about the concept of delegation. The course may also present cases and exercises in delegation, reinforcing knowledge of the concept, but the program does not allow the participants to benefit from real-world applications. The applications, when they become familiar, usually vary to some extent from the class examples and from each other. Consequently, each situation itself becomes a learning process that requires continually adjusting the elements of the concept to suit reality. Do this successfully often enough and the delegation concept will have become a new (and hopefully comfortable) pattern of behavior. Thus a participant will have absorbed a concept in the classroom, carried it onto the job and put it into practice in a number of situations, and developed the hazily defined "skill" called delegation.

The problem with most concept programs (and thus with most management, supervisory, or leadership programs) is that they are treated by many learners and teachers alike as informational programs. As a result, concepts are absorbed but are not applied. Many supervisors can discourse at length about proper delegation, but not nearly as many practice delegation wisely and well.

A concept program that can be tested for retained knowledge can also be done with an informational program. However, the true success of a concept program depends largely on changes in behavior and attitude.

■ The Employee Role

Regardless of how small the group supervised, the manager will often find it difficult to meet all the training needs of every employee. Thus training becomes an ongoing job made necessary by promotions, transfers, and replacements. Unless every employee does precisely the same things (which is a condition encountered by very few supervisors) any of the foregoing actions can create gaps in the capability of the work group. Even if the group remains stable in terms of personnel transactions, there are always the gaps created by illness, vacations, and other time off.

A recommended objective for training in any employee group is the achievement of the maximum possible amount of flexibility, through cross-training among employees working at comparable skill levels. The correctional officer workforce is a prime example of this in most systems—officers are expected to be able to work a variety of posts with minimal loss in efficiency or effectiveness.

For another example, look at the position of a records supervisor who has three records clerks assigned to three different areas. Although all three are classified as records clerks, one is assigned to compiling files on new inmates and filing and retrieval of file contents, a second to computing sentences, and the third to responding to correspondence and information requests. The absence of any one of these workers could mean the development of a disruptive backlog if no one else is able to step into that person's job for a day or two. However, if all three records clerks were trained in all three functions, the two on the job could be shifted around to serve the greatest needs. If all three clerks were present, similar positive action could be taken if one activity was backlogged but the other two were current.

Cross-training of employees is of value to the supervisor because of the versatility of coverage just described. It can also be of appreciable benefit to employees. Many employees welcome the opportunity to learn different tasks and undertake activities new to them, and enjoy breaking away from prolonged periods spent performing the same tasks. The stimulation and thus the motivation provided by task variety is often sufficient to encourage the supervisor to rotate employees through several different assignments at regular intervals. And of course, cross-training increases promotional potential for participants. Take care, however, to assure that cross-training and job rotation are limited to employees and jobs in the same general job classification or pay range, and that no labor agreements are breached by exchanging assignments in this way.

The immediate goal in approaching training with employees is to impart knowledge or skill. However, the long-range goal should be to create a learning attitude among them. Not all employees can be encouraged along these lines, so the creation of a learning attitude remains an ever-present goal that is never completely fulfilled. However, managers will know that this goal is being served when some employees begin to seek out additional knowledge on their own, without further supervisory urging.

Getting Started

Since this chapter provides no more than an overview, the initial suggestion offered for pursuing training in a given department is to look about the institution for advice and assistance from other managers who are already involved in training. Of course, most agencies, and all that are ACA accredited, will have initial and refresher training programs that inform and update staff on basic practices. Beyond that level, if the institution is large enough or sufficiently committed to the concept, there may be a separate training or staff development department. There also may be an in-service training component of the security department. If that is the case, the chief of security and the training officer can help with program structure and general approach. More likely, however, the training officer can provide advice, information, and assistance to another supervisor preparing to begin an in-house training or training program.

Determining Needs

Before launching any training program, however, examine the apparent needs of the work group and select some potentially fruitful starting points. In other words, it is best to begin with some form of <u>learning-needs analysis</u>.

There are a number of ways to accomplish this, starting with the ACA's accreditation standards for training. But as an example of a more individualized approach, **Exhibit 28-1** shows a portion of a simple skills inventory matrix created for the employees in a food service department. All of the employees in a particular job classification are listed in the column on the left side of the form. The remaining columns contain the supervisor's assessment of each employee's capabilities in the jobs identified at the tops of the columns. Note the simplicity of the breakdown: A number 1 indicates that the employee needs complete training in this activity; a number 2 designates the need for some degree of refresher training; and a number 3 indicates that the employee now performs that particular job satisfactorily.

Do not be misled by the simplicity of the matrix. To be sure, each number is a judgment call. In creating the matrix (especially in determining the 2s and 3s),

Employees	Sanitation	Preparation	Dietary Knowledge	Inmate Supervision
P. Abel	2	3	3	3
C. Brown	3	3	1	2
N. Carter	1	1	3	3
J. Davis	3	3	1	2
D. Evans	2	1	3	3

1. Needs complete training
2. Needs refresher
3. Satisfactory

Exhibit 28-1 Skills inventory: food service.

since a 1 is automatically indicated for an employee who has not done a particular job, it is necessary to speak with each employee and observe each at work. When this exercise is completed, the manager will have a fairly good idea of where to focus early training efforts.

Another form of the skills inventory, a variation applicable to the assessment of the learning needs of all employees' capabilities relative to all jobs in the department, is shown for secretarial staff in **Exhibit 28-2**. For each employee and each job you are really asking two questions:

- Is training on this job applicable to this employee?
- Can this employee, working under normal supervision and direction, perform this job satisfactorily?

A "not applicable" answer may be called for because a job is totally unrelated to an employee's general line of work or is of a different pay grade or classification. Be careful, however, of how freely the "N/A" indicator is applied. It may be to both the manager's and the employee's advantage to consider training in completely different areas of activity as long as the jobs are consistent in classification and pay grade.

For the major question regarding the employee's capability of performing the job with normal supervision and direction, the task (again accomplished through individual observation and judgment) is simply to answer yes or no. All "no" answers again indicate potentially productive starting points for staff training.

The Supervisor as a Teacher

The supervisor may do some, although not all, of the teaching in a training program for a department. For some of the topics and skills the supervisor will be the most qualified and most readily available instructor. But chances are that at least a few topics will be better handled by others.

Employees	Transcription	Filing	Phone Contacts	Personal Contacts
R. Baker	Yes	No	Yrs	Yes
P. Fredericks	N/A	Yes	Yes	No
M. George	Yes	Yes	Yes	Yes
N. Lori	N/A	No	No	Yes
D. Quincy	N/A	No	Yes	No

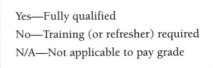

Yes—Fully qualified
No—Training (or refresher) required
N/A—Not applicable to pay grade

Exhibit 28-2 Skills inventory: secretarial.

For example, a records office manager who is interested in cross-training several records office staff in sentence computation might be wise to use the employee who is regularly assigned to this task as an instructor. With proper encouragement and assistance, the person who best knows how to perform the details of a given task can become the best possible resource for teaching that task to others. The supervisor should not presume to do all of the teaching all of the time. But the department's training program will remain the supervisor's responsibility and part of this responsibility can be carried out by helping others develop as instructors.

Teaching a class can loom as a formidable task to a newcomer. If the idea of teaching a small class of employees is frightening, rest assured that that feeling is not novel. Just about everyone involved in teaching or public speaking has experienced similar qualms. It helps to regard early ventures into teaching as learning experiences in themselves, remembering also that most managers are not professional teachers, and that the employees being taught are familiar and not a group of unknown "students." Also recall that instruction in this environment is best accomplished in an informal, friendly atmosphere.

The keys to building effectiveness as an instructor are preparation and practice. By being reasonably comfortable with the subject and taking care to organize the course material for logical presentation, most people can overcome apprehensions about teaching. The more a person teaches a given subject, the better able they are to teach it in the future. The more often an initially inexperienced instructor faces a class of learners, the less bothersome uneasiness about teaching will become.

The Process

Speaking now primarily of the process of imparting skill or knowledge to non-supervisory employees, the person providing the training should be trying to do the following for the trainees:

- Motivate them. Attempt to reach them with reasons why they should want to learn. Let them know what is in it for them—new skills? Something interesting for its own sake? Opportunity for more interesting, challenging, or varied work? Let them know what they stand to gain, and what the department stands to gain by taking advantage of what is to follow.
- Tell and demonstrate. Present the subject in logical order, and when teaching skill-related topics (for instance, how a particular task is performed) actually show them how it is done.
- Check for their understanding. Ask for questions and discussion and generally encourage feedback. If the course content is reflected in those interactions, chances are that effective communication is taking place.
- Let them try it. Again related primarily to skill training, having told them how it is done, shown them how to do it, and once there is a reasonable assurance that they understand, have them actually perform the task. Again, encourage feedback and work with them until they are able to perform the task satisfactorily.

Some Points to Remember

The importance of preparation has been mentioned already. However, preparation is only a part of what is necessary to get a message across. The balance of the task lies in dynamic instruction. Entertaining instruction can at least partially salvage thin material, but dry, lackluster presentation can render the best-prepared material useless.

Combine instructional methods whenever possible. Neither telling, showing, nor doing alone is nearly as effective as combining, for example, telling and showing, or ideally, telling, showing, and doing.

Use multiple modes of presentation. Avoid reliance on pure lecture, 100 percent transparencies, complete tape-recorded presentations, or the predominant use of any other single medium of presentation. Use mixed media, perhaps lecturing for a while, supporting comments with transparencies, showing a brief film now and then, or using short segments of tape-recorded material. Also, use interactive techniques whenever possible—discussion, exercises, case studies, and in general anything that will draw employees into the learning process.

Always be aware that the learners in this context are workers, not students. These sessions are taking place in the middle of or perhaps after a regular workday, and chances are they have not come to the class fresh and relaxed. Never forget that they are full-time workers who probably have job problems on their minds.

Accept the high likelihood that the presentation will rarely reach everyone in every way. People are different; some are receptive to new information and new skills and some are not. But it is important to try. Failure to reach everyone is not necessarily an indication of a teacher's shortcomings, but failure to try to do so is such an indication.

Remember that for the instructor the process of teaching is itself a valuable form of learning. Most supervisors who have done some teaching have discovered that there is no better way to expand one's knowledge of a subject than to teach that subject.

■ Training and You

Formal training programs for the supervisor often deal with matters of supervisory skill or management practice—usually the "concept programs" discussed earlier. There is, of course, difficulty in including practical applications in this kind of program, since even the best-prepared exercise or case problem is at best only a model of reality. Consequently, concept programs deal primarily with ideas and theories.

As many people in management are constantly discovering, the gap between theory and practice can be broad and deep. The principal shortcoming of most management programs lies in their failure to provide any substantial means of bridging the gap between theory and practice. Many managers learn things that they may retain in an informational sense, but they never take the steps necessary to incorporate changes based on this information into their patterns of behavior.

Whenever possible, take advantage of the opportunity to attend managerial training programs. A manager may feel inclined to bypass such opportunities because of workload considerations. But if management training is approached with an open and receptive attitude, participants are practically guaranteed to come away with something that will help them do a better job.

Again, because of the concept-based nature of management training, management programs (at least many of the better ones) are oriented more toward attitude change than toward imparting specific skills. Because of this, a great deal of self-starting is required on the manager's part if maximum benefit is to be obtained from such training. Even the best of programs leave it largely up to the participant to translate what is learned in the classroom into action on the job.

Aside from management and supervisory programs provided by the organization, training can also include:

- Books, journals, and professional bulletins and newsletters and other publications.
- Courses available outside of the institution, such as professional seminars, and workshops and management programs at local colleges.
- Audio- and videotape programs made available for individual self-study.
- Correspondence courses and conferences, seminars, and workshops made available by various professional organizations.

The key difference between training for a line manager and for subordinate employees lies in the matter of who makes such training available. It is up to management to make training available to line employees. However, in what is a somewhat contradictory situation, do not assume that it is up to your superiors to make training available to you as a line manager. The institution, if prompted by its supervisory personnel, may make management-oriented training programs available. However, access to many other forms of supervisor-oriented training is largely controlled by the individual manager.

When a manager has an opportunity to attend such programs because the administration, the personnel department, or the staff development component of the organization has made such training available, take advantage of it. The preferred avenue to such training programs is being given the option of attending. However, sometimes a person is "asked" to attend in such a way that refusal would clearly prejudice their career. And sometimes a manager is ordered to attend a program. Even if attendance is mandated or forced, the participant in such a program will do well to maintain a positive attitude toward the course itself.

Briefly recalling the matter of commitment discussed in the early part of this chapter, it is the individual manager's personal commitment to training that will most influence how that person and their employees develop educationally. If the commitment is verbal only (if, for instance, management simply launches a training program for employees and then backs away) employees are likely to follow that lead and participate only superficially. Employees will take their cues from their supervisor. If they see management as truly "into training," they are much more likely to take any such program seriously.

Training is one of the responsibilities of a manager. The organization will likely expect its managers to stimulate and guide employee training while actively pursuing their own.

EXERCISES

Exercise 28-1: The Skills Inventory

Using the approach of either Exhibit 28-1 or Exhibit 28-2, prepare a preliminary skills inventory for your department or a portion of your department.

The technique of Exhibit 28-2 is adequate for a smaller department, for example, fewer than 10 employees and not more than 10 or 12 major tasks. If you supervise a larger department, concentrate on a single job classification or labor grade and apply the technique of Exhibit 28-1.

Do not simply render all judgments "off the top of your head." When you have doubts concerning a person's ability to perform a certain task, spend a few minutes with the employee and ask some pertinent questions.

Use your skills inventory to develop a listing of learning needs for your employees: who must be trained in certain tasks and who could be of greater benefit to the department (by providing greater flexibility, improved task coverage, for example) by being trained in additional tasks.

Exercise 28-2: Cross-Training

If you supervise two or more persons in equivalent pay grades who perform somewhat different tasks, develop a plan for cross-training at least two (and preferably three) such persons in the others' tasks. If you do not yet supervise others, develop a hypothetical two- or three-person cross-training plan based on positions and tasks in the department in which you work. Your plan should include:

- A brief learning-needs analysis indicating who needs how much training in what functions.
- Who will do the training?
- What training processes will most likely be used?
- How long the training should require?
- How you will assess the results of the training to decide when it is complete?

ENDNOTES

1. *Standards for Adult Correctional Institutions,* 3rd ed., The American Correctional Association: Lanham, MD, 1990.
2. Joint Commission on Accreditation of Healthcare Organizations.

The Supervisor and the Law

29

Chapter Objectives

- Discuss the issue of inmate litigation and basic managerial responses to inmate legal issues.
- Review the essential elements of a supervisor's response to inmate litigation.
- Provide a review of pertinent areas of legislation with which the supervisor should be generally familiar.
- Offer special emphasis on addressing the increasingly visible phenomenon of sexual harassment.

Laws should be like clothes. They should be made to fit the people they are meant to serve.

Clarence Darrow

Our Constitution is color-blind, and neither knows nor tolerates classes among citizens. In respect of civil rights, all citizens are equal before the law. The humblest is the peer of the most powerful.

Justice John Marshall Harlan

FROM THE INSIDE

Inmate and Employee Legal Issues: Two Situations

Situation One: The inmate's crime had garnered him some publicity, but then came the tedium of a life term. In the depths of a maximum-security unit in a maximum-security prison, he turned to litigation as his refuge, or perhaps his recreation.

He sued the institution discipline committee, "individually and in their official capacity," every time the committee upheld an incident report that had been filed against him.

He sued the hospital administrator and other medical staff of the institution, seeking to be allowed to use marijuana to moderate the effect of his glaucoma condition.

He sued the institution commissary supervisor and the company that supplied the institution's peanut butter because he said he found glass in the jar he bought in the commissary.

He sued the food service supervisor because he said the trays his food was served on were giving him hemorrhoids.

Situation Two: The correctional officer had been charged with a murder in the community, and had been acquitted on the grounds of self-defense. He had a history of other problems, including serious domestic violence charges, although he had never been convicted of them either.

The agency fired him, based on his history of violent acts as well as the killing, which to the agency looked quite a bit different than it had to the jury. The employee appealed to the appropriate civil service merit board. The board reinstated him, finding that there was no "nexus" (fancy word for connection) between the man's employment as a correctional officer (who sometimes was required to use firearms on duty) and his undisputed pattern of resorting to violence in his private life.

The agency appealed this ruling (this time to a court) and kept the officer on administrative leave (not permitting him to enter the institution) for the duration of the litigation. Finally, a court upheld the termination—after years of costly wrangling.

This chapter will address two very different types of legal concerns of correctional managers. The first is offender litigation, which can be filed when an offender claims some staff or agency action has denied him or her some right under the U.S. Constitution, or a statute or regulation. The second area is the wide range of mostly employee-related legal issues that have been created by laws relating to the workplace, discrimination, and other matters.

A major disclaimer—the authors are not attorneys. They do not in this chapter advocate any particular course of action or give any legal advice on any specific issue. The content that follows is intended to provide only an introduction to these issues. Any specific case- or fact-related interpretation required by the reader should be sought from qualified legal counsel.

■ Offender Litigation

No employee in corrections, regardless of level or position, is unaware of the fact that inmates sue staff. Regularly. Frivolously. Frustratingly. Expensively. Stressfully. Even when (as in most cases) the agency provides legal representation for employees acting within the scope of employment, it is a concern.

Inmate litigation is a reality in our profession that has to be mentioned, even if just in passing. However, it is hoped that this section will introduce new managers to the idea that offender litigation is not only expectable, but avoidable in many instances.

First, consider the difference between civil and criminal cases. Broadly speaking, a civil case is brought by a supposedly injured party (which in some cases could be an arm of government) and has the potential for monetary, injunctive, or other noncriminal sanctions. A criminal case, on the other hand, is brought by a law enforcement arm of the government, involves an alleged infraction of a criminal statute, and carries specified criminal penalties.

A correctional officer searches and impounds some inmate property, and the property subsequently is inadvertently lost or destroyed. Ordinarily, there is a lower level procedure (an agency-level administrative claim) for the inmate to pursue for redress. But in the extreme, the inmate may sue the employee and/or the agency for the loss. Such an action is a civil complaint, and in all likelihood the agency will provide representation for the employee or employees being sued.

There are cases where an employee seeks independent counsel in such cases. The reasons are somewhat complicated, but usually they involve a lack of confidence in agency counsel, or a potential conflict of interest between the agency and the individual. Another such instance might be that of an employee who engages in illegal conduct (sex with an inmate, for instance) in the workplace. That person almost surely will be the target of a civil suit, and may be prosecuted criminally as well if the victim is an inmate. The agency would not be likely to provide representation in either the civil or criminal matter that could result. That is because it would be hard to construe that the employee was acting within the scope of their official duties, ordinarily an important test for such litigation.

As an example of a criminal matter, suppose an employee is working in the property storage area and loots the valuables safe, taking watches, rings, and other valuables. In such an instance, it is likely that an investigation by the agency will result in criminal charges being filed by an outside law enforcement agency. As an employee, violate a law while on duty and you are on your own in almost every instance.

As a related matter, alleged labor law violations can give rise to civil litigation that may or may not involve individual employees. The agency's legal resources will be brought to bear on such cases because the administration supposedly was seeking to assert legitimate management prerogatives when it took the action in question. This also ordinarily is the case when an employee sues the agency for some personnel action, such as termination.

Inmates threaten civil litigation almost as much as they actually file lawsuits. But staff who are following policy and statutory guidance need not fear this kind of intimidation. Every agency (public or private) can bring legal representation to bear on an inmate-filed complaint, and will provide that representation to employees who are operating within policy, in goodwill, and within the scope of their official duties. In many cases, if an employee is found liable in such an instance and was acting within the scope of employment, the agency may have provisions for indemnifying the employee for any financial sanctions.

The important keys to an individual manager's response to inmate litigation are as follows:

1. Know policy and remain within its bounds.

2. Make informed decisions that reflect a goodwill effort to carry out policy.

3. Carefully document any unusual activities or decisions taken where there is some indication that litigation may result.

4. When actually served with a complaint, immediately notify your supervisor and open a line of communication with agency legal counsel.

5. Preserve any relevant records, and in no case destroy or conceal records or information required by any later stage in the legal process.

6. Never threaten or take reprisal action against an inmate or staff member who threatens to or actually files litigation.

7. If enjoined temporarily by a court order from doing something, don't do it, and don't get clever and think you can find a way around the order.

8. If presented with a final order, comply with it; follow agency counsel advice with respect to any appeals.

Litigation is a legitimate part of our legal system and inmates have a right to sue. Painful as it is to admit, there are abuses in corrections—abuses at the individual and systemic levels. We as professionals often wish the courts were less elastic in the types of cases they entertain. The examples at the start of the chapter include actual cases filed, some of which were clearly not meritorious on their fact. However, instances of overreaching by the courts are numerous, at least from the viewpoint of any experienced correctional manager. But civil litigation also indisputably has led to curtailment of abuses, improvement in conditions of confinement, and an overall upgrade of corrections in the United States in the last four decades.

■ Non-Inmate-Related Supervisory Legal Issues

The remaining portions of this chapter will cover some major areas of legislation that place requirements on organizations and thus influence supervisory behavior.

First, every manager should know that it is possible that an aggrieved employee may file a grievance, appeal, or lawsuit after some adverse personnel action is taken. Such matters often are resolved by a merit board of some kind within the relevant governmental structure. If the outcome of an initial appeal is unsatisfactory to the aggrieved employee, full-scale litigation may result, and often those cases are founded on one or more of the laws discussed next.

To defend such cases, every agency will have at its disposal some legal resources that specialize in labor law. This may be in-agency counsel, or it may be provided by another related agency, such as the state's attorney general's office. Nevertheless, such cases should be taken seriously, and the eight recommendations at the end of the previous section also apply for this kind of litigation.

To be clear, learning the small amount of applicable labor law conveyed in this book will not enable any manager to decide legal issues independently. Rather, this material is meant to increase awareness of some of the things a manager should or should not do because of the existence of various forms of legislation. Most agency policy and procedure in place will already incorporate a basic structure for the manager to operate within. When a true legal question arises, seek the answer through the personnel department or through in-house legal counsel.

The National Labor Relations Act

The National Labor Relations Act (NLRA) of 1935, as amended by the Taft-Hartley Act, provides critical structure for labor management relations in the United States. Numerous references to the NLRA are actually references to this law as amended by Taft-Hartley. The key provisions of this law are incorporated in the labor-management and personnel management structures of every correctional agency. There is little the individual supervisor can do in this area but follow the procedures and guidance laid out by agency policy. A more complete discussion of the issue of union relations follows in Chapter 30.

Wage and Hour Laws

Of primary interest is the 1938 Fair Labor Standards Act (FLSA), which is the federal wage and hour law. As such, it is the model for the wage and hour laws of many states. Occasional points that might not be covered by federal law maybe covered by pertinent state laws. Generally, if the same points are covered by both state and federal laws but differences exist between the two, the more stringent legislation will apply.

Categories of Employees

Exempt Employees

Minimum wage and overtime requirements apply in most correctional settings. Three types of employees remain exempt and are identified as follows:

1. **Executives.** An executive employee must generally spend 50 percent or more of the time in direct management of an enterprise or an organizational subunit such as a department. In addition, an executive employee must direct the activities of two or more persons. The executive definition may also require that a person possess the authority to hire and fire or so recommend; possess discretionary powers, rather than be assigned largely routine work; and from work week to work week spend no more than 40 percent of the time on nonmanagerial work.

2. **Administrative.** An administrative employee must spend 50 percent or more of the time on office or nonmanual work related in some way to policy, general business, or people in general and must be required to exercise discretion and independent judgment. Other tests of the administrative classification may be assisting executive or administrative personnel; handling special assignments with only general supervision; working in a position requiring special training, experience, or knowledge; and spending not more than 40 percent of the time on nonadministrative work.

3. Professional. Professionals in corrections (for example, doctors, nurses, psychologists, teachers) are so classified by virtue of spending 50 percent or more of the time in work that requires advanced specialized knowledge or is original or creative in nature. The definition may also require that the professional be consistently required to exercise discretion and independent judgment, be employed at intellectual and varied work, and be engaged in nonprofessional activities not more than 20 percent of the time.

Decisions on the applicability of a particular definition are generally made on the basis of the percentage of time spent on various activities. The time test applies on a work week to work week basis.

Nonexempt Employees

All employees who do not fall under the executive, administrative, or professional category are considered nonexempt employees. They must be paid at least the prevailing legal minimum wage for each hour worked in a workweek, and they must be paid at a rate of one and one-half times the regular rate for all overtime hours. In addition, they must be given equal pay for equal work unless there are legitimate factors that justify the establishment of different rates. (Sex is not a legitimate factor.) The organization is required to keep detailed records of hours worked and wages paid.

There are a few well-defined exceptions to the payment of the legal minimum wage. Special regulations allow the payment of lower rates to students, interns, and apprentices. Employment of such persons is also subject to additional requirements and restrictions.

What Kind of Employee? Exempt and nonexempt are common organizational terms that are used essentially synonymously with salaried (exempt) and hourly (nonexempt). Being nonexempt is advantageous to the employee because overtime must be paid for all hours in excess of 40 in a week. Being exempt may be seen by some as advantageous to the employee because of the associated rate of pay and flexibility of hours. It is advantageous to the employer because of stability of labor cost and the ability to get additional work accomplished (say beyond 40 hours in a week) without additional payment.

It is important to recognize the difference between the use of "may" and "must" in the preceding paragraph. The law clearly states that any position that does not meet the exempt criteria must be considered nonexempt and be paid overtime. However, a position that meets the exempt criteria is not legally required to be treated as exempt.

While this generally is a personnel office function, it is worthwhile for managers to know that there is some risk in incorrectly classifying employees as exempt when they should in fact be nonexempt. In some organizations certain positions have been treated as exempt simply because they were compensated at or above the minimum exempt salary requirement. For example, a position as "secretary" might be retitled "administrative assistant" and made an exempt "administrative" position simply because of satisfying the salary requirement. However, because the position may not involve a sufficient percentage of true administrative work it may, upon audit by the Labor Department, be ruled nonexempt. If this occurs, the organization will be called on to pay imputed overtime costs for positions incorrectly classified. For a position to be treated as exempt, it must meet the requirement for work content as well as for salary.

Overtime Compensation

Managers who actually control the work schedules of their employees, and at the same time need to control salary expenses, need to understand how overtime is structured under applicable labor law.

The Work Week

The FLSA defines the work week as a fixed, recurring period of 168 hours; that is, seven consecutive 24-hour periods. These 24-hour periods need not be calendar days, and the seven periods together need not be a calendar week. In many organizations, work weeks beginning and ending at midnight Friday or midnight Sunday are not uncommon. The work week may be changed, and many organizations have done so to facilitate payroll accounting, but it cannot be changed in such a way as to avoid payment of overtime.

Time and One-Half

The FLSA requires payment of one and one-half times the regular rate for all overtime hours. Overtime hours are defined as those hours in excess of:

- Forty hours in a 7-day work week, where the usual 7-day work week is used.
- Eight hours per day or 80 hours per 14-day period, when the use of the 14-day period has been approved and posted.

The institution may use either or both methods for certain of its employees but may use only one method at a time for a specific employee group. If the so-called 8 and 80 provision is used, overtime must be paid for all hours worked in excess of 8 in one day or in excess of 80 in the 14-day period, whichever results in the greater number of overtime hours.

In the example below, the employee worked a total of 80 hours. The employee is owed 3 hours of overtime that is derived from the fourth day, when 10 hours were worked, and the eighth day when 9 hours were worked (although on one day the employee worked only 5 hours).

Look at this example.

Day	Hours
1	8
2	8
3	5
4	10
5	8
6	0
7	0
8	9
9	8
10	8
11	8
12	8
13	0
14	0
14 days	80 hours

Overtime owed = 3 hours (2 from day 4; 1 from day 8)

In this case the employee worked more than 8 hours on one or more days and more than 80 hours for the 14-day period. This example assumes that the employee worked 8 hours in each of the 9 days and 10 hours on the 10th day and thus is due 2 hours of overtime. Note that the employee has worked 2 hours in excess of both the 8 hours per day and 80 hours per work period provisions. However, this does not mean that the employee is owed overtime for 4 hours (based on 2 hours in excess of 8 and 2 hours in excess of 80). The employee is owed but 2 hours of overtime pay.

Hours are not double counted. Rather, when the totals of daily overtime and over-80 differ it is the higher that must apply. The FLSA also specifies that only hours actually worked need to be counted toward determining overtime. That is, the institution is not required to count nonworked time such as vacation days, sick leave, holidays, and personal time as part of the 80 hours.

The "Regular Rate"

The so-called regular rate referred to in the FLSA includes the scheduled hourly rate plus on-call pay, call-in pay, and shift differential. Here follows an example of the effects of these additions on the rate.

Assume the employee in the example is paid overtime under the 7-day, 40-hour work week. The employee receives the following amounts: an hourly rate of $9.00, a flat rate of $15.00 for on-call time, $50.00 for 4 hours work on call-in, and 60 cents per hour shift differential. Assume the employee actually worked a total of 50 hours including the 4 hours of call-in time, and that shift differential was not used for the 4 hours of call-in time.

Illustration of "Regular Rate"

Overtime period: 7 days, 40 hours

Employee worked 50 hours, including 4 hours of call-in time

Rates paid:	Basic: $9.00/hour
	Shift differential: $0.60/hour
	Call-in: $50.00 (4 hours)
	On-call: $15.00 (flat)

Calculation:	$9.00 × 46 hours	= $414.00
	0.60 × 46 hours	= 27.60
	50.00 call-in	= 50.00
	15.00 on-call	= 15.00
	Subtotal	$506.60

$$\frac{\$506.60}{50} = \$10.132/\text{hour "regular rate"}$$

$10.132 × 1/2 = $5.066/hour overtime premium

Basic Earnings (from above)	$506.60
Premium ($5.066 × 10 hours)	50.66
Total Earned	$557.26

Since the employee worked a total of 50 hours, divide 50 into $506.60 to arrive at a "regular rate" of $10.132 per hour. Therefore, for the 10 hours of excess time the employee must be paid time and one-half this regular rate, or 10 hours at $15.198 per hour. Having already been paid the regular rate for each of the 50 hours, the employee is owed only the difference between that and $15.198 for the 10 excess hours, that is, $5.066 × 10 hours or $50.66. The total owed the employee in this example is $557.26.

Generally, hours spent at home "on call" are not counted as hours worked. This treatment depends on the employee's freedom of movement while on call. Pay received for such time, however, is counted in determining the regular rate. Note also, however, that when an employee who is "on call" is actually called to perform work, the hours actually worked are counted in the total hours worked. In determining whether on-call time must be counted as hours worked, the government will generally look to determine whether the employee must remain on the employer's premises or be sufficiently close that the time cannot be used as the individual chooses. If this is the case, the hours will be treated as working time for purposes of both minimum wage and over time requirements.

Equal Pay

A key section of the FLSA prohibits discrimination among employees on the basis of sex when the employees are doing equal work on jobs requiring equal skill, effort, and responsibility and performed under similar working conditions. In correcting unlawful differences in rates of pay, the act requires that the lower rate be increased; it is not permissible to decrease the higher rate. The act does make provision, however, for unequal pay if the inequality is directly attributable to a bona fide seniority system, merit system, incentive compensation system, or any other plan calling for a differential in pay based on any factor other than sex.

Affirmative Action and Equal Employment Opportunity

This is an area of the law that has certain fundamental underpinnings, but which is constantly changing as courts rule on various cases. Rather than try to recapitulate case law, which certainly will change over time, this section will discuss the important underlying statutes that directly impact a manager's day-to-day activities.

Title VII of the Civil Rights Act of 1964

As amended by the Equal Employment Opportunity Act of 1972, this legislation prohibits discrimination because of race, color, religion, sex, or national origin in any term, condition, or privilege of employment. The Equal Employment Opportunity Act of 1972 greatly strengthened the powers and expanded the jurisdiction of the Equal Employment Opportunity Commission (EEOC) in enforcement of this law.

As amended, Title VII now covers the following:

- All private employers of 15 or more persons.
- All educational institutions, public as well as private.
- State and local governments.
- Public and private employment agencies.

- Labor unions with 15 or more members.
- Joint labor-management committees for apprenticeship and training.

The EEOC investigates job discrimination complaints, and when it finds reasonable cause that the charges are justified, attempts, through conciliation, to reach agreement by eliminating all aspects of discrimination revealed by the investigation. If conciliation fails, the 1972 amendments give EEOC the power to go directly to court to enforce the law. Among other important provisions, the 1972 act also provides that discrimination charges may be filed by organizations on behalf of aggrieved individuals, as well as by employees and job applicants themselves. Applicants may also go to court directly to sue employers for alleged discrimination.

The Equal Pay Act of 1963

The Equal Pay Act of 1963 requires all employers subject to the Fair Labor Standards Act (FLSA) to provide equal pay for men and women performing similar work. In 1972, coverage of this act was extended beyond employees covered by FLSA to an estimated 15 million additional executive, administrative, and professional employees (including academic, administrative personnel and teachers in elementary and secondary schools) and to outside salespeople.

The Age Discrimination in Employment Act of 1967

The Age Discrimination in Employment Act (ADEA), covering age discrimination in essentially all aspects of employment, was first passed in 1967.

Amended a number of times since its initial passage, ADEA applies to private employers and state and local governments having 20 or more employees and to labor unions having at least 25 members. The original act prohibited such employers from discriminating against persons in the 40- to 70-year-old age range in any area of employment because of age. The Age Discrimination in Employment Amendments Act of 1986, effective for most employers on January 1, 1987, removed the age 70 limitation on ADEA protection. An employer can neither place an age limit on candidates for employment (except for those occupations for which it has been established that age is a bona fide occupational qualification) nor establish a mandatory retirement age for most employees. The amended ADEA has also necessitated the amendment of numerous insurance plans and other employee benefits plans to permit their continued provision to all active employees regardless of their age. Essentially, ADEA in its present state requires employers to provide the same terms, conditions, and privileges of employment to all employees regardless of age.

Title VI of the Civil Rights Act of 1964

Title VI of the Civil Rights Act of 1964 prohibits discrimination based on race, color, or national origin in all programs or activities receiving federal financial aid. Employment discrimination is prohibited because a primary purpose of federal assistance is the provision of employment, such as apprenticeship, training, work-study, or similar programs. Revised guidelines adopted in 1973 by 25 federal agencies prohibit discriminatory employment practices in all programs if such practices cause discrimination in services provided to program beneficiaries.

The Americans with Disabilities Act

Passed in 1990 and largely effective in 1992, the Americans with Disabilities Act (ADA) affirmed the rights of persons with disabilities to equal access to employment, services, and facilities available to the public (whether under public or private auspices), transportation, and telecommunications. Covered disabilities are defined in the law. This legislation provides a comprehensive mandate for barring discrimination against persons with disabilities and provides enforceable standards addressing such discrimination.

The ADA requires employers to provide reasonable accommodation for disabled individuals who are capable of performing the essential functions of the positions for which they apply. This may include altering physical facilities to make them usable by individuals with disabilities, restructuring jobs around their essential functions, and altering or eliminating nonessential activities so that disabled persons can perform the work. Regulations implementing ADA were issued by EEOC, the agency responsible for dealing with complaints of discrimination under all major federal antidiscrimination laws.

Case law is accumulating on this topic, as the act defines "disabilities" extremely broadly, to include hearing and vision impairments, paraplegia and epilepsy, HIV or AIDS, and dozens—possibly hundreds—of other conditions. The list of disabilities continues to grow as legal disputes continue concerning what is or is not a disability. There continue to be disagreements over what constitutes a "reasonable accommodation" in any particular case. "Reasonable" has never been very well defined.

Some elements of the law have been clarified. In January 2002, the Supreme Court narrowed the number of people covered by the ADA, ruling that merely having an impairment does not make one disabled for purposes of ADA; that a person's ailment must extend beyond the workplace and affect everyday life. In other words, the Court has said that a person who can function normally in daily living cannot claim disability status because of physical problems that limit the person's ability to perform certain manual tasks on the job. The Court also ruled that disabled workers are not always entitled to premium assignments intended for more senior workers; so in many instances seniority can take precedence over disability. In June 2002, the Supreme Court ruled that disabled workers cannot demand jobs that would threaten their health or safety. This came from a case in which a worker with a particular medical condition wanted to return to his original position even though it was adjudged medically risky for him to do so.

A number of additional cases are pending, so the continuing implementation of ADA is likely to be the source of more case law for some time to come.

Civil Rights Act of 1991

The Civil Rights Act of 1991 was essentially passed to reverse several Supreme Court decisions that had the effect of weakening existing law. It provided for the most extensive modification of Title VII (of the Civil Rights Act of 1964) in more than 20 years. This newer law relies on jury trials, along with statutorily limited compensatory and punitive damages, as the basic litigation scenario under Title VII of the Civil Rights Act of 1964 and the Americans with Disabilities Act. This act introduces jury trials into employment law to determine liability and compensatory and punitive damages for violations that are found to constitute intentional discrimination. The net effect of this legislation on employers is to increase the likelihood of legal action and increase legal costs associated with trials since potential plaintiffs and their attorneys are attracted by the prospect of damage awards and attorneys' fees (rather than simply compensation for losses, as under Title VII).

The National Labor Relations Act and Related Laws

In addition to the previous requirements, discrimination on the basis of race, religion, or national origin may arise under these laws. It may be unlawful for employers to participate with unions in the commission of discriminatory practices unlawful under these acts, or to practice discrimination in a way that gives rise to racial or other divisions among employees to the detriment of organized union activity. It may also be unlawful for unions to exclude individuals from union membership, thereby causing them to lose job opportunities; to discriminate in the representation of members or nonmembers in collective bargaining or in the processing of grievances; or to cause or attempt to cause employers to enter into discriminatory agreements or otherwise discriminate against union members or nonmembers.

The Rehabilitation Act of 1973

As amended in 1974, this law requires affected employers to maintain affirmative action programs to ensure the hiring and promotion of qualified handicapped persons.

The Vietnam Era Veterans Readjustment Assistance Act of 1974

This act extends the protection of affirmative action to disabled veterans and veterans of the Vietnam period employed by contractors holding federal contracts of $10,000 or more.

Other Discrimination Laws

Employment discrimination has also been ruled by the courts to be prohibited by the Civil Rights Acts of 1866 and 1870 and the Equal Protection Clause of the Fourteenth Amendment to the Constitution. Action under these laws on behalf of individuals or groups may be taken by individuals, private organizations, trade unions, and other groups.

Many state and local government laws also prohibit employment discrimination. When EEOC receives discrimination charges, it defers them for a limited time to various state and local agencies having comparable jurisdiction and enforcement status. Determination of which agencies meet this deferral standard is a continuing process. These agencies' procedures and their requirements for affirmative action vary, but if satisfactory remedies are not achieved the charges will revert to EEOC for resolution.

Other Applicable Laws

The Family and Medical Leave Act (FMLA) 1993

This law makes it possible for an eligible employee (one who has been employed at least 1 year and has worked at least 1,250 hours) to take up to 12 weeks of unpaid leave in a 12-month period for certain specific reasons without loss of employment. The qualifying reasons are: for the birth of the employee's child or the care of that child up to 12 months of age; for the placement of a child with the employee for adoption or foster care; for the employee to care for spouse, child, or parent having a serious health condition; and for the employee's own serious health condition involving the employee's inability to perform the essential functions of the job. An employee returning to work within the 12-week limit must be returned to his or her original position or to a fully equivalent position in terms of pay and benefits and overall working conditions.

Leave taken under FMLA must often be coordinated with short-term disability and other time-off plans. Also, since certain forms of leave may be taken intermittently or on a reduced day or hours schedule and since there are rules governing the treatment of employee benefits while on such leave, this act has created additional work for department management and human resources.

The FMLA has created a number of problems relative to other laws governing employment. There is overlap in the treatment of sick leave between FMLA and the ADA, and rarely is it clear which law's provisions take precedence. Also, portions of the Fair Labor Standards Act (FLSA) and various state workers' compensation laws conflict with provisions of both ADA and FMLA. When confronted with any but the simplest questions raised by FMLA, the individual supervisor is advised to seek answers through human resources or in-house legal counsel if available.

As a result of these interactions, the FMLA has made managing and staffing considerably more difficult for some managers, and litigation has resulted in a few clarifications in a few areas. In January 2003, the Supreme Court heard arguments that Congress overstepped its bounds when it extended a guaranteed 12 weeks of family leave to state employees. At issue was Congress's power to impose leave requirements on state governments and the power of state workers to sue if they believe their leave rights have been violated. The Court essentially upheld the rights of all workers under FMLA. In May 2003, the Supreme Court ruled that states can be sued for violating workers' federally guaranteed right to take time off for family emergencies. This occurred in response to one state's claim of constitutional immunity from suit under the FMLA.

At present, leave granted under FMLA is unpaid leave. There have been frequent proposals to expand coverage of the FMLA to more people and to grant at least partial pay for such leaves. These proposals are quite controversial, and to date no single proposal has enjoyed a great deal of support.

Title IX, Education Amendments Act of 1972

This law extends coverage of the Equal Pay Act, prohibiting discrimination on the basis of sex against employees or students of any educational institution receiving federal financial aid. Provisions covering students are similar to those of Title VI of the Civil Rights Act of 1964.

Health Insurance Portability and Accountability Act (HIPAA), 1996

The portion of HIPAA implied by its title—portability of health insurance for workers who change organizations—was implemented relatively smoothly in 1996. It was of little concern to anyone other than the human resources department and the organization's health insurance carriers.

However, HIPAA also addressed some issues of privacy and confidentiality. Primarily governing the ways in which health information is gathered and used and disclosed, HIPAA affects organizations involved in health care in any way. The HIPAA Privacy Rule took effect in mid-April 2003; the Transactions and

Code Sets (TCS) Rule, having to do with consistent transmission of information, in October 2003; and the HIPAA Security Rule is scheduled for implementation in April 2005. These portions of HIPAA represent a fair amount of work for any function concerned with delivering health care or handling health information.

■ Sexual Harassment

Because of its prominence in today's society a few separate comments on the subject of sexual harassment are in order. The number of sexual harassment complaints filed with the EEOC and various state agencies continues to increase, as does the number of employers involved and the extent of punitive monetary damages. In recent years sexual harassment has been one of the two leading causes of legal complaints against employers (the other being age discrimination).

The legal basis for defining and addressing sexual harassment has been in place for some time. Sexual harassment is in fact a form of sex discrimination under Title VII of the Civil Rights Act of 1964. Sexual harassment consists of unwelcome sexual advances, requests for sexual favors, or other conduct of a sexual nature if submission is either an actual or implied condition of employment, submission or rejection is used as a basis for making employment-related decisions, or the conduct interferes with work performance or creates an offensive work environment.

A key concern in the foregoing paragraph lies in the word "unwelcome." Conduct is considered unwelcome if the employee neither solicited nor invited it, and regarded it as undesirable or offensive. To a considerable extent, whether a particular occurrence is or is not sexual harassment may depend largely on the perception of the victim.

Sexual harassment can take a number of forms. Sexually explicit pictures, calendars, or other materials; offensive sexually related language (including sexual humor) or other sexual conduct that creates a hostile environment; sexually explicit behavior; indecent exposure; sexual propositions or intimidation; offensive touching; and participation in or observation of sexual activity are all examples of sexual harassment.

So also is something as seemingly innocent (to some) as repeatedly asking a coworker or subordinate for a date after having been turned down. This latter situation adds the dimension of repetition to some harassing behavior. Asking a time or two might be considered reasonable, but asking repeatedly (especially after having been turned down) may be considered harassing.

Sexual harassment is not limited strictly to the workplace. Similar acts also constitute sexual harassment if they occur off-premises at employer-sponsored social events. It can even occur off-premises at private sites if it involves people who have an employment relationship with each other. In addition to involving employees, sexual harassment can involve visitors, vendors, patients, and others as potential perpetrators or victims.

To limit or avoid liability for sexual harassment it is necessary for the employer to promptly and confidentially investigate all complaints, take appropriate remedial action, and create and retain complete and accurate documentation.

The importance of a sound prevention program cannot be overstated as far as sexual harassment is concerned. At a minimum such a program should include a published sexual harassment policy and a detailed procedure for investigating complaints. Ideally, all employees (and most certainly all managers) should be educated in the recognition and prevention of sexual harassment.

The most important thing for a manager to remember about sexual harassment (after not engaging in it personally) is to not tolerate any form of it in your area of responsibility. If managers are aware of and ignore or condone any form of sexual harassment, they are doing at least three things. First, they are missing an opportunity to intervene and stop a subordinate employee from committing further, possibly even more serious, acts. Second, they are placing themselves into a potentially liable situation, since if they know of the act and do not intervene, they become in a sense complicit. And third, they are placing the organization at increased risk of a sexual harassment lawsuit, for the manager is seen as representing the organization in this matter.

A special note about sexual contact with offenders as a subset of this type of conduct is warranted. A person in a position of authority over an inmate who has sexual contact with that inmate is committing a grossly improper act at a minimum. It is unethical, it is unprofessional, it poses immense civil liability exposure for the individual and the agency, and in many jurisdictions it is illegal.

■ Who Needs More Rules?

This discussion has barely scratched the surface of the collection of laws, rules, and regulations with which correctional managers must comply, looking only at the major areas that are likely to be of concern. There are many additional regulations bearing on other issues. These primarily are the day-to-day concerns of the personnel department, and many of them will generally not be immediately visible to the individual supervisor. However, managers are likely to see many of the effects of new legislation as regularly occurring change in the features and facets of issues such as employee benefits programs.

Who needs more rules? Certainly not supervisors. Remember, however, supervisors are also employees, and the protection afforded to line employees under legislation such as equal pay and affirmative action also extends to the managerial ranks. Supervisors should be willing to recognize that certain laws represent a well-defined set of boundaries—the outside limits within which managers must learn to work in fulfilling their responsibilities.

EXERCISES

Exercise 29-1: Asking to Be Sued

Inmates had chosen a moment when the two cell house officers were outside the range grills and had barricaded themselves inside. Inmates could be seen going in and out of cells, and broken table and chair legs were being brandished as makeshift weapons. Response squads were assembled, and a sufficient supply of chemical agents was on hand. The deputy warden, captain, unit manager, and other managerial personnel were at the entry to the unit.

As per standard procedure for any disturbance or forced cell move (when there was time to procure one), a video camera was filming the scene. The video tape, played in federal court in connection with an inmate lawsuit over the incident, revealed the following:

> The unit manager called down the range, saying, "Come on out men, surrender now." Various vulgar endearments were returned by those inmates who had heard him.

> The captain (not known for his quiet ways) yelled out a similar order for the inmates to give up. He, too, was greeted with a less than polite series of catcalls.

> The deputy warden, a man who was known to border on hating inmates, then called downrange, "Let's go fellows, come on out." He, too, was greeted by a less-than-polite response—perhaps a bit more personal in nature because of his reputation among the inmate population.

> Then, in a voice heard clear as a bell over the tape, the associate warden says, "Thank you."

> His hand waves, gas fills the cell house, you see the backs of the squad members as they go in, and soon you see inmates being dragged struggling from the cell house.

Think about the following issues:

- Did staff demonstrate a reasonable attempt to resolve this problem before resorting to forcible means?
- Were staff acting within the scope of their employment in taking this action?
- Were there any obvious actions that could have triggered an inmate lawsuit?
- What did the associate warden's remark say about his approach to his official duties? Did it suggest that he was looking for an excuse to gas the inmates? Would such a remark hamper a "goodwill" defense?
- What impact should the impact of possible court review of staff conduct have on that conduct?

[In the real-world case from which this exercise is drawn, the tape proved to be embarrassing when played for the judge, but no liability was found against individual staff or the institution.]

Exercise 29-2: What Kind of Employee?

Relatively new to department management in the agency, you have heard the terms *exempt* employee and *nonexempt* employee used with some regularity. Only recently the department of corrections reissued its employee handbook to include two sections of supposed wage-and-hour rules identified only by the headings "Exempt Employees" and "Nonexempt Employees." The handbook provides no defining distinction between the two, and as a result you find you have been the target of numerous clarification requests from your employees. Beyond a simple definition, a number of your employees are asking you "Just what does this exempt and nonexempt stuff mean, anyway?"

In preparation for a staff meeting at which you will be expected to address this issue with your employees, review all of the apparent advantages and disadvantages of being classified in each of these categories. This should include knowing the difference between those employees who must, under law, be treated as nonexempt and those employees who, although they might qualify as exempt, may be treated as nonexempt for payroll purposes at the organization's option.

Exercise 29-3: Rates, Hours, and Overtime

Your institution operates on the "8 and 80" basis for overtime. One of your employees worked the following days and hours (mostly on the 11:00 P.M. to 7:00 A.M. shift):

Day	Hours
1	8
2	8
3	6
4	9
5	7
6	0
7	0
8	10
9	10
10	12
11	6
12	8
13	0
14	0

The employee's base rate is $8.60 per hour. Shift differential is $0.65 per hour. Day 10 included 4 hours of call-in, paid at a flat $40.00. (The employee was asked to come in at 7:00 P.M., 4 hours early.)

Determine the following:

- The hours of overtime due the employee.
- The "regular rate" for determining overtime premium.
- The employee's total earnings for the 2-week period.

Unions: Building Constructive Relationships

30

To protect the workers in their inalienable rights to a higher and better life; to protect them, not only as equals before the law, but also in their health, their homes, their firesides, their liberties as men, as workers, and as citizens; to overcome and conquer prejudices and antagonism; to secure to them the right to life, and the opportunity to maintain that life; the right to be full sharers in the abundance which is the result of their brain and brawn, and the civilization of which they are the founders and the mainstay . . . The attainment of these is the glorious mission of the trade unions.

Samuel Gompers

If you don't have them, the best way to avoid them is to create a Theory Y environment where your people have a chance to realize their potential. If you already have unions, then deal with them openly and honestly.

Robert Townsend

FROM THE INSIDE   Learning to Admit and Correct Your Mistakes

It's a low-security institution with a new associate warden, a crusty old captain, and a union president who is pressed by "radical" elements in the union to take a firmer stand on major issues when they arise. The institution had a history of contentious labor-management relations, generally characterized by filing formal complaints at the drop of a hat. The new associate warden was trying to intervene in that pattern.

One of the lieutenants had done something really stupid. He had reprimanded a correctional officer in front of a group of inmates for something that wasn't even close to serious. The officer went to the captain to complain, and basically was told to forget it. The next place he went was to the union president, who came calling on the captain, only to hear the same thing.

The associate warden's office was the next stop for the union president, advising that the employee was about to file a formal grievance. After hearing the story second-hand, the associate warden separately talked to each of the principals involved. He then called in the captain and told him that the lieutenant was to apologize personally to the officer for upbraiding him inappropriately, and for doing so in front of inmates. The captain went ballistic, but eventually made it happen. The lieutenant wasn't particularly happy either, but he made the apology. The employee didn't file a grievance.

In the aftermath, the union president came to the associate warden and said, "There's SOBs, and then there's SOBs. You're my kind of SOB." With that demonstration of top management willingness to admit and correct mistakes, labor management relations began to turn around in the institution.

■ The Setting

Most correctional agencies operate in a unionized environment. Public employee unions are important functional forces in the workplace, as well as potent political forces in the wider arena of public policy and funding. In the private sector, it is less common to work in a facility that has a recognized bargaining unit, but it is the view of the authors that over time unionization will be a factor for private corrections to contend with as well.

The lofty ideals expressed by Samuel Gompers years ago still motivate organized labor in this country and elsewhere. But even if those goals had not been formulated thusly, employees will object to management decisions from time to time. They will have grievances over a variety of issues. They will experience morale problems. They will be motivated toward increased pay and benefits. They will see what other organized workers receive in their negotiated pay and benefits packages, and want those gains duplicated in their own pockets.

How managers deal with these issues can go a long way toward ensuring harmony and productivity in the workplace. Doing so brings into play many of the other topics covered in this book, particularly that of communications. Making good decisions, communicating well with staff, admitting mistakes and correcting them, management-by-walking-around—all these and many other management practices will help keep peace in the workforce.

Labor relations in the generic sense (that is, not referring specifically to union relations) are an unavoidable part of managing in the correctional setting. The question is whether or not management is operating within a contractual or collective bargaining structure, and to a certain extent this is a divide that boils down to public versus private corrections. Regardless of the situation however, managers have to deal with employees on matters that either do or could involve a union.

The next two sentences are absolutely critical for managers to remember. **Failing to adequately deal with labor relations issues in a non-union setting can lead to a work environment that is ripe for union organization. Failing to deal fairly and legally with a union in a facility that already is organized can not only bring about serious operational problems, but be quite costly as well.**

■ Can Unionization Be Avoided?

For current managers who are working in a non-unionized setting, this section will be of particular interest. If the agency or facility is already formally organized, then it will be important in setting the stage for good relations with the union and its members.

For background, one can look to the early 1970s when a major study was made of 379 union elections. These elections extended over a period of 3 years and involved approximately 30 industries in nearly as many states. Of these 379 elections, 281, or 74 percent, were won by the unions. It seemed reasonable for the researchers to ask why the unions had such success. Three conclusions were offered:

1. Most of the organizing campaigns focused on wages or other economic issues. Most of the initial demands were unreasonable, and most reflected ignorance about, or indifference to, the organizations' financial positions. Although in many cases wages appeared to be a good reason, it was often not the real reason. In many cases the unions won because of **apparent management indifference to complaints, no response or effort to respond to employees' problems, and the organizations' lack of credibility** with employees in regard to costs and true operating circumstances.

2. In most of the cases in which the unions won, **antimanagement sentiment was initially brought about by poor working conditions such as substandard facilities, poor organizational communications, or arbitrary or seemingly uncaring management.**

3. In many instances the anxiety produced by **generally widespread lack of knowledge about what was truly taking place** was helpful to the union cause.

In all cases of lost elections covered in the survey there were serious morale problems among employees. Also in all cases, management was operating in somewhat of a vacuum as to true employee feelings and opinions. Communications issues are a common theme as well. These are the very issues that this book has returned to repeatedly in describing the attributes of a good manager. And if management does not listen to the employees, union organizers will.

In testing the potential for unionization it is simply not enough for top management to have supervisors talk to employees and report back on their feelings. True antimanagement sentiment, potentially beneficial to a union organizer, is determined only through effective listening. Many employees (quite likely the majority) would prefer to be loyal to the organization, but the organization's seeming indifference to upward communication can discourage such loyalty. Also, the price of management indifference can be extremely high. If a union loses a bargaining election it may try again after 1 year has elapsed, and again and again until it succeeds. Management need lose only once.

There are three basic errors commonly committed by management in assessing the potential for success of union organizing efforts.

1. There is a widely prevalent management notion that most elements of worker dissatisfaction are due to wages, fringe benefits, and other economic items. However, initial organizing activity usually springs from noneconomic matters involving issues that are not nearly as quantifiable as dollars. Employee dissatisfaction will ultimately be expressed in the form of financial demands. A specific financial package can be obtained by contract, but there is no contractual way to obtain less tangible items such as sympathetic listening, open communications, and humane and respectful treatment.

2. Many top managers automatically assume that all supervisors are on the side of management. However, most supervisors came "up from the ranks" within the functions they supervise. As such, they are an integral part of the work group. Also, in many institutions first-line supervisors have been kept out of participation in real management decision making.

3. Frequently nobody at the top of the organization has any solid idea of what is really troubling the ranks of nonmanagerial employees.

Committing any of these basic management errors in one form or another can pave the way for a successful union organizing drive. More specifically, an institution can push its employees closer to a union by:

- Introducing major changes in organization structure, job content, equipment, or operating practices without advance notice or subsequent explanation.
- Giving employees little or no information about the status of important events at the institution or about its plans, goals, or achievements.
- Making key decisions in ignorance of the employees' true wants, needs, and feelings.
- Using pressure (authoritarian or autocratic leadership) rather than true leadership (consultative or participative leadership) to obtain employee performance.
- Disregarding or downplaying instances of employee dissatisfaction.

■ Corrections and Unions

Unionization of government employees is not a new phenomenon. As a result, public sector correctional managers almost certainly are dealing with one or more unions already, while private correctional managers probably are not.

But it is an open question as to how long private prisons will remain unorganized. As the total number of jobs available in the manufacturing sector of the U.S. economy has been steadily declining for a number of years, unions have correspondingly lost a source of most union membership. As this has taken place, unions have shifted a great deal of their focus (which already had included government employees at all levels) to the service sector, which now employs 75 percent or more of the U.S. workforce. This suggests that private corrections could be fertile grounds for union organizing in the future, particularly as they grow in number. Already, some private corrections firms have encountered union organizing issues in the United States and Puerto Rico.

With the economy of the new century displaying a great deal of turmoil and uncertainty, employees are feeling more and more uneasy about the future. As many organizations lay off staff, including mid-level managers, even employees in the public sector perceive a threat to their continued livelihood. And privatization itself has been cause for alarm among public sector employees as they feel threatened by the possibility of losing jobs or job security to a nongovernment entity.

One would think the continuing expansion of the correctional segment of the criminal justice system in the 1990s and into the next decade would reassure correctional staff of their job security. One would think that private corrections would present no real threat to the public sector. But that has not been the case. There has been continued resistance on the part of government employee unions to the trend toward privatization. Union workers remain concerned for their jobs and their futures.

As noted earlier, tangibles such as pay and benefits are always prominent in the presence of labor unrest. But uncertainty in the United States workplace drives far more than these economic issues. When employees feel that their concerns are not being adequately addressed by management (and corrections is faced with some nearly overwhelming concerns, many of which seem to defy all attempts at resolution because of outside pressures and restrictions) these employees will turn to someone else who will seem to listen. Often this "someone else" is a union.

On the other hand, it may not be too strong a statement to say that the formation of a union is (at least in part) evidence of an inability on the part of managers to properly balance the equally legitimate needs of the workplace and the worker. Managerial insensitivity to proper concerns, use of inappropriate supervisory techniques, and the failure to regard employees as an important and valuable resource that requires tending—all of these factors can create a fertile environment for a union.

■ The Supervisor's Position

It is possible, however, for the non-unionized institution to remain that way. As might be expected, the individual supervisor is critical in such a situation.

Throughout this book there have been repeated references to the importance of the supervisor-employee relationship. Supervisors are the members of management whom employees know best. The first line manager indeed may be the only member of this mysterious entity called management whom most employees know on a first-name basis or even know on speaking terms at all. Thus as employees see their immediate supervisor, so are they likely to see all of management and the organization itself. If they see their immediate supervisor as unconcerned, uncaring, distant, or indifferent, they are likely to view the organization as a whole in that manner also. Supervisors are the key communications link upon which so many employee issues hinge, as demonstrated by the previously cited study.

It follows, then, that the supervisor is in a key position when it comes to dealing with the threat of unionization. Line supervisors are the link that ties employees to higher management and thus to the organization. A manager's long-term behavior will have a great deal to do with whether the department or institution is a fertile ground for union organizing activity. And in particular, a manager's conduct and actions during an organizing campaign will exert a significant influence on employees' reaction to the organizing drive.

■ The Organizing Approach

Ordinarily, this section will not apply to public sector correctional managers. However, it could be applicable in situations where one union tried to usurp representation of bargaining unit members from another labor organization, which is possible in the public sector as well.

What do the first stages of a union organizing campaign look like? One of the first signs visible may be "leafleting" or the distribution of union literature to employees at walkways, driveways, and parking lot entrances. However, although serious leafleting is an undeniable indication of union activity it is ordinarily not the first step in an organizing campaign. Chances are the union has been studying the institution for weeks or even months, to judge its organizing potential, before the first literature appears.

When organizing activity actually begins, management may well know nothing about it. In fact, during the earliest stages the union may take great precaution to prevent management from learning about their interest. The controlled access features of a correctional institution reduce the options to nonemployee union organizers for activity of this type. But any area where employees congregate informally can be used for this purpose. The union may send organizers to locations (restaurants or bars frequented by off-duty staff) where they can loiter and listen and pick up what they can from conversations.

The organizers will try to learn as much as possible about the institution before revealing themselves. They will also attempt to pinpoint employees who have the potential to serve as internal organizers. They will look especially for those employees who are popular, knowledgeable, reasonably articulate, and in some way unhappy with the organization.

Should their silent assessment of the institution raise serious doubts that it could possibly be unionized, the organizers might simply withdraw without ever announcing their presence. However, if they believe the union stands a chance of succeeding they will likely identify themselves to a few selected employees. This will be the start of preparations to carry their message to others. Leafleting is likely to begin at about this stage.

The major exception to the usual significance of leafleting occurs in a practice sometimes referred to as a "pass-through." This refers to devoting a day or two to distributing literature at perhaps several targeted installations in the same general area. These are generally "cold" visits, with little or no advance work. The union will simply "pass through" the area and drop off as much literature as possible with employees (usually at shift change time) and follow up only if they receive expressions of interest from employees. (The pass-through literature usually includes a reply card to be returned for more information.)

When the organizers are out in the open and their purpose is generally known, they will step up their activities in meeting with employees and contacting them in other ways. Somewhere along the way, possibly through sympathetic employees, they will attempt to obtain a list of the names and addresses of all the institution's nonmanagerial employees. The union will most certainly be contacting many individual employees by telephone and may seek to visit the homes of others.

In talking with employees, the union will attempt to uncover issues to use as rallying points for employee sympathy and support. The organizers will attempt to identify martyrs and victims of "the system" and will effectively play on emotions in spotlighting incidents of alleged unfair treatment and discrimination.

The organizers will go to great lengths to impress on employees their right to be treated as individuals. This may seem elementary, since most people will express strong belief in the rights of the individual. If, however, in the face of seemingly indifferent management the union organizer is the first person to tell them this, then the grounds for union credibility may exist. The organizers will make every effort to develop a supportive, communicating relationship with employees. This should sound familiar, since the development of such a relationship is part of a supervisor's role.

You can be sure that most issues and incidents surfaced by the union are specially selected to make the agency's management look bad. Lacking sufficient factual material, organizers may stage incidents intended to make the union look good and make management look foolish.

The supervisor's awareness of this particular organizing tactic is critical. It is all too easy to make an inappropriate statement or incorrect decision when confronted with a trumped-up grievance or problem at an inconvenient time. These incidents often will take place under awkward circumstances (which usually includes the presence of some employee witnesses). Such matters would be rightly dealt with by the administration, the personnel officer, or whoever else may be coordinating the institution's counter-organizing activities. However, there is a need for the supervisor to react on the spot, without making promises or commitments and without seeming to be refusing to listen to an employee. Afterward the incident can be promptly reported to the proper persons.

■ Unequal Positions

Under the National Labor Relations Act (NRLA), unions and employers are not on an equal footing in the organizing process. In many respects the union enjoys the upper hand. Under the act, an employer can commit an unfair labor practice and such charges can be brought against the organization by the union. If the National Labor Relations Board (NLRB) rules and upholds the union's claim, then the union may be automatically certified as a recognized bargaining agent without the necessity of a representation election.

The law, however, does not work the other way around; generally, there is no such thing as an unfair labor practice committed by a union. Also, as noted earlier, if the union should lose a bargaining election it may petition for another election after 1 year has elapsed. The employer may well have to win year after year to remain non-unionized. However, the employer need lose only once and the union is in, permanently, for all practical purposes. As a practical matter, de-certification of a union is difficult to achieve and occurs infrequently.

■ The Manager's Active Role

The guidelines relevant to supervisory behavior during a union organizing campaign make up a sizeable collection of do's and don'ts. These are important issues for a manager because the stakes are so high for the parent agency. Individual managers should not "wing it" in the labor-management arena. They should, instead, get as much input as possible from the institution or agency's personnel office or labor management department before taking any action.

What the Supervisor Can Do When a Union Beckons

- Campaign against a union seeking to represent employees, and reply to union attacks on the institution's practices or policies.
- Give employees your opinions about unions, union policies, and union leaders.
- Advise employees of their legal rights during and after the organizing campaign, and supply them with the institution's legal position on matters that may arise.
- Keep outside organizers off institution premises.
- Tell employees of the disadvantages of belonging to a union, such as strikes (for nongovernment employees) and picket-line duty; dues, fines, and assessments; rule by a single person or small group; and possible domination of a local by its international union.
- Remind employees of the benefits they enjoy without a union, and tell them how their wages and benefits compare with those at other facilities (both union and nonunion).
- Let employees know that signing a union authorization card is not a commitment to vote for the union if there is an election.
- Tell employees that you would rather deal directly with them than attempt to settle differences through a union or any other outsiders.
- Give employees factual information concerning the union and its officials, even if such information is uncomplimentary.
- Remind employees that no union can obtain more for them than the institution is able to give.

- Correct any untrue or misleading claims or statements made by the union organizers.
- Inform employees that the institution may legally hire a new employee to replace any employee who strikes for economic reasons.
- Declare a fixed position against compulsory union membership contracts.
- Insist that all organizing be conducted outside of working time.
- Question open and active union supporters about their union sentiments, as long as you do so without direct or implied threats or promises (see "Shifting Ground Rules" later in this chapter).
- State that you do not like to deal with unions.

What the Supervisor Cannot Do When a Union Beckons

- Ask employees about their union sentiments in a manner that includes or implies threats, promises, or intimidation in any form. Employees may volunteer any such information and you may listen, but you may ask only with caution (see "Shifting Ground Rules").
- Attend union meetings or participate in any undercover activities to find out who is or is not participating in union activities.
- Attempt to prevent internal organizers from soliciting memberships during nonworking time.
- Grant pay raises or make special concessions or promises to keep the union out.
- Discriminate against pro-union employees in granting pay increases, apportioning overtime, making work assignments, promotions, layoffs, or demotions, or in the application of disciplinary action.
- Intimidate, threaten, or punish employees who engage in union activity.
- Suggest in any way that unionization will force the institution to close up, move, lay off employees, or reduce benefits.
- Deviate from known institution policies for the primary purpose of eliminating a pro-union employee.
- Provide financial support or other assistance to employees who oppose the union, or be a party to a petition or such action encouraging employees to organize to reject the union.
- Visit employees at home to urge them to oppose the union.
- Question prospective employees about past union affiliation.
- Make statements to the effect that the institution "will not deal with a union."
- Use a third party to threaten, coerce, or attempt to influence employees in exercising their right to vote concerning union representation.
- Question employees on whether they have or have not signed a union authorization card.
- Use the word "never" in any statements or predictions about dealings with the union.

It is vitally important to observe the limitations these requirements place on managerial actions and comments when dealing with employees. Ideally, the supervisors in an institution undergoing organizing pressure should receive formal training in these guidelines from a labor attorney or a labor relations expert within the agency.

It is also to the manager's advantage (at all times, but especially during a union organizing campaign) to know employees as individuals, and know them well. People cannot be stereotyped and there are few reliable generalizations concerning employees' receptiveness to a union. However, it is nevertheless possible for a manager to make some reasonable judgments as to how certain employees might react under organizing pressure. Often the employee sympathetic to the union's cause may:

- Feel that the organization is not communicating well on important workplace issues.
- Feel unfairly treated by the organization and believe that reasonable work opportunities have been denied.
- Feel that the organization has been unsympathetic regarding personal problems and pressures.
- Express a lack of confidence in supervision or individual managers.
- Feel he or she cannot talk openly with members of management.
- Feel unequally treated in terms of pay and other economic benefits.
- Take no apparent pride in affiliation with the institution.
- Exhibit career-path problems, having either changed jobs frequently or having reached the top in pay and classification while still having a significant number of working years remaining.
- Be a source of complaints or grievances more often than most other employees.
- Exhibit a poor overall attitude.

As a supervisor it is extremely important to know employees' attitudes toward the institution, and to be able to develop a sense for how well communications are working. Ultimately, a labor union has little to offer if employees already feel that the organization is responding to their needs.

Shifting Ground Rules

The previous lists of what the supervisor can and cannot do in the presence of union organizing are based on interpretations of the NLRA by the NLRB. Many of these interpretations are clear cut and have stood the test of time regardless of the composition of the NLRB. Some, however, are not clear cut and are likely to change as the board's composition changes.

By way of illustrating the shifting ground managers must negotiate, there is the matter of management's questioning of employees about union sentiments and activities. From 1980 it was relatively accurate to cite the so-called "TIPS Rule" in summarizing the most important elements of what a member of management could not do during union organizing. A manager could not Threaten, Interrogate, Promise, or Spy. (You may have encountered "TIPS" as "SPIT" or "PITS," depending on the arrangement of the four prohibitions.)

In the middle 1980s the NLRB loosened its interpretation of interrogation to suggest that it is lawful for an employer to question union supporters about their union sentiments as long as the questioning carries with it no threats or promises and in no way interferes with or restrains the employees in the exercise of their rights under the National Labor Relations Act.

The present posture on interrogation is hardly new; it had been an applied principle of labor relations for 30 years until 1980. In 1980, however, when the NLRB was dominated by one particular political party, the stricter interpretation of the interrogation prohibitions of the law was imposed. This stricter interpretation was reversed in 1984, when the NLRB composition changed again.

However, even the principle as currently applied does not mean that most questioning of employees about union involvement is necessarily "safe." It essentially means that an unfair labor practice charge concerning interrogation will not automatically go against management. Instead, it will likely be subjected to searching analysis to determine whether any aspect of the questioning may have been coercive. The "TIPS Rule" remains valid in that interrogation still carries with it a fair amount of risk.

It is absolutely critical that if involved in any aspect of a union organizing drive, the individual manager pay strict attention to guidance provided by the agency's legal counsel or labor-management team. Although the actions enumerated in the do's and don'ts lists should remain largely valid, some of them may vary in content or emphasis from one national administration to another depending on the makeup of the NLRB.

■ The Bargaining Election

The Mechanics of an Election

Working through both outside and inside organizers, the union will go about the business of securing sufficient employee interest to allow it to petition the NLRB for a bargaining election. Generally the indications of such support will take the form of simple cards that employees sign to indicate interest in having an election. Employees should be aware that signing a card is not an automatic "yes" vote for union representation, but rather simply an expression of interest in having an election.

When sufficient signatures are gathered (usually half or more of the number of employees in the unit that the union is seeking to represent) the union will petition the NLRB. After a preliminary investigation, the board usually will sanction an election and a date for voting will be set.

Election is by secret ballot, and all employees who work in the unit the union is seeking to represent are eligible to vote. If the union receives a simple majority of the vote, it will then be certified by the board as the legal bargaining agent for all persons who work in the unit. Compulsory union membership is not required by law. However, in all likelihood everyone working in the unit will eventually join the union, since this particular right of the union is usually bargained for in the initial contract.

If the union fails to achieve a simple majority, various possible legal challenges are possible. If they do not upset the results of the election, the union will withdraw, at least for a while if the vote was close. It may wait for a longer period if the results were clearly one-sided.

Keep in mind, however, that some elections are little more than formalities—many elections are lost long before the organizers ever show up. In an organization that is low-performing in the many areas noted as critical, employees very likely will be strongly disposed to organize. If the trend in relations between employees and management is clearly in the direction of a union, this can be difficult to reverse. Reversal may, however, be accomplished through hard work and plenty of open and honest communication.

Even if a single unit of employees is lost to a union (for instance, a union representing correctional officers), new steps aimed at creating positive communicating relationships can still pay off. A new atmosphere can make contract bargaining easier, smooth out day-to-day labor relations matters, and help keep other bargaining units out of the institution.

■ If the Union Wins

Shortly after the union is certified as a recognized bargaining unit, an initial contract will be negotiated. This gives supervisors a whole new set of rules and regulations to live with.

Whether a private correctional manager learning to work within a unionized environment or a public correctional manager who deals with a well-established union, every supervisor should know the union contract inside out. Know what it says, know what it does not say, and know why it says what it says. More importantly, comply with it faithfully. Some contracts seem top-heavy with numerous details and exacting requirements, but the manager may find that some parts of his or her job are actually easier because there now are hard and fast rules for situations that were previously subject to interpretation and judgment.

No doubt management will have a formal labor-management committee or other structure to work through with the union on contractual and interim issues. In many institutions, monthly meetings between local union officials and designated management staff are used as a forum for working out problems and concerns as they develop. The union will no doubt have some input, comment, or opportunity for key recurring events, like formulation of quarterly rosters or other assignment-related matters. But those meetings do not mean that individual managers are relieved from their responsibility to communicate with their staff, nor to work on an individual department level on with union officials on matters of mutual concern. Above all, as advised by Robert Townsend in the opening quotation, be open and honest in your dealings with the union.

The presence of a union does not mean a manager can back off in communications with employees and simply wave the contract at them. In fact, that probably is the worst thing a manager could do. Complete two-way communication remains essential in establishing and maintaining sound relationships with every employee—whether or not there is a union involved. After all, employees work for the institution, not for the union. Generally the union will be the employees' voice only if the employees feel they are not recognized as individuals and are not being heard by management.

Personal Managerial Qualities

Finally, a manager who deals with a union must display several key qualities to be successful.

Honesty is always the best policy. If you don't know, or know but can't tell, say so. Credibility is hard to gain and almost impossible to regain. If your union (and by extension your employees) learn that they can't trust you, your ability to get things done will be immensely more complicated.

If you or your organization are wrong, admit it and go about correcting the mistake in the quickest time possible. Sometimes this will mean that a subordinate's decision must be overturned, creating problems with that person. The vignette at the beginning of the chapter shows just such a situation. But managers are not infallible, and if the union (and again by extension your line employees) sees that a manager is willing to admit and correct mistakes, a far more constructive working relationship will result.

EXERCISES

Exercise 30-1: The Organizer

You are the powerhouse supervisor in a correctional institution presently under union organizing pressure. The power and steam plant, supporting shops, and supply areas you supervise are outside the secure perimeter of the institution.

The union's drive has reached the stage of signature cards. You are passing through the power house's machine shop when you observe an individual who you believe is a union organizer backing one of the millwrights into a corner. He is waving what appears to be a union authorization card. The millwright looks worried and in considerable distress, and also appears to be physically trapped in the corner by the other party. You cannot hear what the person with the card is saying, but you believe you recognize the kind of card this person is waving and you can tell this person is speaking quite forcefully.

Describe what you would do under the following two sets of circumstances:

1. You recognize the probable organizer as an employee of the correctional facility, but belonging to a department other than your own.

2. You are reasonably certain the probable organizer is not an employee of the institution.

Exercise 30-2: The Confrontation

You are unit manager of a fairly large, private drug treatment facility for low-security offenders. The program has been operating understaffed for a number of months. Times have been hectic, so you have been pitching in yourself much more than used to be necessary. On days when you have been short-staffed with floor officers you have taken to providing lunch reliefs personally. This practice has caused you to change your own lunchtime to the time when the staff lounge/dining area is most crowded.

Today you have just gotten your lunch and are about to take a seat when you are approached and very nearly circled by three of your staff members. One of them says to you, "We've been meaning to talk with you, but we're all so much on the run that we haven't gotten to you. Things have got to change around here. We can't keep going the way we're going. We're thinking of asking a union to come in, and we want to talk with you about it. Now."

There you stand in the middle of the lounge, tray in both hands, feeling surrounded.

- How do you believe you would handle this incident?
- What discussion areas should you avoid?
- Should this discussion take place in such a public location?

Suggested Reading

Adams, Scott. *The Dilbert Principle.* New York: Harper Business Publications, 1996.

Appropriately at the head of the list is this classic takeoff on corporate and bureaucratic behavior. Mixing text commentary and comic strip-based commentary, the author pokes fun at organizational behavior that is common in the business world, but also will look very familiar to anyone working in government. Read and refresh your sense of humor about your chosen line of work, serious as it may be.

Addeo, Edmond G. and Robert E. Burger. *Egospeak.* Radnor, PA: Chilton Book Company, 1973.

The main point of *Egospeak,* subtitled *Why No One Listens to You,* is: We don't listen because we're too busy talking—either actively speaking or thinking of what to say next—to listen effectively. Talking and listening, it is suggested, cannot happen at the same time. Of special interest to managers are the chapters concerning "JobSpeak" and "BusinessSpeak."

American Correctional Association. *Correctional Officer Resource Guide.* Laurel, MD: 1989.

This publication provides a broad range of information on practical issues faced by the correctional manager, not only in the security department of a correctional facility, but in all departments. It begins with an overview of the criminal justice system and proceeds to cover all typical areas of institution operations, many of which are part of the day-to-day concerns of managers in other correctional departments.

Readers are referred generally to ACA's Communications and Publication Division, which publishes a wide variety of corrections-related books as well as a periodical that covers, in a timely manner, contemporary issues in the field. Many important management-related books and articles are available from this source.

American Correctional Association. *Guidelines for the Development of Policies and Procedures*. Laurel, MD: 1991.

This book provides an inside view of the kinds of policies and procedures that a correctional manager must deal with on a daily basis. It keys model correctional policies—covering virtually all institutional operations and departments—to professional standards promulgated by the American Correctional Association and thus provides the reader with a clear picture of the basic operational factors that a manager must incorporate into the administration of a correctional department.

Banta, William F. *AIDS in the Workplace*. New York: Lexington Books, 1993.

AIDS is a topic of considerable concern in the correctional workplace, and not just in relation to inmates. Banta provides a very complete discussion of the supervisory implications of HIV. This book is an excellent starting point for anyone who wishes to learn more about the multifaceted impact that this disease is having in the U.S. workplace.

Barnes, Ralph M. *Motion and Time Study,* 7th ed. New York: John Wiley & Sons, 1980.

This book is probably the definitive work on the analysis of work motions and the establishment of time or performance standards. It is easily the most comprehensive volume available on the tools and techniques of methods analysis. Although the book is clearly intended as a text for industrial engineers or management engineers, a number of its sections, such as those on the general problem-solving process and human engineering, are useful to anyone interested in improving work methods. This volume clearly and understandably details all the known and proven tools applied in methods improvement.

Bartollas, Clemens, Ph.D. *Becoming a Model Warden: Striving for Excellence*. Laurel, MD: American Correctional Association, 2003.

This publication covers a wide range of practical topics such as staff development, integrity issues, proactive management, applying experience to management situations, and leadership development.

Berne, Eric. *Games People Play*. New York: Grove Press, 1964.

A great deal of Berne's work presented in *Games People Play* formed the basis of the later work by Thomas Harris, *I'm O.K.—You're O.K.,* essentially making Berne, a psychiatrist, one of the founders of transactional analysis. *Games People Play* is an interesting and important book in the study of the psychology of human relationships.

Berne, Eric. *The Structure and Dynamics of Organizations and Groups*. New York: Grove Press, 1963.

This book is about groups—from small, informal groups to large, formal organizations—and what can go wrong with them. It is not a handy leadership manual but rather a scientific work, and it is moderately difficult to read and absorb. There is a great deal to be learned from this book concerning the psychology of the work group, but the approach and presentation are primarily academic.

Block, Peter. *The Empowered Manager: Positive Political Skills at Work*. San Francisco, CA: Jossey-Bass, 1990.

While this book is something of an inspirational, rather than objective, work, Block provides some interesting views on the entrepreneurial and visionary aspects of management and leadership. For the manager caught in a bureaucracy or a bureaucratic mindset, his development of the issues of personal responsibility and initiative is important.

Cribben, James J. *Leadership: Strategies for Organizational Effectiveness*. New York: AMACOM, American Management Association, 1981.

This is a clear and readable treatment of leadership that takes a fairly down-to-earth look at the essentials of leadership from the individual manager's point of view. Especially helpful are chapters concerned with organizational approaches to motivation and leadership values.

Davis, Brian L., et al. *Successful Manager's Handbook*. Minneapolis: Personnel Decisions, Inc., 1989.

A well-organized general work on a number of issues related to management development, this book contains a great deal of information on honing administrative skills, organizing, developing leadership skills, conflict management, and other important management tasks.

United States Department of the Interior. *Gobbledy Gook Has Gotta Go*. Washington, DC: Government Printing Office, 1966.

A small (112 pages) but extremely interesting book that uses actual examples of some of the worst government writing ever seen to illustrate how official prose can be improved. This book is fun reading on its own—just to see the many ways the English language can be butchered in the course of a government agency's operation—but it also contains many useful ideas on improving writing skills in general.

DiIulio, John, Jr. *Courts, Corrections, and the Constitution*. New York: Oxford University Press, 1990.

Dealing with legal issues is a reality for today's correctional manager. This excellent work summarizes the impact the courts have had on corrections in recent decades. It provides an insightful picture of how court interventions have actually worked out in several major correctional systems, both for good and bad.

DiIulio, John, Jr. *Governing Prisons*. New York: The Free Press, 1987.

This book highlights some of the broader issues of managing prisons, through a comparative study of correctional management in three correctional systems. DiIulio uses the categories of order, amenity, and service as bases for evaluating the quality of correctional life (and therefore governance), and also delves into the issues of internal controls and bureaucracies that typify public correctional operations in the United States.

DiIulio, John, Jr., et al. *Improving Government Performance.* Washington, DC: The Brookings Institution, 1993.

This book charts a course of thought parallel to that of the National Performance Review undertaken in the early 1990s by the federal government. It promotes the view that evolutionary rather than revolutionary change is a more effective way to streamline the bureaucracy and manage more efficiently. Geared more toward a macro-view of management, and directed toward the federal level of government, it still makes for informative reading by the individual manager seeking to work effectively within the larger organization.

DiIulio, John, Jr. *No Escape: The Future of American Corrections.* New York: Basic Books, 1991.

In this book, DiIulio combines scholarly information, on-site observation in numerous prisons, and a thoughtful practicality to produce an excellent picture of where the United States may be heading in the field of corrections. This work continues to emphasize the theme that DiIulio has developed in many of his writings—that the quality of management has a direct effect on the quality of imprisonment. This book is a must for any correctional manager concerned with how his or her chosen career is shaped and will continue to be affected by a variety of forces.

Drucker, Peter F. *The Practice of Management.* New York: Harper & Row, 1954.

Along with Drucker's *Managing for Results* (Harper and Row, 1964), *The Practice of Management* stands out as one of the highlights of this particular author's productive output. This is recommended reading for any supervisor who truly enjoys management and aspires to rise higher in the organization. One chapter of particular value is "Management by Objectives and Self-Control." (This portion of the book alone provided a whole "new" approach to management—management by objectives.)

Ewing, David W. *Writing for Results in Business, Government, the Sciences and the Professions,* 2nd ed. New York: John Wiley & Sons, 1979.

This is an excellent guidebook for persons who do the bulk of their writing in the institutional setting. The "situational approach" to writing—first examining the situation to determine what the writer wants to accomplish, with what readers, and under what circumstances—is promoted. It suggests strongly that effective writing means analyzing and planning as well as doing, and it is especially helpful in "targeting" written communications according to audience and purpose and in dealing with the often complex terminology of government, the sciences, and the professions.

Fast, Julius. *Body Language.* New York: M. Evans and Company, 1970.

This interesting and entertaining book deals with kinesics, the study of nonverbal human communication, or "body language," which can include any reflexive or nonreflexive movement of all or a part of the body used by a person to communicate an emotional message. The implications for interpersonal communication are significant; for example, the book cites studies that reveal the extent to which body language can actually contradict verbal communications. *Body Language* will provide some interesting and potentially helpful insights into the actions that surround or accompany a person's words.

Fournies, Ferdinand F. *Coaching for Improved Work Performance.* Blue Ridge Summit, PA: Liberty House, 1978.

This book focuses on analytical and practical techniques for managers who want to improve their ability to motivate employees. It is filled with many examples and case studies, as well as problem-solving techniques. While somewhat oriented toward a sales or marketing environment, many of its suggested approaches are quite applicable to the face-to-face interactions that characterize the correctional setting.

Gold, Michael Evan. *An Introduction to the Law of Employment Discrimination.* Ithaca, NY: ILR Press, 1993.

A small, well-organized book that covers key areas of employment discrimination law in a succinct, useful manner. A good "pocket guide" to the whys and wherefores of a complex field.

Gordon, Thomas. *Leader Effectiveness Training (L.E.T.).* New York: Wyden Books, 1977.

One of the truly indispensable books recommended here is *Leader Effectiveness Training.* Written in a clear, easily readable style, it takes a commonsense, humanistic approach to the task of getting things done through people. This book stresses cultivation and maintenance of open and honest interpersonal relationships. The chapter titles speak for themselves: "Doing It Yourself— Or with the Group's Help," "Making Everyday Use of Your Listening Skills," and "The No-Lose Method: Turning Conflict into Cooperation."

Harris, Thomas A. *I'm O.K—You're O.K.* New York: Harper & Row, 1967.

This book is most appropriately read for improved understanding of human behavior—your own as well as others—and for deeper insight into the problems of interpersonal communication. Although a great deal of the book consists of the definitive presentation of the field we call transactional analysis (TA), it can help managers become more attuned to "where someone is coming from" in interpersonal dealings.

Heckmann, I. L., Jr., and S.G. Huneryager. *Human Relations in Management.* Cincinnati: South-Western Publishing Co., 1960.

Intended as a college text, this book is largely a collection of readings, questions, and bibliographies. Although moderately difficult to read in places, it is a goldmine of information for managers at all levels. Included, for instance, are A. H. Maslow's "A Theory of Human Motivation," Douglas McGregor's "The Human Side of Enterprise," and Gordon Allport's "The Psychology of Participation."

Henderson, James, et al. *Guidelines for the Development of a Security Program,* 2nd ed. Laurel, MD: American Correctional Association, 1997.

A key resource work for the correctional manager, this book outlines the core functions around which every correctional operates. Every section will not only apply to every manager of a noncustodial department, but taken as a whole, the body of information contained in this publication would be essential to any manager who aspired to head an institution some day.

Hersey, Paul. *The Situational Leader.* New York: Warner Books, Inc., 1984.

This is a short, highly readable book based on a simple model that has been used to train managers at more than five hundred corporations. Using clear, interesting examples, this book reminds us that it is not enough to simply describe your leadership style or communicate your intentions. Rather, a "situational leader" assesses the performance of others and takes responsibility for making things happen.

Hersey, Paul and Kenneth H. Blanchard. *Management of Organizational Behavior: Utilizing Human Resources.* Englewood Cliffs, NJ: Prentice Hall, 1977.

This wide-ranging book focuses on linking management theory with practice, and relies on a number of behavioral science frameworks. Drawing on examples from many areas of organizational life, it is well written and amply annotated, and provides a great deal of useful information for new and current managers.

Hickman, Craig and Michael Silva. *Creating Excellence: Managing Corporate Culture, Strategy, and Change in the New Age.* Nightingale-Conant, 1984.

As suggested by the title, this book uses case studies and various exercises to inform the reader of methods for adapting and improving organizations in the face of modern change. While quite business oriented, its fundamentals are applicable to the correctional setting as well.

Jay, Antony. *Management and Machiavelli.* New York: Holt, Rinehart & Winston, 1967.

Subtitled *An Inquiry into the Politics of Corporate Life,* this book is decidedly slanted toward the overall management of entire organizations. It has its basis in Jay's interpretation of the psychology and conduct of modern corporations paralleling the principles of management employed in the medieval political state. Readers with an interest in the rights and wrongs of corporate management will get something from *Management and Machiavelli* that applies to all types of organizations, especially those of the bureaucratic and institutional form (large hospitals, various associations, and with state and federal agencies).

Kepner-Tregoe. *Problem Analysis and Decision Making.* Princeton: Princeton Research Press, 1973.

This is a tersely written little book used in connection with management seminars conducted by Kepner-Tregoe, Inc. It is oriented toward a very rational, no-nonsense approach to the problem-solving and decision-making side of management.

Likert, Rensis. *New Patterns of Management.* New York: McGraw-Hill, 1961.

This is a most valuable work in promoting understanding of the "systems of management" that exist in work organizations. The basic organizational differences attributable to correctional institutions are not the differences of "prison" versus "industry" but rather the differences between organizations doing repetitive work and those doing varied work. *New Patterns of Management* sheds considerable light on the matter of understanding the varying styles of supervision related to the different kinds of work the organization does.

Maxwell, John C. *Thinking for a Change*. New York: Warner Business Books, 2003.

Another easy-to-read volume that contains a variety of useful strategies for personal intellectual development, and includes well-designed exercises that apply chapter principles.

Maynard, H.B., ed. *Industrial Engineering Handbook*, 3rd ed. New York: McGraw-Hill, 1971.

This weighty volume, consisting of more than 1,500 pages containing the written contributions of more than 100 authors, is primarily a reference book for practicing industrial engineers. However, it is also especially useful to supervisors and managers whose work is affected by that of industrial or management engineers or who become actively involved in methods improvement projects. It makes an especially handy reference for a methods improvement project team involved in the analysis of manual tasks.

McCarthy, Edward, *Speechwriting: A Professional Step-by-Step Guide for Executives*. Dayton: The Executive Speaker Company, 1989.

Inevitably, managers and executives are called upon to speak before groups, a task that (at least initially) strikes fear in the hearts of many. This book provides a clear and functional approach to public speaking. While not capable of dispelling actual butterflies, it can be quite useful in learning how to prepare and present information to groups through speeches.

McConnell, Charles R. *The Health Care Manager's Guide to Performance Appraisal*. Gaithersburg, MD: Aspen Publishers, Inc., 1993.

This book is suggested as an adjunct to Chapter 14, "Performance Appraisal: The Supervisor's Darkest Hour," for those who wish to further pursue the topic of performance appraisal. Appraisal is briefly examined in philosophy and principle. Guidance is provided for handling the individual elements of appraisal and for designing and developing an appraisal system. Appraisal's relationship to job descriptions and other source documents is established, and basic variations in appraisal practices are examined.

McGuigan, Patrick B. and Jon S. Pascale. *Crime and Punishment in America*. Washington, DC: The Institute for Government and Politics, 1986.

This collection of contributed works covers the entire spectrum of criminal justice, from street enforcement to imprisonment. It provides the reader with a full-fledged primer that encompasses family and juvenile crime issues as well as judicial and correctional management concerns, such as alternatives to incarceration.

Nierenberg, Gerard and Henry Calero. *Meta-Talk*. New York: Simon and Schuster, 1973.

For those seriously interested in learning more about oral communication, this is a book to be reread and studied. Subtitled *Guide to Hidden Meanings in Conversations,* the book deals with the kinds of things people say and why they probably say them—and not with the exact words used, but rather with the messages and meanings "between the words." One reading of *Meta-Talk*—or two or three readings, for that matter—will not make you an expert on hidden meanings. However, the book should provide insights that cannot help but enhance your ability to understand others better. The book's message is essentially that no "meaning" is ever absolute, and that true meaning lies in the combination of speaker, listener, and circumstances.

Oakley, Andy. *Issues Confronting City and State Governments*. Skokie, IL: P.O. Publishers, 1994.

This small volume contains a great deal of interesting data and commentary relating to important issues of practical governance. Of particular interest is the section on prisons and jails, but the author also touches on crime prevention, outsourcing, automation, and other issues of relevance to the correctional manager.

Odiorne, George S. *How Managers Make Things Happen*. Englewood Cliffs, NJ: Prentice Hall, 1982.

An excellent, easily readable rendering of the basics of management in modern work organizations, this work covers the likes of how to change poor work habits into good ones, develop a leadership approach that is solid and yet flexible when necessary, know when to criticize and when to praise, and combat carelessness and indifference. Odiorne's chapter on decision making is especially well done.

Osborn, Alex F. *Applied Imagination: Principles and Procedures of Creative Thinking*. New York: Charles Scribner's Sons, 1953.

This is one of the best books possible for help in "getting your mind in gear." It may well stand yet as the definitive work on creativity; certainly it is among the most entertaining and readable works on the subject. A great deal of what the book contains may strike you as old, familiar stuff, but it is presented appealingly and most of it still holds true today.

Parkinson, C. Northcote. *Parkinson's Law*. Boston: Houghton-Mifflin Company, 1957.

This book is essentially the forerunner of the many volumes that take a humor-in-the-bitter-truth approach to the problems of management and administration. Parkinson's elaborate and pompous style is exactly suited to the sometimes outlandish ideas he presents. The reader may be inclined to believe that the author rarely entertained a serious thought—for example, he concedes that serious books on public or business administration have their place "provided only that these works are classified as fiction." But a single reading also provides a great deal of insight into organization behavior. The best chapter is the first chapter, "Parkinson's Law or the Rising Pyramid." The "law" itself is stated in the book's opening sentence: Work expands so as to fill the time available for its completion.

Peter, Laurence J., and Raymond Hull. *The Peter Principle*. New York: William Morrow and Company, 1969.

This highly successful book presents some deadly serious messages in a wholly entertaining manner. The "Principle" states simply: In a hierarchy every employee tends to rise to his or her level of incompetence. The author is suggesting that people rise just so high, and no higher, in any organization, and that the level at which they stop is just a bit over their heads. It follows, Peter suggests, that eventually every position tends to be occupied by someone who is incompetent to carry out its duties, and that the real work is accomplished by people who are still on the way up to their levels.

Peter, Laurence J. *The Peter Prescription*. New York: William Morrow and Company, 1972.

Whereas *The Peter Principle* is subtitled *Why Things Always Go Wrong,* this natural sequel is subtitled *How to Make Things Go Right.* As sequels often go, *The Peter Prescription* has neither the freshness of humor nor the fullness of insight that the first book has. Yet this is a valuable book since it is focused much more clearly on what the individual can do in the way of self-improvement. Especially recommended for reading and reflection is "Part Two: Protect Your Competence."

Phillips, Richard and John Roberts. *Quick Reference to Correctional Administration*. Sudbury, MA: Jones and Bartlett Publishers, 2000.

This volume of model policies and procedures embodies many of the concepts described throughout this book. The authors provide an excellent means of quickly researching a body of information on key correctional management topics, including both inmate- and staff-management issues.

Pinchot, Gifford and Elizabeth Pinchot. *The End of the Bureaucracy and Rise of the Intelligent Organization*. San Francisco: Berrett-Koehler, 1997.

These authors present an interesting perspective on workplace issues and the future, as organizations adapt to societal change. Paradigm shifts and leadership issues are a focus, along with a not-often-found emphasis on fairness in the workplace.

Repa, Barbara. *Your Rights in the Workplace*. Berkeley: NOLO Publishing, 2002.

This comprehensive and well-organized book is written for employees, and covers virtually all of the central human resources issues from a non-management perspective. It is recommended because of its excellent summaries of such issues as immigration, discrimination, and worker privacy, as well as a fine appendix containing a wide range of resource information on topics such as HIV/AIDS and mediation/arbitration.

Ring, Charles R. *Contracting for the Operation of Private Prisons: Pros and Cons.* College Park, MD: The American Correctional Association, 1987.

Ring's small, easy-to-read analysis of the "pros and cons" of private corrections is essential reading for corrections professionals as privatization continues to grow. Although private prisons are much more prevalent now than when this book was written and some material thus is dated, this still is a well-written exposition of the central issues involved.

Sack, Steven. *From Hiring to Firing.* Merrick, NY: Legal Strategies, Inc. 1995.

Sack is an attorney who writes well and comprehensively about both the mechanics and legal implications of personnel administration. His chapters on employee privacy, nondiscrimination, workplace policies, and how to properly terminate employees are particularly useful. While written for the business world, it provides a great deal of useful information for managers in general, often in the form of "Tips" and "Counsel Comments" that provide succinct advice on specific points he is discussing.

St. James, Elaine. *Simplify Your Work Life.* New York: Hyperion, 2001.

This compact book is full of easy-to-read hints for streamlining your personal life and professional management techniques. It covers areas such as time, effectiveness, productivity, and working with others.

Schichor, David. *Punishment for Profit: Private Prisons/Public Concerns.* Thousand Oaks, CA: SAGE Publications, 1995.

Schichor makes no attempt to conceal his antiprivatization bias, but still presents a very complete picture of the issues surrounding privatization in corrections (and other social services as well). This book, read in connection with Charles Ring's work cited earlier, provides a great deal of thought-provoking information on this issue, which is of increasing importance in the field of corrections.

Shave, Gordon A. *Nuts, Bolts, and Gut-Level Management.* West Nyack, NY: Parker Publishing Company, 1974.

This book evolved from a series of interviews with a man described as "an extremely successful manager." Although the focus is primarily on the problems and responsibilities of middle and upper management in the manufacturing industry, there is plenty of common-sense advice applicable to managers at all levels in all lines of work. Supervisors who feel that there do not seem to be enough hours in the day would do well to read Chapter 4, "How to Vastly Increase Your Discretionary Time."

Sherman, Michael and Gordon Hawkins. *Imprisonment in America.* Chicago: The University of Chicago Press, 1981.

This book is policy oriented and within that context provides an overview of the evolution of correctional policies and practices in the United States. It concentrates on the issues of who should be locked up, for how long, and under what conditions, and concludes with a proposal for an integrated approach to imprisonment in the United States.

Stockard, James G. *Rethinking People Management.* New York: AMACOM, American Management Association, 1980.

    This book generally lives up to its claim of being "constructively critical of personnel administration." In addition to providing the supervisor with insight into what an appropriately focused personnel function should be doing, it also suggests how the supervisor can act to enhance constructive responses from personnel.

Strunk, William, and E.B. White. *The Elements of Style,* 2nd ed. New York: Macmillan, 1972.

    According to E.B. White, who in 1957 was commissioned to revise Strunk's original 1919 publication for the college market and the general trade, *The Elements of Style* was Strunk's attempt to "cut the vast tangle of English rhetoric down to size and write its rules and principles on the head of a pin." The attempt was successful; the book is a tight summation of the case for cleanliness, accuracy, and brevity in the use of written language. If limited to only a single book about writing, *The Elements of Style* should be the choice.

Terry, George R. *Supervisory Management.* Homewood, IL: Richard D. Irwin, 1974.

    This book takes the view that behavioral objectives of the personnel involved are inseparable from getting the work out, so the goal of supervision is not a satisfactory work group but also a satisfied work group. The book's focus is primarily the first-line supervisor, making it one of the few works available to the lower levels of management. *Supervisory Management* touches upon all aspects of the supervisor's job, although some are necessarily treated once-over-lightly. Questions and cases accompany each of its 18 chapters, making it appropriate as a course text.

Timm, Paul R. *Managerial Communication.* Englewood Cliffs, NJ: Prentice Hall, 1980.

    This book is a refreshing exception to an old unwritten rule; it is a college textbook that reads as clearly and easily as a work of popular nonfiction. Although almost every chapter could well be a book or educational program in its own right, coverage of each aspect of organizational communication is more than adequate for the working supervisor. Chapters of special value to working managers at all levels include "Speaking Before Groups" and "Letters and Memos."

Toffler, Alvin. *Future Shock.* New York: Random House, 1970.

    An immensely popular book, *Future Shock* is most valuable in creating a full appreciation of the rate at which all aspects of life and living are changing around us. Also of note are the author's observations concerning the ways people cope—or fail to cope—with accelerating change. Anyone seeking insight into the impact of change and the origins of resistance to change would do well to examine the lengthy table of contents and read a few selected sections, especially those relating to health and work and work organizations.

Townsend, Robert. *Up the Organization*. New York: Alfred A. Knopf, 1970.

Up the Organization is an entertaining book. It is arranged in alphabetical order by topic, and since no topic requires more than a few minutes reading time it is a book that can be read in bits and pieces with no loss of impact. Townsend's approach is energetic and people centered; his is a management style that depends heavily on a basic belief in the willingness of most people to produce, given the proper environment. Although aimed largely at top management, there is a great deal in the book's pages for first-line supervisors. Sections on "Delegation of Authority" and "People" are recommended. Read this book with some caution, however; a great deal of its gutsy leadership style is personality based, and not everyone is a Robert Townsend. Although based largely on humanity and common sense, in practical terms the Townsend style will reform an organization only when applied from the top down.

Trachtman, Michael G. *What Every Executive Better Know About the Law*. New York: Simon and Schuster. 1987.

While oriented toward the business executive, this well-written book has important sections on common sense and the law, documentation, labor law, and what happens if you actually go to court. It also contains a very complete glossary that is written in English, not Legalese.

Ward, David A. and Kenneth F. Schoen. *Confinement in Maximum Custody*. Lexington, MA: Lexington Books, D.C. Heath and Company, 1981.

This book is a collection of presentations made at a conference held at the Spring Hill Center in Wayzata, Minnesota in June 1978. While rather narrowly focused on the issues of maximum-security confinement, it provides the reader with an excellent feel for the complexity of legal, psychological, and human issues that challenge managers in this very specific kind of correctional environment.

Weiss, Donald H. *Fair, Square, and Legal*. New York: AMACOM, American Management Association, 1991.

An excellent summary of applicable legal issues that managers must take into account to assure that they are engaged in lawful hiring, day-to-day supervision, and discharge practices. This work is particularly well organized and easy to read.

Weiss, W. H. *Supervisor's Standard Reference Handbook*. Englewood Cliffs, NJ: Prentice Hall, 1980.

This is an excellent book for first-line supervisors in any field. Its presentation is straightforward and very nearly person-to-person conversational. Despite its title it can be used effectively as a topic text as well as an occasional reference. Its emphasis is decidedly on dealing with day-to-day problems, and as such it does not delve deeply into longer-term processes such as planning.

Zinsser, William. *On Writing Well: An Informal Guide to Writing Nonfiction,* 2nd ed. New York: Harper & Row, 1980.

Lively and readable in its own right, this is also a helpful reference for people who are seriously interested in learning how to improve their writing in general. It is not a textbook, and certainly not an English grammar lesson. This book will provide valuable overall guidelines for writing for simplicity and clarity in today's world. The book's main theme is that there is no subject that cannot be made accessible if the writer writes with humanity and cares enough to write well.

Index

(Note: Page numbers in italics indicate material found in tables, figures, or exhibits.)